Lecture Notes in Business Information Processing 486

LNBIP reports state-of-the-art results in areas related to business information systems and industrial application software development – timely, at a high level, and in both printed and electronic form.

The type of material published includes

- Proceedings (published in time for the respective event)
- Postproceedings (consisting of thoroughly revised and/or extended final papers)
- Other edited monographs (such as, for example, project reports or invited volumes)
- Tutorials (coherently integrated collections of lectures given at advanced courses, seminars, schools, etc.)
- Award-winning or exceptional theses

LNBIP is abstracted/indexed in DBLP, EI and Scopus. LNBIP volumes are also submitted for the inclusion in ISI Proceedings.

Inès Saad · Camille Rosenthal-Sabroux ·
Faiez Gargouri · Salem Chakhar ·
Nigel Williams · Ella Haig
Editors

Advances in Information Systems, Artificial Intelligence and Knowledge Management

6th International Conference
on Information and Knowledge Systems, ICIKS 2023
Portsmouth, UK, June 22–23, 2023
Proceedings

Springer

Editors
Inès Saad
ESC Amiens and University of Picardie Jules Verne
Amiens, France

Faiez Gargouri
University of Sfax
Sfax, Tunisia

Nigel Williams
University of Portsmouth
Portsmouth, UK

Camille Rosenthal-Sabroux
University of Paris Dauphine-PSL
Paris, France

Salem Chakhar
University of Portsmouth
Portsmouth, UK

Ella Haig
University of Portsmouth
Portsmouth, UK

ISSN 1865-1348 ISSN 1865-1356 (electronic)
Lecture Notes in Business Information Processing
ISBN 978-3-031-51663-4 ISBN 978-3-031-51664-1 (eBook)
https://doi.org/10.1007/978-3-031-51664-1

This Springer imprint is published by the registered company Springer Nature Switzerland AG
The registered company address is: Gewerbestrasse 11, 6330 Cham, Switzerland

Paper in this product is recyclable.

Preface

A large amount of data, information, and knowledge is being created and used in organizations. With the evolution of information and communications technology platforms, organizations in the process of digital transformation constantly search for innovative concepts and creative applications. The International Conference on Information and Knowledge Systems (ICIKS) looks to bring together academics, researchers, practitioners and other professionals from around the world and provide opportunities to highlight recent advances in the fields of Information Systems, Artificial Intelligence, Knowledge Management, Decision Support and Digital Decision Tools. ICIKS also provides opportunities to network and to share ideas and expertise and promote discussions on various organizational, technological and socio-cultural aspects of research in the design and use of information and knowledge systems in organizations.

The 6th International Conference on Information and Knowledge Systems (ICIKS 2023) was held in Portsmouth during June 22–23, 2023. Portsmouth is a port city and naval base on England's south coast, known for its important maritime heritage. ICIKS 2023 was organized in cooperation with the Association of Computing Machinery (ACM) Special Interest Group on Artificial Intelligence (ACM SIGAI) and Special Interest Group on Management Information Systems (SIGMIS) and received the support of the University of Portsmouth (UK), Amiens Business School (France), University of Picardie Jules Verne (France), Paris Dauphine University (France), and University of Sfax (Tunisia). It is with great pleasure that we present to you the proceedings of ICIKS 2023.

The ICIKS 2023 received 58 submissions from the UK and the following countries: Algeria, Brazil, Canada, Egypt, France, Ireland, Italy, Nigeria, Saudi Arabia, Tunisia and Ukraine. Each submission was carefully reviewed by at least three members of the Program Committee. In addition to the General Conference track, ICIKS 2023 included the following four special tracks: (1) Learning Paradigms for Intelligent Systems; (2) Blockchain Technology: Latest Developments, Applications and Challenges; (3) Risk and Uncertainly in Strategic Competitive Intelligence; and (4) Natural Language Processing for Decision Systems and Knowledge Management. For the final conference program, 30 papers were accepted for oral presentation, from which 18 (31.03%) full papers and 6 (10.34%) short papers were selected for inclusion in these proceedings. These papers cover a wide range of Information Systems, Knowledge Management and Artificial Intelligence topics, including Decision Support Systems, Machine Learning, Intelligent Systems, Natural Language Processing, Recommender Systems, Cybersecurity, Modelling and Simulation, and Ontology based Systems. The papers included in these proceedings are organised into five parts: (1) Decision Making, Recommender Systems and Information Support Systems; (2) Information Systems and Machine Learning; (3) Knowledge Management, Context and Ontology; (4) Cybersecurity and Intelligent Systems; and (5) Natural Language Processing for Decision Systems.

The conference program was enriched by three keynote presentations. The keynote speakers were Mark Xu from University of Portsmouth, UK, with a talk on "Managerial Information Processing in the Era of Big Data and AI - An Evolutionary Review"; Sajid Siraj from Leeds University Business School, UK, with a talk on "Explainability and Fairness in Machine Learning and Data Mining"; and Adiel T. de Almeida and Lucia Reis Peixoto Roselli from Federal University of Pernambuco (Brazil) with a joint talk on "Advances in Preference Modelling with Improvement of Multicriteria Decision Support Systems Using Behavioural and Neuroscience Studies".

ICIKS 2023 also introduced a Best Paper Award to recognize and promote quality contributions to academic research. A special scientific committee assessed, scored and ranked all accepted and presented papers. The Best Paper Award selection process went through four phases: (1) An initial score was assigned to each accepted paper based on innovation, significance to the research community, scientific quality and technical excellence. (2) A short list of seven top-ranked papers was chosen based on the initial scores. (3) The scientific committee members scored the shortlisted papers based on communication and presentation capabilities. (4) The shortlisted papers were rank ordered based on the sum of the review and presentation scores. The Best Paper Award was given to the paper with the highest overall score.

The ICIKS 2023 shortlisted papers (randomly ordered) are:

- FITradeoff Decision Support System applied to solve a Supplier Selection Problem by Lucia Reis Peixoto Roselli and Adiel Teixeira de Almeida
- ACTIVE SMOTE for Imbalanced Medical Data Classification by Sana Ben Hamida and Raul Sena Rojas
- Epistemology for Cyber Security: A Controlled Natural Language Approach by Leigh Chase, Alaa Mohasseb and Benjamin Aziz
- Moving towards explainable AI using Fuzzy Networks in decision making process by Farzad Arabikhan, Alexander Gegov, Rahim Taheri and Negar Akbari
- Decision and Information Support System to implement the Framework of Twelve steps for building decision models by Adiel Teixeira de Almeida and Lucia Reis Peixoto Roselli
- Topic Modelling of Legal Texts using bidirectional encoder representations from Sentence Transformers by Eya Hammami and Rim Faiz
- Models-simulators in business decision-making processes for pharmaceutical enterprises by Zoia Sokolovska, Oksana Klepikova, Iryna Ivchenko and Oleg Ivchenko

The ICIKS 2023 Best Paper Award was given to the following paper:

- ACTIVE SMOTE for Imbalanced Medical Data Classification by Sana Ben Hamida and Raul Sena Rojas

The 2023 version of ICIKS also introduced a Best Review Award and Local Organisation Award. The ICIKS 2023 Best Review Award this year was given to John Edwards (Aston University, UK) while ICIKS 2023 Local Organisation Awards were given to Inès Saad (General Chair) and to Camille Rosenthal-Sabroux and Faiez Gargouri (Program Chairs).

We would like to thank the ICIKS Steering Committee, Program Committee members, Track Chairs, and Session Chairs for their time, dedication and effort providing valuable reviews. We also would like to thank all the authors for submitting their contributions. Finally, we are grateful to the editors of the LNBIP series from Springer for their help in the preparation of these proceedings.

October 2023

Inès Saad
Camille Rosenthal-Sabroux
Faïez Gargouri
Salem Chakhar
Nigel Williams
Ella Haig

Organization

General Chair

Inès Saad	Amiens Business School & UPJV, France

Program Committee Chairs

Camille Rosenthal-Sabroux	Paris Dauphine University – PSL, France
Faiez Gargouri	University of Sfax, Tunisia

Local Organisation Committee

Salem Chakhar (Chair)	University of Portsmouth, UK
Nigel Williams	University of Portsmouth, UK
Ella Haig	University of Portsmouth, UK
Louisa Wood	University of Portsmouth, UK
Amine Belmejdoub Quotb	University of Portsmouth, UK
Asad Khan	University of Portsmouth, UK
Farok Bin Iqdara	University of Portsmouth, UK

Steering Committee

Marie-Hélène Abel	University of Technology of Compiègne, France
Pierre-Emmanuel Arduin	Paris Dauphine University, France
Markus Bick	ESCP Business School, Germany
Tung Bui	University of Hawaii at Manoa, USA
Fatíma Dargam	SimTech Simulation Technology, Austria
John Edwards	Aston University, UK
Rim Faiz	IHEC, University of Carthage, Tunisia
Michel Grundstein	Paris Dauphine University, France
Gilles Kassel	University of Picardie Jules Verne, France
Shaofeng Liu	Plymouth University, UK
Jean-Charles Pomerol	Sorbonne University, France

Program Committee

Marie-Hélène Abel	University of Technology of Compiègne, France
Zeeshan Ahmed Bhatti	University of Portsmouth, UK
Frederic Andres	National Institute of Informatics, Japan
Pierre-Emmanuel Arduin	Paris Dauphine University, France
Muhammad Awais Shakir Goraya	University of Portsmouth, UK
Jean-Paul Barthes	University of Technology of Compiègne, France
Sana Ben Hamida	Paris Dauphine University, France
Markus Bick	ESCP Europe, Germany
Amel Borgi	University of Tunis El Manar, Tunisia
Wafa Bouaynaya	University of Picardie Jules Verne, France
Sarra Bouzayane	Audensiel, France
Tung Bui	University of Hawaii at Manoa, USA
Frada Burstein	Monash University, Australia
Salem Chakhar	University of Portsmouth, UK
Fatima Dargam	SimTech Simulation Technology, Austria
Claudia Diamantini	Marche Polytechnic University, Italy
John Edwards	Aston University, UK
Rim Faiz	IHEC, University of Carthage, Tunisia
Faïez Gargouri	University of Sfax, Tunisia
Mila Gascó-Hernandez	University at Albany, USA
Susan Gauch	University of Arkansas, USA
Sahar Ghrab	University of Sfax, Tunisia
Michel Grundstein	Paris Dauphine University, France
Ella Haig	University of Portsmouth, UK
María Isabel Sánchez Segura	Universidad Carlos III de Madrid, Spain
Céline Joiron	University of Picardie Jules Verne, France
Dimitris Karagiannis	University of Vienna, Austria
Gilles Kassel	University of Picardie Jules Verne, France
Asad Khan	University of Portsmouth, UK
Ashraf Labib	University of Portsmouth, UK
Shaofeng Liu	Plymouth University, UK
Maria Manuela Cruz Cunha	Polytechnic Institute of Cávado, Portugal
Florinda Matos	ISCTE – Instituto Universitário de Lisboa, Portugal
Nada Matta	UTT-Troyes, France
Brice Mayag	Paris Dauphine University, France
Seif Mechti	ISSEPS, University of Sfax, Tunisia
Elsa Negre	Paris Dauphine University, France
Renata Paola Dameri	University of Genoa, Italy
Camille Rosenthal-Sabroux	Paris Dauphine University, France

Inès Saad	Amiens Business School & UPJV, France
Sajid Siraj	University of Leeds, UK
Eric Tsui	Hong Kong Polytechnic University, China
Mohamed Turki	University of Sfax, Tunisia
Takahira Yamaguchi	Keio University, Japan
Sam Zaza	Middle Tennessee State University, USA

Sessions and Keynote Talks Chairs

Marie-Hélène Abel	University of Technology of Compiègne, France
Farzad Arabikhan	University of Portsmouth, UK
Pierre-Emmanuel Arduin	Paris Dauphine University, France
Muhammad Awais Shakir Goraya	University of Portsmouth, UK
Alexander Gegov	University of Portsmouth, UK
Sana Ben Hamida	Paris Dauphine University, France
Sarfaraz Hashemkhani	Universidad Católica del Norte, Chile
Olumuyiwa Matthew	University of Portsmouth, UK
Salem Chakhar	University of Portsmouth, UK
Inès Saad	Amiens Business School & UPJV, France
Sajid Siraj	University of Leeds, UK
Adiel Teixeira de Almeida	Federal University of Pernambuco, Brazil
Mark Xu	University of Portsmouth, UK

Contents

Decision Making, Recommender Systems, and Information Support Systems

Models-Simulators in Business Decision-Making Processes for Pharmaceutical Enterprises

Zoia Sokolovska(✉), Oksana Klepikova, Iryna Ivchenko, and Oleg Ivchenko

Odessa Polytechnic National University, Shevchenko Ave., 1, Odessa 65044, Ukraine
{szn,o.a.klepikova,i.y.ivchenko}@op.edu.ua

Abstract. This article focuses on computer simulation and shows the practical application models in the decision-making processes for industrial enterprises. The key points of the research are the demonstration of a hybrid paradigm of computer simulations based on the system dynamics method and agent approaches using the AnyLogic software platform. The proposed simulation models and experiments are aimed at situational testing of pharmaceutical production sales and distribution in the uncertainty of complex socio-economic systems in order to control a competitive market because traditional analysis and forecasting methods do not always provide the de-sired results in such conditions. The model's modularity and openness provide the possibility of adaptability according to the specifics of the industrial enterprises. The proposed approach allows pharmaceutical enterprises to develop a business plan; analyze and pick up management factors for achieving sustainable and profitable development; study the indicators reflecting the internal and external factors that affect the stability of the enterprise and the pharmacy market of Ukraine.

Keywords: Pharmaceutical company · Decision-making process · Simulation model · Situational simulation experiment

1 Problem Definition

The process of decision-making management in the modern economy is becoming more complicated. Intensive economic changes are accompanied by a significant number of participants interacting in an uncertain environment. This demands automated Decision Support Systems (DSS) from both government and businesses.

The need to track the dynamics of enterprise development in consideration of the complex interrelationships and influences of many stochastic factors leads to an increase in the role of computer simulation modeling.

Simulation models are often integrated into the Decision Support System. The most typical scenarios are the following:

- a model is built into the current business processes for the object of study to be automatically launched after the specific operations;
- a model is aimed at developing options for the object's business strategies in a situational model;

I. Saad et al. (Eds.): ICIKS 2023, LNBIP 486, pp. 3–18, 2024.
https://doi.org/10.1007/978-3-031-51664-1_1

– a model is launched by a user with specific setting parameters in the decision-making process.

The main purpose of using simulation models as the DSS built-in blocks (components of the model base) is determined by mathematical methods. The dynamic simulation of the studied processes is accompanied by the implementation of many situational experiments with the possibility of their operational regulation (due to the flexible user interface) and the update of the input data.

The relevant level of simulation modeling is the micro level, particularly in enterprises with a high level of product market competition, and the rapid changes in the categories and sales.

The pharmaceutical industry is one of the most important sectors of the world industry saturating the consumer markets with pharmaceutical products. In Ukraine, the pharmaceutical industry is one of the most dynamic and profitable sectors of the country's economy, during this difficult socio-economic situation. According to the State Statistics Services of Ukraine [1], the average annual growth of the national market over the last 5 years remains at the level of 15–20%. In 2019, the volume of the pharmaceutical market was 3.3 billion dollars USA; in 2020, it grew to 3.4 billion dollars USA; in 2021, it grew to 4.1 billion dollars USA. In 2022, the volume of pharmacy sales decreased by 6% and is expected to stay at about 4 billion dollars [2].

This sector is an important part of the market, which determines the national and defense security of the country. The market's economic characteristics are the high level of competition, significant segmentation, import dependence, and sufficiently significant state regulation with little funding. Systemic problems of the industry are the impossibility of ensuring a full cycle of production (the absence of active pharmaceutical production resources); the insufficient technological and innovative level; the provision of raw materials, and the financial restrictions.

Consequently, the results of pharmaceutical production entirely depend on the effectiveness of strategic and operational business decisions. This study proposes the computer simulation models developed for the relevant situations in the pharmaceutical products market, with the corresponding adjustment of business plans for production and sale. The article shows the possibility of making business decisions based on simulation experiments, using the nonlinear dynamic processes reproduced on these models and the relevant feedback.

2 Analysis of Recent Research and Publications

A significant number of literary sources are devoted to the problems of implementing decision support systems in business [3–8].

However, it is necessary to highlight the articles which investigate the direct connectivity between the implementation of the DSS and business decisions efficiency. The obtained research results from 881 companies are presented in the article [9]. The paper shows that in the digital economy, companies' investments in IT resources are considered part of their capital investments. Therefore, companies need efficient management of IT sources to achieve the necessary business advantages. The study contributes to the development of IT systems management processes through their conceptualization, evaluation, and implementation of effective designs. The authors propose a concept and model of IT resource management that shows the capabilities of companies to identify, design, implement, and use IT management processes to select the allocation of IT resources and IT metrics. It is shown that increasing the efficiency of automated decision support systems due to the effective management of IT resources contributes to a significant increase in business productivity.

The influence of DSS on marketing decision-making in the pharmaceutical industry of Pakistan is studied in the article [10]. The authors analyzed the materials of 96 pharmaceutical companies in Pakistan. It has been shown using of information technologies in decision-making processes significantly affects market indicators in the pharmaceutical industry and facilitates strategic planning, organizational support, knowledge management, and technological progress.

The impact of the marketing information system on decision-making in Indian pharmaceutical companies is studied in the article [11]. The sustainable influence is confirmed by the exponential growth of the marketing information system productivity. It is shown that this was facilitated by the implementation of such initiatives as the digitalization of company activities, marketing intelligence, and improvement of the internal accounting system. The authors research the most influential system factors in decision-making processes in the Indian pharmaceutical sector.

The development of management decision-making methodology was researched in article [12]. It is noted that the decision-making process involves criteria with a high level of uncertainty. However, in practice, the multi-attribute theory of utility (proposed as the main method in the course of decision-making in uncertainty) is overloaded. The authors propose a new approach (SURE). The modeling is based on triangular distributions to create a graph that visualizes the preferences and the crossing alternative's uncertainty in decision-making. Approbation of the method was carried out based on a large pharmaceutical company. The SURE can correctly identify the alternative ultimately chosen by the company under study. The research showed that the method can be used at the same time as technological design in decision-making in uncertainty with different criteria. The approach is recommended for use in increased uncertainty.

Modern achievements in the development of methodology and applied applications of simulation modeling are regularly presented in software [13–15].

A wide range of articles presented using paradigms of simulation modeling in specific fields.

Agent and hybrid paradigms have become widespread in recent years.

The article [16] is devoted to the use of the agent paradigm. The authors show the popularity of the method in business and management and decision-making. A positive point is the review of the literature sources of the agent paradigm in business and management using bibliometric and content analysis. The analysis showed an increase in agent modeling applications in management; marketing, operations, supply chains; financial management, and risk management.

The articles [16] and [17] compare the current results with 2014 in marketing. The analysis of the use of the agent paradigm showed: decision-making in marketing; structures of simulated systems; goals of simulation experiments and models. The most popular field of using this method is marketing especially for the comparison of retrospective and current data.

The articles [16] and [18] show the development of agent modeling in management. The last analysis showed that agent modeling can study complex organizational structures and the systems' process features.

The article [19] is devoted to the hybrid modeling paradigm implementation issues. The advantages of using the hybrid paradigm in planning and decision-making in various sectors were identified. At the same time, it is noted that the time to create the hybrid paradigm models is significantly longer than using one of the simulation paradigms. Determination of specific components of the hybrid paradigm at the early stages and model development can make this process easier. The authors propose five system characteristics as decision-making points that should help developers make the right decision.

Using a hybrid modeling paradigm in the pharmaceutical industry started research many years ago considering the development of multi-approach modeling software. This article [20] is devoted to the problems of creating a complex model for the pharmaceutical enterprise investment portfolio. The optimal portfolio management strategies and tracking of the dynamics of investment resources are revealed. It was emphasized that the AnyLogic software provides support for basic simulation methods.

The AnyLogic software platform in the implementation of the hybrid approach is discussed in the article [21]. A comprehensive review of implementing a hybrid methodology in the AnyLogic software is presented by the developers in the book [22].

Pharmaceuticals are related to health care, so simulation software often implements various aspects of these related industries. The most developed field is the organization and management of pharmaceutical supply chains. There are numerous examples of the application of various simulation technologies/paradigms in this field.

In the article [23], multi-agent modeling of supply chains was carried out in the pharmaceutical industry. The model was tested based on a thousand runs, each lasting 39 years.

The system dynamics method was used for the development of the model in the article [24]. A simulation model is used to determine the characteristics of the pharmaceutical supply chain in a hospital (Colombian hospitals). One of the advantages of the model is that, as a result, the hospital can determine the real cost of drugs, considering various processes: from taking drugs to administering them to patients. A cause-and-effect diagram allows for identifying the interactions between model variables to develop a flow chart of the inventory with the determination of the final results. The model allows decision-makers in hospitals to identify variables that affect the final cost of drugs. This determines the amount of compensation allowed by national institutions.

The article [25] is devoted to the supply chains in pharmaceutical companies using the system dynamics methodology.

The hybrid paradigm was used in the article [26]. The purpose of the proposed hybrid structure is to obtain a more detailed view of drug supply chains to ensure the rhythm of supply and strengthen the chain's sustainability. The model is tested on generic drugs. The positive aspect is using various scenarios of the supply chain using a wide range of disruptive factors.

The article [27] presents a model developed for determining the elements affecting the sale of generic drugs throughout their life cycle. This helps to improve the efficiency of planning and predict the stage of decline in the product life cycle. The model is used to identify the demand, supply, and competition as three main subsystems in generic pharmaceutical products. A positive point of simulating the life cycle of generics is that manufacturers get more information about sales processes. It becomes possible to predict the decline stage, even when the overall demand for the generic drug falls in the market.

The system dynamics paradigm was used in the model in the article [28]. The main focus of the model is forecasting drug prices and production volume, taking into account the manufacturer's profit, raw material prices, and drug quality. The simulation was carried out on the Vensim software for a specific drug. The simulation experiments showed that with an increase in the percentage of maximum profit for manufacturers and with the use of high-quality raw materials, the profits of companies, the volume of production, and the quality of medicines will increase.

The modern direction is the introduction of the Internet of Things (IOT) in the pharmaceutical industry - which was reproduced in the development of a system-dynamic model [29]. The research aims to evaluate alternative investment strategies for the implementation of IOT. The results of simulation experiments are the basis for the development of cost-effective solutions by managers.

Thus, it should be noted that the use of simulation modeling in the management of enterprises in the pharmaceutical industry has found concrete implementation but is still quite limited. The main drawback is the lack of a unified approach to the creation of complexes of simulation models, which would cover various aspects of enterprise activity. The field of business solutions of pharmaceutical enterprises flexibly changes over time and is unlimited. This determines the relevance of the chosen issue.

The aim of the study is to explore the possibilities of using hybrid simulation technologies in the processes of developing business solutions for the production and sale of specific product items in the competitive market of pharmaceutical products.

3 The Proposed Approach

We will consider the possibilities of using simulator models in decision-making processes at the micro level using the example of the competitive market of pharmaceutical products.

For simplicity, let's examine the two product markets of pharmaceutical products, which are saturated with two supply chains of one pharmaceutical enterprise ("competing drugs") or supply chains of two competing pharmaceutical enterprises. Conventional nomenclature positions – drug A and drug B.

Orders from the pharmacy network form the composition of orders for the production of products by the pharmaceutical enterprise (enterprises). Three types of orders are studied:

- on drug A;
- on drug B;
- to drug A or drug B.

The forecast output is limited only by the production capacity of the enterprise (enterprises). Finished products are delivered to pharmacy warehouses within the set time. The duration of delivery processes is considered.

Orders for the production of drugs are formed as a result of market demand research in pharmacy chains. Demand simulation is possible under various scenarios (Fig. 1): 1) according to the specific intensity; 2) randomly (according to the specified law of random variable distribution); 3) as a result of communication between customers of pharmacy chains (as a result of receiving verbal information); 4) as a result of promotions; 5) considering the seasonality of pharmaceutical drugs; 6) considering the intensity of specific diseases in specific regions of the study, associated, for example, with negative environmental consequences, etc.

The variable parameters of the model are:

- the production capacity of the pharmaceutical enterprise (enterprises) related to the production of the researched types of drugs;
- duration of supply of various drugs (one or different manufacturers) to pharmacy warehouses (to the pharmacy chain);
- parameters of market demand formation for specific drugs;
- coefficients for adjusting the volume of products launched into production (the forecast coefficients are intended for adjusting the received volume of orders from the pharmacy network). Coefficients can be formed based on the possibility of the occurrence of various production situations of rejection of orders of some pharmacy chains due to their failure to observe financial discipline; as a result of the expected market entry of an analogue drug shortly and the predicted decrease in market demand for the existing drug; due to the need to create additional insurance stocks of drugs (within their expiration dates) and due to any other situations dictated by specific circumstances; a combination of agent and system-dynamic approaches was chosen for implementation.

System dynamics approach used for modeling production and sales. The agent paradigm was used for the simulation of market demand for pharmaceutical products of the enterprise (enterprises). Agent simulation was chosen for the possibility of modeling the emerging behavior of buyer agents in the formation of the need and sale in the pharmacy network.

The model is implemented on the AnyLogic Simulation Software [30].

The main elements of model are (Fig. 1):

- Funds: FactoryStockA, FactoryStockB (production results - and stocks in the warehouses of drug manufacturers A and B); RetailerStockA, RetailerStockB (stocks of drugs A and B in the pharmacy network).

- Parameters: ForecastA, ForecastB (forecast coefficients for adjusting the demand for drugs A and B). It is implied that the available production capacity of the respective enterprise cannot always satisfy the market demand for specific products. It is

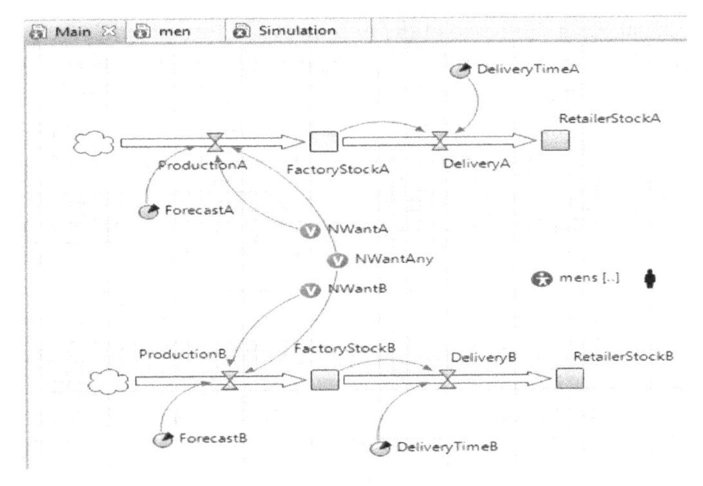

Fig. 1. Competitive market model screenshot

implied that the available production capacity of the respective enterprise cannot always satisfy the market demand for specific products. The variation of the values of the coefficients helps to reflect the real production situation. The coefficients are used in the process of simulating production volumes of drugs A and B, respectively.

– Dynamic variables, the values are formed during simulation experiments: NWantA (market demand for drug A); NWantB (market demand for drug B); NWantAny (customer orders for any in-stock device).

The state diagram (state chart) for determining agents' behavior is shown on Fig. 2.

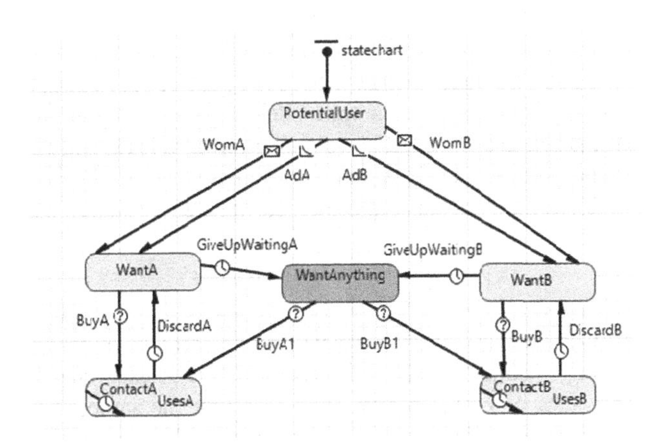

Fig. 2. Behavior diagram (state chart) of consumer agents

The diagram shows that the consumer agent can be in several states.
The first state – of PotentialUser (potential consumers).

Next, the agent can go into one of the two states – of wanting to purchase drug A (WantA) or B (WantB). The transition is carried out due to the influence of advertising of the corresponding drug – AdA and AdB.

In addition to the influence of advertising, the transition to the WantA or WantB states can also be carried out due to consumer communication, that is when receiving relevant messages.

From the states WantA or WantB, agents move to the states of actual consumers of – UsesA or UsesB products. This is done due to the BuyA or BuyB transitions (the warehouses of the pharmacy chain must have the corresponding products).

Purchases of drugs A or B can be canceled by – DiscardA, DiscardB transitions (they are carried out randomly according to the given law of distribution). Contacts between consumers are due to the transitions ContactA and ContactB. If the consumer cannot buy drug A or B because it is not in stock, he goes to the state of wanting to buy any drug, – the WantAnything state. Transitions to the WantAnything state are made differently for potential consumers of drug A (transition GiveUpWaitingA) and potential consumers of drug B (transition GiveUpWaitingB). If the consumer agent is in the state of wanting any drug (WantAnything), it can go to both the state of consumers of drug A (UsesA) and the state of consumers of drug B (UsesB), depending on which drug is in stock.

In the model, the space of movement of consumer agents is also determined, and with AnyLogic graphic elements, it is possible to reproduce the interaction of agents during the implementation of simulation experiments.

4 Results

The technology of conducting simulation experiments on the model will be illustrated using the example of a number of situations that arise during the production and sales activities of a pharmaceutical company. Materials used by FARMAK company. The experiments are aimed at the target audience of the pharmaceutical market with a capacity of 1,000 people. It is assumed that one agent buys one package of the drug. On all-time charts with accumulation, here is X – the time intervals of the simulation.

according to the selected term; here Y are quantitative indicators in natural units of measurement (number of drug packages, number of buyers, etc.).

The simulation experiment results are presented for specific situations.

4.1 Situation 1

The company analyzes the feasibility of releasing two alternative drugs of the same pharmacological group – conditionally, drugs A and B. It is considered that the duration of the production cycles of both drugs is approximately the same; delivery time to pharmacy warehouses is as well (Fig. 3).

At the same time, the setting parameters can be the intensity of advertising, the intensity of the transition of potential customers to the category of customers (users who are interested in buying the drug), and the intensity of drug purchases in the pharmacy network. Drug A has been on the market for some time, drug B has just been released. The

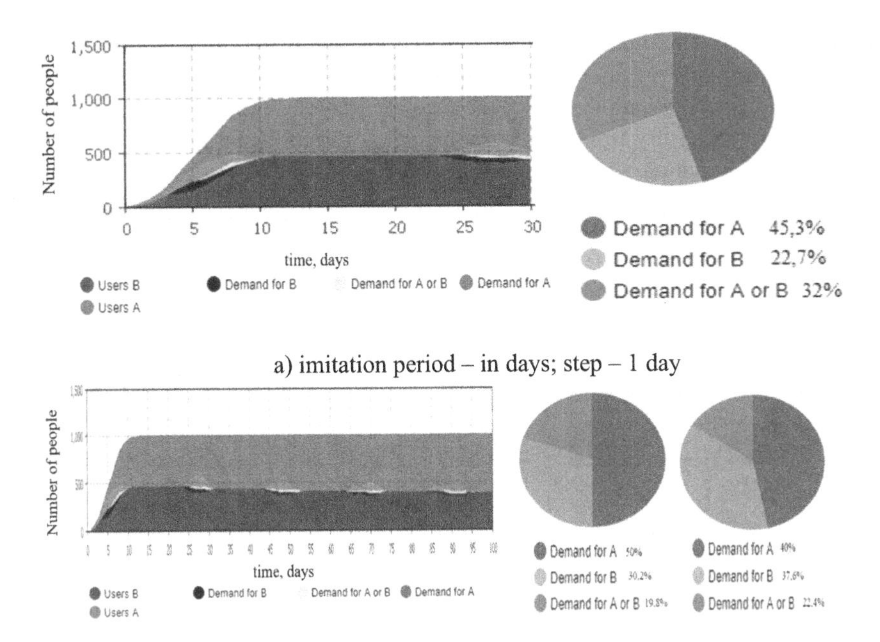

a) imitation period – in days; step – 1 day

b) simulation period of – 100 days, step – day c) increase in advertising intensity

Fig. 3. Experiment 1: launch of production alternative products of one pharmacological group

situation of two food markets in one research region is considered. The initial population of agents – is 1000 (=Y).

The results of the experiment with the simulation period of 30 days, and a step of – 1 day, are presented in Fig. 3a). The greatest demand for drug A is 45.3% of the total demand; demand for drug B – 22.7%; demand for drug A or B – 32%.

When the simulation period increases to 100 days, the given trend persists – Fig. 3b).

The situation is improving significantly with the increase in the intensity of advertising and the extension of the drug's shelf life on the pharmaceutical market – Fig. 3 c): the demand for drug A is 40% of the total demand; the demand for a new analog drug is increasing to 37.6%.

Further experiments showed that the situation improves with an increase of the simulation time, which confirms the expediency of releasing drug B.

4.2 Situation 2

The state of the two analog drugs is studied in the conditions of different regions.

The conducted experiments showed that the feasibility of launching and effective coexistence in one regional market of analog drugs depends on the group of drugs and the spread of diseases in a region.

The intensity of the pharmaceutical market in different regions of Ukraine was evaluated in FARMAK company by stimulating the demand for drugs from different pharmacological groups. Means for the treatment of infectious diseases and blood diseases were subject to research. The situation in Kyiv, Odesa, Lviv, and Ivano-Frankivsk regions was analyzed. The variable parameters of the model were the intensity of customer requests (the intensity of generation of agents), the receipt orders from the pharmacy network; the term of drugs delivery from production warehouses to the pharmacy network. The general term of imitation was also variable. The simulation step – is one day. In the Odesa region, drugs for the infectious diseases are constantly in high demand. This trend is observed throughout the simulation period – Fig. 4

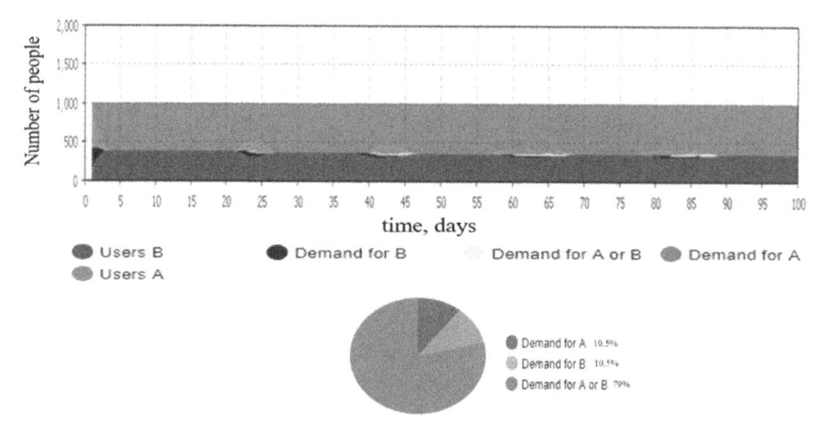

Fig. 4. Experiment 2: research of the Odesa pharmaceutical market of two analogue drugs for the treatment of infectious diseases.

According to these results (Fig. 4) drug A (which has been on the market for some time) has the largest percentage of demand in the situation when users do not care which drug to buy (A or B).

At the same time, the demand for researched analog drugs in the Lviv region is not permanent (temporary) and is specific, and targeted – either drug A or drug B. There is almost no demand of the type "drug A or drug B".

Regarding the market of pharmaceuticals for infectious diseases in the Kyiv region, the results of the research of this market are presented in Fig. 5. The demand for these drugs is much smaller than in the Odesa region. Buyers are mostly focused on buying drug A (64.5%), the sector of drug B is much smaller (32.2%). The sector of "indifference" to a particular drug is rather insignificant (3.3%). According to these results (Fig. 4) drug A (which has been on the market for some time) has the largest percentage of demand in the situation when users do not care which drug to buy (A or B).

Regarding the market of pharmaceuticals for infectious diseases in the Kyiv region, the results of the research of this market are presented in Fig. 5.

At the same time, the demand for researched analog drugs in the Lviv region is not permanent (temporary) and is specific, and targeted – either drug A or drug B. There is almost no demand of the type "drug A or drug B".

The demand for these drugs is much smaller than in the Odesa region. Buyers are mostly focused on buying drug A (64.5%), the sector of drug B is much smaller (32.2%). The sector of "indifference" to a particular drug is rather insignificant (3.3%).

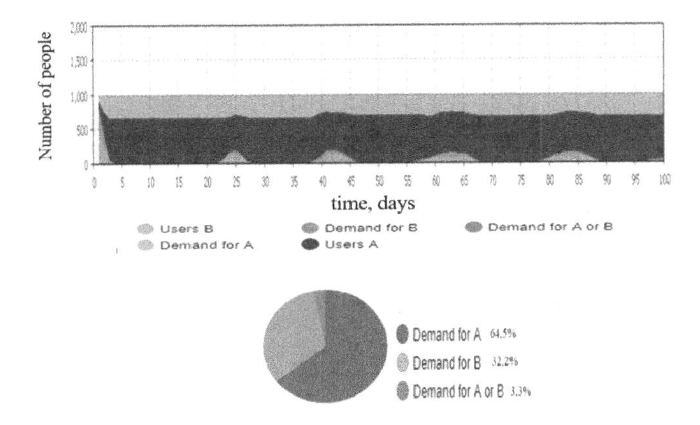

Fig. 5. Experiment 2: research of the pharmaceutical market of the Kyiv region for analogue drugs for blood diseases

As for drugs for the treatment of blood diseases, in this case, the opposite pattern is observed by region.

The highest level of blood disease is in the Ivano-Frankivsk region. Although the sales results indicate a higher sales volume of drug A, sales of the new analog drug B are stable. The demand sectors for drugs A and B are approximately the same.

The sector of "indifference" of those wishing to buy drug A or drug B is quite large. The results are presented in – Fig. 5. The results showed that the demand for these drugs is much smaller than in the Odesa region. Buyers are mostly focused on buying drug A (64.5%), and the sector of drug B is much smaller (32.2%). The sector of "indifference" to a particular drug is rather insignificant (3.3%) – Fig. 5.

The results showed that the demand for these drugs is much smaller than in the Odesa region. Buyers are mostly focused on buying drug A (64.5%), and the sector of drug B is much smaller (32.2%). The sector of "indifference" to a particular drug is rather insignificant (3.3%) – Fig. 5.

As for drugs for the treatment of blood diseases, in this case, the opposite pattern is observed by region.

The highest level of blood disease is observed in the Ivano-Frankivsk region. Although the sales results indicate a higher sales volume of drug A, sales of the new analog drug B are stable. The demand sectors for drugs A and B are approximately the same. The sector of "indifference" of those wishing to buy drug A or drug B is quite large. The research results are presented in – Fig. 6.

As for the Odesa and Kyiv regions, the obtained results are approximately the same compared to the Ivano-Frankivsk region, the specific weight of diseases and the demand for the corresponding drugs is much lower – Fig. 7.

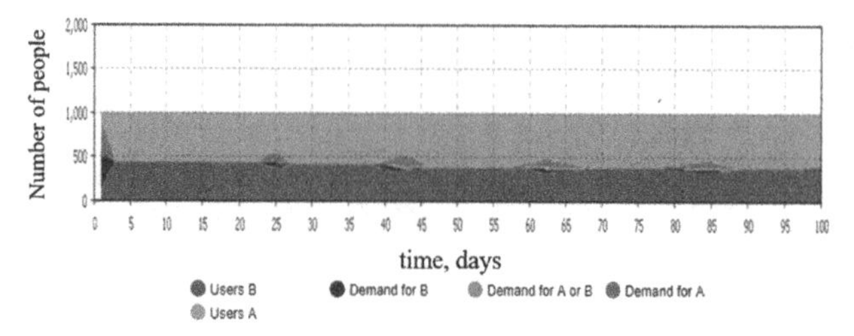

Fig. 6. Experiment 2: research of the pharmaceutical market of the Ivano-Frankivsk region for analogue drugs for the treatment of blood diseases

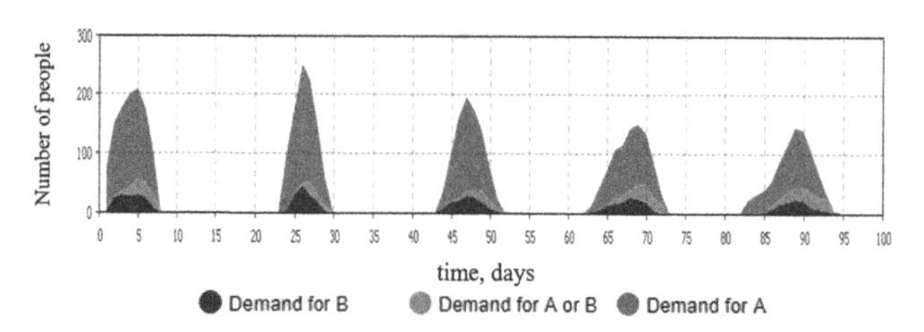

Fig. 7. Experiment 2: research the pharmaceutical market of the Kyiv region for analogue drugs for the treatment of blood diseases.

4.3 Situation 3

The effect of reducing the time of delivery of pharmaceuticals to the pharmacy chain on their sales volume was shown in this research (Fig. 8).

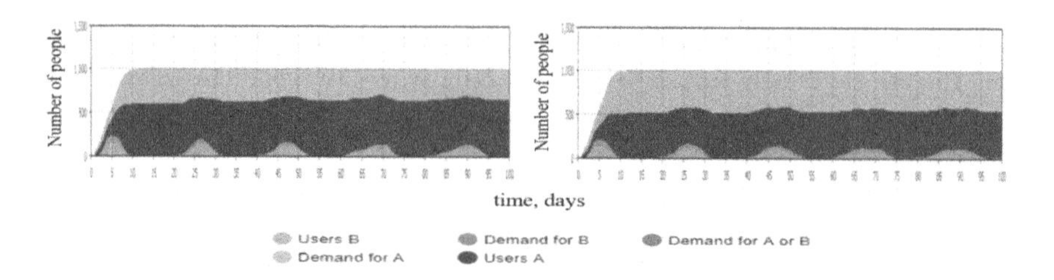

(delivery time to the pharmacy network both drugs) (reduced delivery time to the same for pharmacy drug network B)

Fig. 8. Experiment 3: study of the influence of the delivery time to the pharmacy network on the volume of drug sales

The previous experiments showed that the release of analog drugs into the market is accompanied by a period of their "adaptation" and subsequently, under favorable conditions (the impact of advertising), a gradual increase in their sales. However, the increase in the specific weight of new drug sales is also influenced by their prompt and rhythmic supply to the pharmacy network, which was proven by the results of simulation experiments (the intensity of advertising is the same for both drugs).

4.4 Situation 4

The influence of the level of drug stocks in the pharmacy network on sales is shown on the Fig. 9.

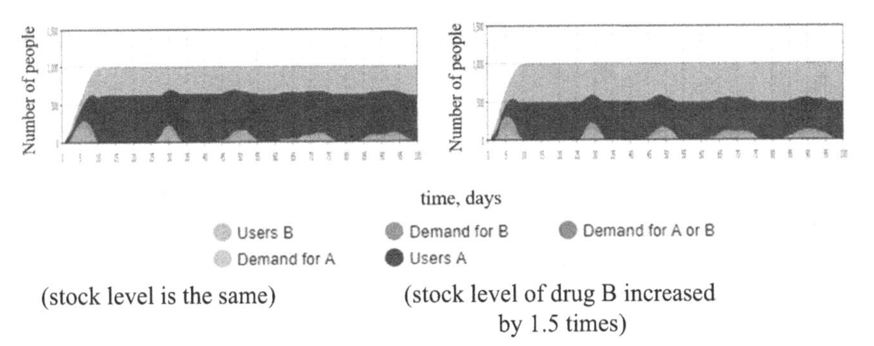

Fig. 9. Experiment 4: study of the influence of the drug stock level in the pharmacy network on sales volumes

The experiments were carried out considering the following assumptions:

- The initial stock levels of both drugs in the pharmacy network are the same; the intensity of advertising and the delivery time from warehouses of finished products to the pharmacy network is the same.
- The level of stocks of drug B in the pharmacy network increases significantly.

The simulation experiments show that the increasing level of stocks plays a positive role. The competitiveness (in terms of sales volumes) of the drug is increasing.

If the company simultaneously increases the intensity of advertising, we can expect a significant reduction in the period of "adaptation" of a new analog drug on the market.

4.5 Situation 5

The actions of pharmaceutical enterprise competitors were studied.

The company has been producing pharmaceuticals for a long time, has well-established advertising, and has determined the necessary level of stocks in the pharmacy network. However, a competing company launches a drug on the market that may become more attractive to buyers. According to this, the following situations are possible:

- The quality of the drug is higher than the existing one.
- The drug is a complete analog but has a lower price than competitors.

In the model, similar situations can be simulated using a variation of.

the ForecastA and ForecastB parameters. The market situation is presented in Fig. 10.

Figure 10 shows that the competitor's drug has a much higher specific weight in the total volume of sales compared to the drug of the investigated enterprise (drug A).

If the company continues to manufacture the drug in the same volume, without consideration the actions of competitors, this leads to overstocking of its warehouses,

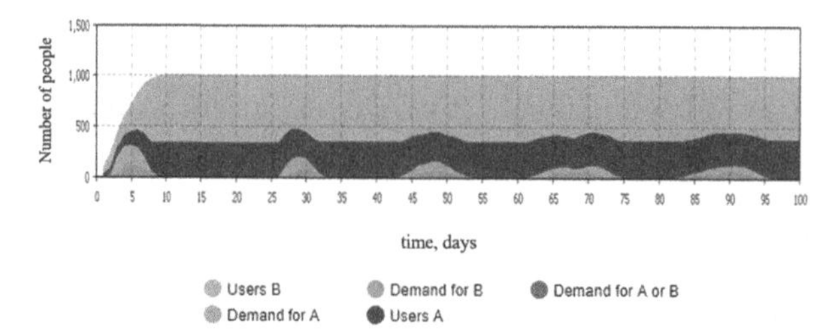

Fig. 10. Experiment 5: research of competition two product markets of pharmaceuticals

as well as, considering the specifics of the products (expiry dates of the drugs), to the possibility of significant financial losses. Considering the actions of competitors using the parameters ForecastA and ForecastB, we get a different picture.

Thus, considering the actions of competitors, the company can adjust production plans on time. In this example, – it is advisable to reduce the production of excess products (preparation A), at the same time, redistribute the released capacity to the production of more in-demand products.

5 Conclusion

In ever changing environment, caused by the war and the uncertainty of business, the survival of enterprises depends on the promptness and efficiency of business decision-making. This is especially true for life-sustaining industries, including the pharmaceutical industry. The specifics of the production and sale of pharmaceutical products create explicit tasks that must be considered in the management of companies.

The modern pharmaceutical market of Ukraine is characterized by a large number of generic drugs, in particular, a certain number of analog pharmaceutical drugs with the same effect on health. As a result, pharmaceutical companies need to understand the pattern of consumer behavior in the process of deciding to purchase specific over-the-counter drugs, as well as their potential to influence this process. Launching into production or supporting the release of specific drugs depends on the market's estimation, production capacities, and investments in advertising companies.

The use of simulation models helps in making appropriate preventive business decisions. The presented simulation experiments are conducted using the data of FARMAK, one of the leading Ukrainian pharmaceutical companies, demonstrate some capabilities of the simulator model. The model is quite typical due to its openness and modularity. AnyLogic software opens up opportunities for further expansion of research:

- Improvement of the behavior of consumers of the pharmaceutical market using the new range of factors.
- Development of the factors influencing the drugs life cycle with the forecast of their life on the market.
- Development of a model module that will reinforce the research on the pricing policy of pharmaceutical companies.
- Improvement of scenarios of situational experiments.

– Expanding the range of the simulation experiments implementation such as optimization, sensitivity assessment, runs comparison, and creating the individual types of experiments.

It has been concluded that the most important parameters of the model are the intensity of advertising of a specific drug, the timely delivery of drugs to the pharmacy network, the maintenance of pharmacy stocks, as well as the consideration and operational response to the actions of competitors. The dependence of business decisions on the delivery of the varied range categories of drugs to different regional market segments is also demonstrated.

References

1. Official website of the State Statistics Service of Ukraine. https://www.ukrstat.gov.ua/
2. Pharmacy sales according to the results of 2022, Weekly Pharmacy, No. 04 (1375), 30 January, pp. 12–13 (2022). https://www.apteka.ua/article/656982
3. Kozioł-Nadolna, K., Beyer, K.: Determinants of the decision-making process in organizations. Procedia Comput. Sci. **192**(2021), 2375–2384 (2021)
4. Santos, J.A., Sousa, J.M., Vieira, S.M., Ferreira, A.F.: Many-objective optimization of a three-echelon supply chain: a case study in the pharmaceutical industry. Comput. Ind. Eng. **173**, 108729 (2022). https://doi.org/10.1016/j.cie.2022.108729
5. Oliveira, F.R., Neto, F.B.L.: Lima neto: flexible dialogues in decision support systems. In: Devlin, G., (eds.) Decision Support Systems Advances in IntechOpen, pp. 278–290 (2010).https://www.intechopen.com/chapters/10957
6. Albers, S.: Optimizable and implementable aggregate response modeling for marketing decision support. Int. J. Res. Market. **29**(2), 111–122 (2012)
7. Dutta, G., Fourer, R., Majumdar, A., Dutta, D.: An optimization-based decision support system for strategic planning in a process industry: the case of a pharmaceutical company in India. Int. J. Prod. Econ. **106**(1), 92–103 (2007)
8. Athreye, S., Kale, D., Ramani, S.V.: Experimentation with strategy and the evolution of dynamic capability in the Indian pharmaceutical sector. Ind. Corporate Change **18**(4), 729–759 (2009)
9. Joshi, A., Benitez, J., Huygh, T., Ruiz, L., De Haes, S.: Impact of IT governance process capability on business performance: Theory and empirical evidence. Decis. Support. Syst. **153**, 113668 (2022)
10. Imran, M., Tanveer, A.: Decision support systems: creating value for marketing decisions in the pharmaceutical industry. Eur. J. Bus. Innov. Res. **3**(4), 46–65 (2015)
11. Saini, G.S., Sharma, S.: Importance of marketing information system in pharmaceutical decision making. Int. J. Manage. Technol. Eng. **IX**(II), 2266–2275 (2019)
12. Hodgett, R.E., Siraj, S.: SURE: a method for decision-making under uncertainty. Expert Syst. Appl. **115**, 684–694 (2019)
13. Proceedings of Winter Simulation Conference (The premier international forum for disseminating recent advances in the field of system simulation).http://meetings2.informs.org/wordpress/wsc2022/ (2021)
14. Proceedings of EUROSIM (Federation of European Simulation Societies (EUROSIM) 1992–2022).https://www.eurosim.info/publications
15. Proceedings of ASIM (Arbeitsgemeinschaft Simulation). https://www.asim-gi.org/asim
16. Onggo, B.S., Foramitti, J.: Agent-based modeling and simulation for management decisions: a review and tutorial. In: Proceedings of the 2021 Winter Simulation Conference (2021).https://www.informs-sim.org/wsc21papers/115.pdf

17. Negahban, A., Yilmaz, L.: Agent-based simulation applications in marketing research: an integrated review. J. Simul. **8**(2), 129–142 (2014)
18. Wall, F.: Agent-based modeling in managerial science: an illustrative survey and study. Rev. Manag. Sci. **10**, 135–193 (2016)
19. Eldabi, T.: Systemic: characteristics to support hybrid simulation modeling. In: Proceedings of the 2021 Winter Simulation Conference (2021). https://www.informs-sim.org/wsc21papers/109.pdf
20. Solo, K., Paich, M.: A modern simulation approach for pharmaceutical portfolio management. In: Proceedings of the International Conference on Health Sciences Simulation (ICHSS 2004), 18–21 January 2004, San Diego California (2004).https://www.anylogic.com/upload/iblock/6dc/6dcedcab3df82604d93a2dafd8e968f2.pdf
21. Grigoryev, I.: AnyLogic in Three Days: Modeling and Simulation (2018). https://www.anylogic.com/resources/books/free-simulation-book-and-modeling-tutorials/
22. Borshchev, A.: The Big Book of Simulation Modeling. https://www.anylogic.ru/resources/books/big-book-of-simulation-modeling/
23. Jetly, G., Rossetti, C.L., Handfield, R.: A Multi-Agent Simulation of the Pharmaceutical Supply Chain, vol. 6 no. 4, pp. 215–226 (2012)
24. Franco, C.: A simulation model to evaluate pharmaceutical supply chain costs in hospitals: the case of a Colombian hospital. DARU J. Pharm. Sci. **28**, 1–12 (2020)
25. Moosivand, A., Ghatari, A.R., Rasekh, H.R.: Supply chain challenges in pharmaceutical manufacturing companies: using qualitative system dynamics methodology. Iran. J. Pharm. Res. **18**, 1103–1116 (2019)
26. Viana, J., Van Oorschot, K., Årdal, C.: Assessing resilience of medicine supply chain networks to disruptions: a proposed hybrid simulation modeling framework. In: Proceedings of the 2021 Winter Simulation Conference (2021)https://www.anylogic.com/upload/iblock/818/6dardcykf1ns5r2cskrmv8lqz5h2tqza.pdf
27. Mousavi, A., Mohammadzadeh, M., Zare, H.: Developing a system dynamic model for product life cycle management of generic pharmaceutical products: its relation with open innovation. J. Open Innov.: Technol. Mark. Complex. **8**(1), 14 (2022)
28. Zare Mehrjerdi, Y., Khani, R., Hajimoradi, A.: System dynamics approach to model the interaction between production and pricing factors in pharmaceutical industry. J. Qual. Eng. Prod. Optim. **5**(2), 163–188 (2020)
29. Akhlaghinia, N., Ghatari, A.R., Moghbel, A., Yazdian, A.: Developing a system dynamic model for pharmacy industry. Ind. Eng. Manag. Syst. **2018**(17), 662–668 (2018)
30. Website The AnyLogic company. http://www.anylogic.com

Decision and Information Support System for a Framework to Building Multicriteria Decision Models

Adiel Teixeira de Almeida and Lucia Reis Peixoto Roselli(✉)

Center for Decision Systems and Information Development (CDSID), Universidade Federal de Pernambuco, Recife, PE, Brazil
{almeida,lrpr}@cdsid.org.br

Abstract. Several Decision Support Systems (DSS) have been presented in literature. This paper presents an Information and Decision System that has been performed to implement a Framework for building decision models in the context of Multi-Criteria Decision Making/Aiding (MCDM/A) approach. This system incorporates the steps of the framework in order to support DMs to build decision models. This system does not intend to suggest a specific MCDM/A method but can be used to structure the model with rigor following the framework steps. In addition, this system can be used to document important steps of the process. The DSS also provides flexibility for users since they can return to previous steps to review or adjust some points. Also, the DSS is interactive since the users present preferential information in all steps. An important step presented in the system regards the preference modeling process. During this step, Decision-Makers (DMs) should contextualize the rationality presented in the MCDM/A problem. Moreover, several tests of consistency are done in order to improve the system and to understand how DMs define the rationality of those problems and if they stay consistent with their rationality during the process. In addition, the DSS included several aspects of Problem Structuring Methods, considering important aspects that are not considered in the previous version of the framework. In special, concepts of the Value Focus Thinking method have been included in the system to support the definition of objectives, criteria, and alternatives. This feature improves the new version of the framework. This DSS presents the most important steps of the framework and also provides insights for the development of behavioral studies, which can bring advances in the MCDM/A approach.

Keywords: FITradeoff method · Decision and Information Support System · Framework · Multi-Criteria Decision Making/Aiding (MCDM/A)

1 Introduction

Several Information Systems (IS) and Decision Support Systems (DSS) have been developed in different contexts to support organizations to deal with decision problems faced in their routines [1–5].

I. Saad et al. (Eds.): ICIKS 2023, LNBIP 486, pp. 19–32, 2024.
https://doi.org/10.1007/978-3-031-51664-1_2

In the context of the Multi-Criteria Decision Making/Aiding (MCDM/A) approach, which involves problems that pretend to attend to more than one objective and have several alternatives (actions) as solutions, several DSSs have been constructed to support Decision-Makers (DMs) to solve them [6–9].

Moreover, in the context of the MCDM/A approach, frameworks have been proposed to support DMs to build decision models. These frameworks present some steps which should be followed by DMs to support the structuring of the problem and modeling preferences expressed by them. These frameworks in general present the same objective but differ in the number and configuration of the steps [7–11]. For instance, De Almeida et al. (2015) [9] proposed a framework that had twelve steps to build an MCDM/A decision model. This framework aims to structure the decision process, permitting DMs to follow sequential steps to build a decision model that incorporates their preferences and suggests a solution that really represents their preferences.

According to Box and Draper [12], models are always wrong since they do not represent reality in a complete way. However, models are useful to support decision processes. Hence, in this view, frameworks are not a perfect feature to construct a decision model, but they are very useful to support DMs to conduct a rational decision process.

A new version of De Almeida et al., (2015) framework has been developed [13]. In special, this new version includes important aspects concerning Problem Structuring Methods (PSM) [14]. PSM is considered the soft part of Operational Research. The aim of these methods has been to support DMs in understanding their problems before trying to solve them [14–17].

In special, concepts of the Value Focus Thinking (VFT) method [15] have been considered in the new framework. These aspects have been included in the preliminary phase, which aims to structure the problem. Steps two, three, and four are those which included concepts of the VFT method.

This framework has been implemented in a Decision and Information System (DIS). In this context, this study aims to discuss this DIS, presenting its steps and showing its usability. The aim of this study has been to discuss how this DIS can be used to build a multicriteria decision model based on the new framework of de Almeida et al. [13]. Therefore, the contribution of this study remains in discussing a system to build multicriteria decision models with rigor based on the framework proposed by [13].

This study shows the usability of this system considering a decision problem in the context of remote work, which is already presented in the literature [18]. This decision problem has been modeled using this system. In addition, this system can be used for documentation of important steps of the process, and support DMs to conduct a rational process.

This study is divided as follows. Section 2 briefly presents the steps of the framework proposed by [13] which is based on the previous one but incorporates new features [15]. Section 3 shows the usability of the Decision and Information System (DIS) to model a multicriteria problem concerning to evaluate initiatives to optimize remote work. This section is the focus of the study, aiming to show its usability. Finally, Sect. 4 presents the conclusion of this study.

2 Framework to Construct Decision Models

A framework for building multicriteria decision models has been presented by de Almeida et al., (2015) [9]. This framework presented twelve steps, which have been separated into three phases: Preliminary phase, Preferential modeling phase, and Finalization.

A new version of de Almeida et al., (2015) framework has been developed [13]. This new framework presents features that have been included to improve the first one. For instance, this framework includes devices of the VFT method in order to support DMs to structure the problem in the Preliminary phase. The VFT. This framework is described below.

- Phase 1 - Preliminary phase
    - Step 1: Contextualization
        - A: Describe the Problem
        - B: Define DM and other actors
- Step 2: Identify objectives using VFT
- Step 3: Define criteria
- Step 4: Establish alternatives
    - A - DM
    - B – Nature
- Step 5: Construct the Consequence Matrix and Define the Consequence Function
- Phase 2 - Preference Modeling and choice of MCDM/A method phase
    - Step 6: Preference modeling
    - Step 7 Conducting an Intra-Criteria evaluation
    - Step 8 Conducting an Inter-Criteria evaluation
- Phase 3 – Finalization phase
    - Step 9: Evaluate alternatives
    - Step 10: Conduct a sensitivity analysis
    - Step 11: Draw up recommendations
    - Step 12: Implement the solution

MCDM/A methods, such as the FITradeoff method [19, 20], are used to evaluate alternatives following a mathematical model with incorporates preferential information of DMs. The MCDM/A methods are applied in a specific step of the framework (Step 9).

The difference between multicriteria model and method regards originality. Decision models are unique, constructed for each MCDM/A problem, following DMs preferences. On the other hand, MCDM/A methods are general and can be applied to solve several problems presented in the literature [9]. For instance, the FITradeoff method [19, 20] has been applied to solve several problems in literature [21–28].

The next section describes the problem used to illustrate the Decision and Information System, which is already presented in the literature [18]. Until now, the DSS presents the six steps of the framework of twelve steps [13]. In this context, the most important phases to construct a decision model have been already presented in the DIS discussed in this study. Moreover, the implementation of the framework in a DSS represents an important advance of the studies in this field.

3 Decision Support System Developed to Implement the Framework

In this section, the Decision and Information System (DIS) has been described. This system has been constructed to implement the framework of twelve steps [13]. A decision problem has been modeled considering this DIS [18].

Considering the context of the COVID-19 pandemic, some companies, that had not already implemented home office jobs, had a short time to adapt to the new reality and started this process in an unstructured way. The issue was that, in many cases, there was not enough time to prepare the team of employees to provide a good service in a productive method. Besides, as many employees had their functions modified, since the original function required face-to-face assistance, they started working remotely without adequate training or equipment. Consequently, these companies faced a direct impact on employees' productivity and on organizational results.

Although the home office has spread with the pandemic, it was possible to notice that it can contribute to a quick response and reduce the companies' fixed costs. With these advantages, highlighted in recent times, the home office has gained space in the market and has become a trend.

In this context, a company wants to perform a study on what are the possible initiatives to optimize the productivity of employees in the home office regime [18]. This company intends to model this decision problem, defining all the conflicting objectives, identifying the criteria to measure the objectives, and defining the alternatives using problem structuring methods, in special the Value Focus Thinking [15].

Therefore, considering this decision problem, in the first step of the DSS, DMs should contextualize the problem. Hence, the DSS presents several "boxes" in which DMs should descript the problem. Moreover, DMs should present the actors involved in the problem.

In the MCDM/A field, the Decision-Maker (DM) or Decision-Makers (DMs) in a group decision problem is the most important actor of the problem. The DM is who presents responsibility for the solution obtained, and who deals with the future benefits or penalties received from the solution. Problems can be solved by only one DM or by more than one DM in Multi-Criteria Group Decision Making (MCGDM) problems. The company problem presents only one Decision-Maker, who is the owner of the company.

After that, the DM proceeds to Step 2 in order to identify his/her objectives. This step is the one that most use the VFT concepts [15]. Step 2 is also an important step in the framework since it will direct efforts to the next steps.

In this context, at this part of the framework, users can use the VFT devices to understand their values. Several devices have been described in the VFT [15]. More than one device can be used by DMs, and each one of them will generate insights about the objectives involved in the problem.

One of the most popular devices is the *Wish List*. Hence, using this device the DM list every element/aspect/or even objective that he/she wishes to obtain in solving the problem. Figure 1 illustrates the use of devices in the DSS.

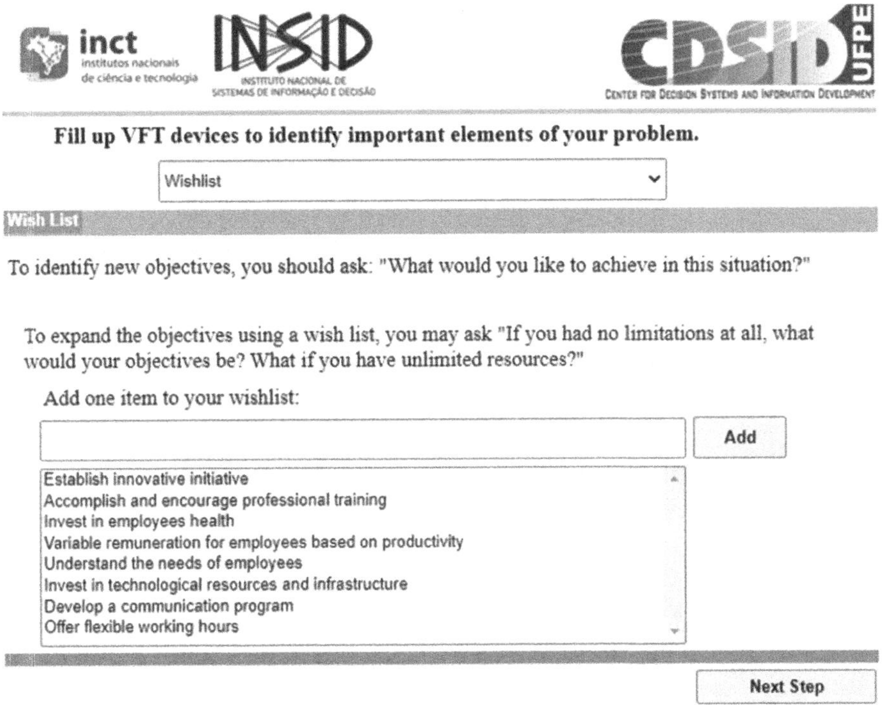

Fig. 1. Step 2 – Devices

For instance, for the DM, some desires for possible initiatives to optimize the productivity of employees are: Establish innovative initiatives; Accomplish and encourage professional training; Invest in employees' health; Variable remuneration for employees based on productivity; Understand the needs of employees; Invest in technological resources and infrastructure; Develop a communication program; and Offer flexible working hours.

After using the devices, users can transform their values into objectives format, review their values or objectives, remove those with are redundant, and obtain a final list of objectives.

When a final list is obtained, the next task has been to separate the objectives into categories. According to Keeney (1992) [15], three categories can be used to classify the objectives: Strategic Objectives, Fundamental Objectives, and Mean Objectives. Strategic objectives are the most general objectives, fundamental objectives are those that are the essential reason for solving the problem, and mean objectives are those that are means to achieve the fundamental ones.

This DIS focused on strategic and fundamental objectives. The fundamental objectives are those that users really aim to achieve when the problem is solved. Moreover, in several problems, the fundamental objectives can be decomposed into lower levels, presenting more details about the fundamental ones in upper levels [15]. Hence, the DSS focused on strategic objectives, fundamental objectives at a superior level, and fundamental objectives at an inferior level, as illustrated in Fig. 2.

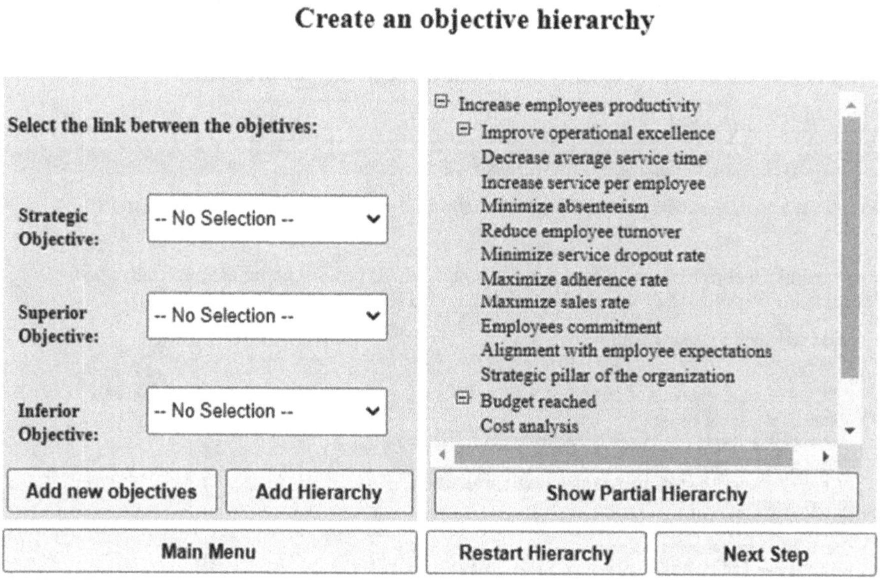

Create an objective hierarchy
Select the link between the objetives:
Strategic Objective: -- No Selection --
Superior Objective: -- No Selection --
Inferior Objective: -- No Selection --
Add new objectives
Add Hierarchy
Increase employees productivity
Improve operational excellence
Decrease average service time
Increase service per employee
Minimize absenteeism
Reduce employee turnover
Minimize service dropout rate
Maximize adherence rate
Maximize sales rate
Employees commitment
Alignment with employee expectations
Strategic pillar of the organization
Budget reached
Cost analysis
Show Partial Hierarchy
Main Menu
Restart Hierarchy
Next Step

Fig. 2. Step 2 – Classification of Objectives

Based on Fig. 2, the Strategic objective is "Increase employee productivity". Based on it, two fundamental objectives are defined: "Improve operational excellence" and "Budget reached". From the first fundamental objective, ten inferior objectives have been considered, namely: "Decreasing average service time", "Increase service per employee", "Minimize absenteeism", "Reduce employee turnover", "Minimize service dropout rate", "Maximize adherence rate", "Maximize sales rate", "Employees commitment"; "Alignment with employee expectation"; and "Strategic pillar of the organization".

After that, with the objectives defined and categorized, the next step of the decision process has been to define the criteria or attributes. Three categories are used to classify the criteria or attributes. Natural criterion is those that is a common variable used by everyone. Constructed criterion is those that should be constructed to measure a specific objective of the problem, using verbal scales or levels, in general. Proxy criterion should be used when a natural and a constructed criterion are not available. It is an indirect variable to measure the objective [15].

Using the DSS, DMs should define a criterion to represent each one of the fundamental objectives. After that, DMs should classify the criterion, inform that unit of measurement, the preference direction (maximization or minimization), and the nature of the consequence measured by the criterion (deterministic or probabilistic consequences). Figure 3 illustrates this step. Figure 4 illustrates all criteria considered in the problem.

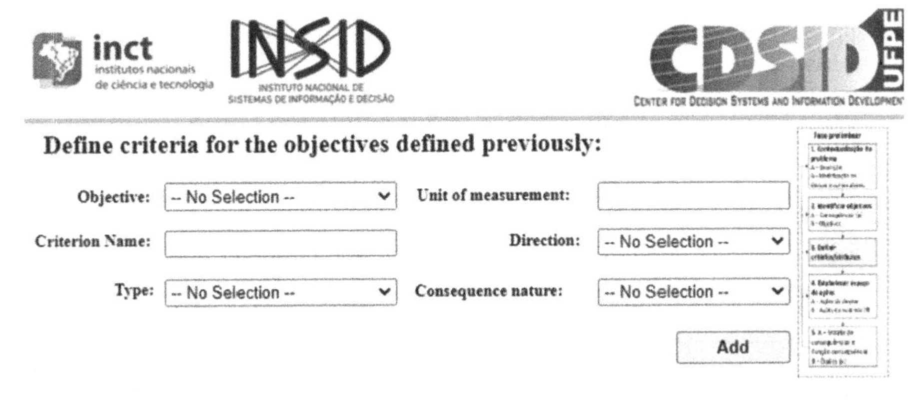

Fig. 3. Step 3 – Criteria Definition

Criterias

Objective	Criteria	Type	Levels	Measurement Unit	Direction	Conse
Alignment with employee expectations	Satisfaction research	Natural		Grade	Maximization	Deten
Increase service per employee	Service time	Natural		%	Maximization	Deten
Innovation rate	Innovation rate	Constructed	3	Levels	Maximization	Deten
Maximize adherence rate	Service dropout rate	Natural		%	Minimization	Deten
Maximize sales rate	Sales rate	Natural		%	Maximization	Deten
Cost analysis	Initiative cost	Natural		Reais	Minimization	Deten
Increase service per employee	Customer rate	Natural		Number	Maximization	Deten
Maximize adherence rate	Adherence rate	Natural		%	Maximization	Deten
Minimize absenteeism	Absenteeism	Constructed	3	Levels	Maximization	Deten

Main Menu | Export criteria sheet | Next Step

Fig. 4. Criteria Sheet

The fourth step concerns listing the alternatives (actions) which can solve the problem. For this problem, the alternatives are some initiatives to optimize the productivity of employees, such as Ideas programs; In-company courses; External courses; Health insurance; Variable salaries, New software.

After the alternatives and criteria have been defined, the consequence matrix should be constructed. The consequences represent the performance of the alternatives evaluated in the criteria. For this problem, all consequences are deterministic, as illustrated in Fig. 5.

Until now, the problem has been structured using concepts of PSM [14, 15]. It is worth mentioning that DMs can return to previous steps to adjust any topic defined before. For instance, DMs can return to include more objectives, criteria, or alternatives.

The next step is the preference modeling step, which starts the second phase of the framework. The preference modeling phase is one of the most important phases in the framework [9]. This step can be considered as the main step of the framework. Based on the preference modeling phase the most suitable MCDM/A method should be indicated to be applied. The method should consider the structure of preference expressed by DMs in this phase.

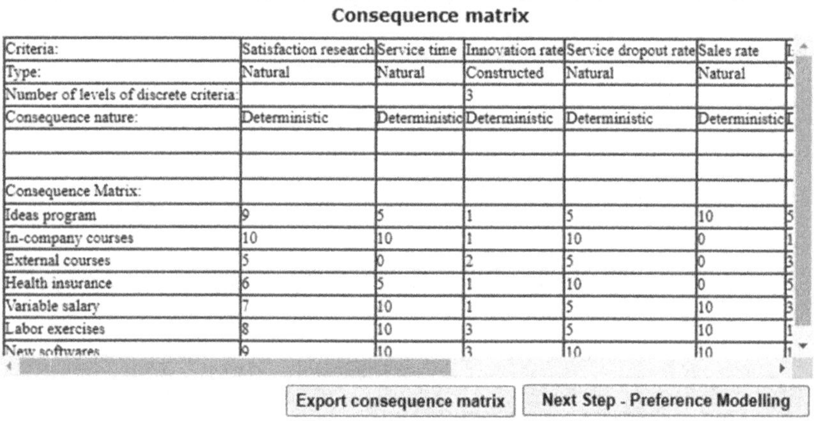

Consequence matrix

Criteria:	Satisfaction research	Service time	Innovation rate	Service dropout rate	Sales rate	I
Type:	Natural	Natural	Constructed	Natural	Natural	N
Number of levels of discrete criteria:			3			
Consequence nature:	Deterministic	Deterministic	Deterministic	Deterministic	Deterministic	I
Consequence Matrix:						
Ideas program	9	5	1	5	10	5
In-company courses	10	10	1	10	0	1
External courses	5	0	2	5	0	3
Health insurance	6	5	1	10	0	5
Variable salary	7	10	1	5	10	3
Labor exercises	8	10	3	5	10	1
New softwares	9	10	3	10	10	1

Export consequence matrix | Next Step - Preference Modelling

Fig. 5. Step 5 – Insert the decision matrix

One of the most important preferential information concerns the rationality involved in the problem [6–9]. DMs can present a compensatory or non-compensatory rationality for the consequences in the decision matrix. If the DM presents a compensatory rationality, for him/her it is acceptable to compensate a lower performance of an alternative in criterion A by a higher performance of the same alternative in another criterion B [5, 9].

For instance, consider a Supplier Selection problem [29], supposing that the problem presents the criteria Cost, Delivery Time, and Quality. If, for the DM, a supplier which had a lower performance in terms of cost (i.e., a high price), compensates its higher performance in terms of quality and delivery time (i.e., shorter delivery time and high quality), the DM presents a compensatory rationality.

On the other hand, if it is unacceptable for DM to make compensations between the consequences, he/she presents a non-compensatory rationality. A famous example of non-compensatory rationality is the U.S. election. [6–9].

The rationality separates MCDM/A methods into two groups: Methods that deal with compensatory rationality and methods that deal with non-compensatory rationality [6–9]. Moreover, Roy (1996) [6] classifies the methods into three categories. Methods of Unique Criterion of Synthesis, which includes methods that consider compensatory rationality and aggregates all the criteria in only one. In this category, the additive aggregation model is the most used technique; and the main representative of this model is the Multi-Attribute Value Theory (MAVT – [5]), and Multi-Attribute Utility Theory (MAUT – [5]). The second category consists basically of Outranking Methods, which considers non-compensatory rationality [6], as methods in the family of ELECTRE and PROMETHEE are the most representative methods. Finally, some methods are considered Interactive methods.

Conjointly with that evaluation, the DM should evaluate his/her preference relation structure. If only strict preferences and indifferences are adequate to represent DMs' preferences between the consequences, it reinforces the Methods of Unique Criterion of Synthesis should be applied. On the other hand, Outranking Methods allow DMs to have hesitations between consequences comparisons, considering different kinds of preference structure relations [6, 9].

Hence, in step number six, DMs will interact with the DSS in order to define the rationality involved in the problem. For this problem, the DM affirms having compensatory rationality, as illustrated in Fig. 6.

In addition, when users start Step 6 they should test the consistency of the rationality declared. Hence, three behavioral tasks should be performed. These tasks require that DMs do holistic evaluations to compare alternatives supported by bar graphs. These tasks have been created from behavioral studies performed previously by the authors. This is an important feature of the DSS to investigate and test the rationality declared by DMs. This study can bring important advances to the MCDM/A approach [20, 30–34].

The first task concerns selecting the best alternative following the rationality declared above. The two alternatives illustrated in the bar graph are alternatives listed by the DM in Step 4 (Fig. 6). The height of the bars represents the performance of that alternative in each criterion. The bar graph considers a ratio scale [9].

After that, the second task concerns testing if users follow the same rationality to select the best alternative in a hypothetical problem (Fig. 7). Finally, the third task involves explaining how the best alternative has been selected in both graphs (the graph with alternatives of the problem and the hypothetical one (Fig. 8).

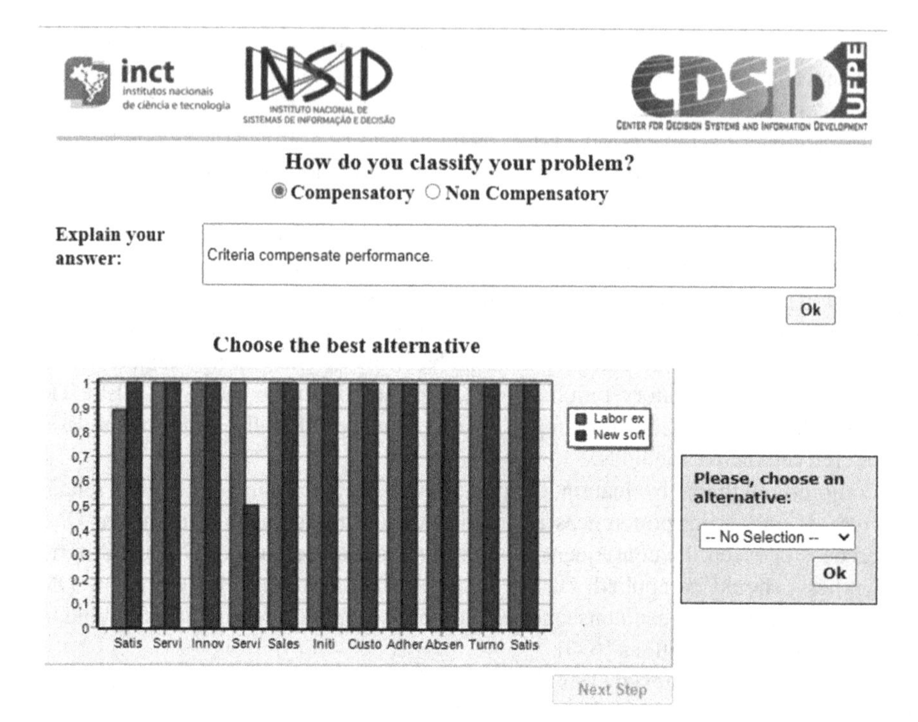

Fig. 6. Step 6 – First task

Choose in this case with the same rationale as you did in the previous:

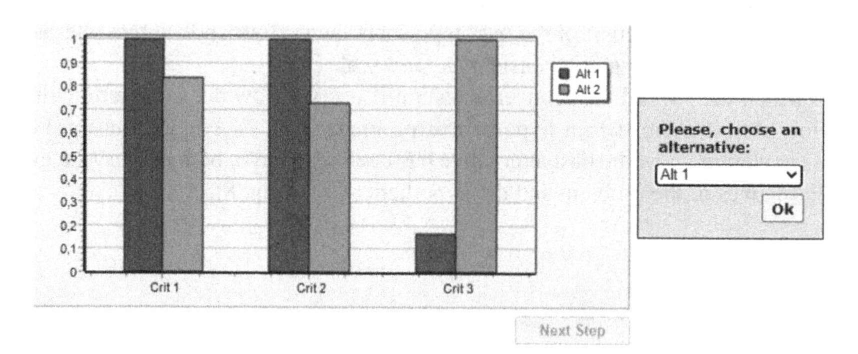

Fig. 7. Step 6 – Second task

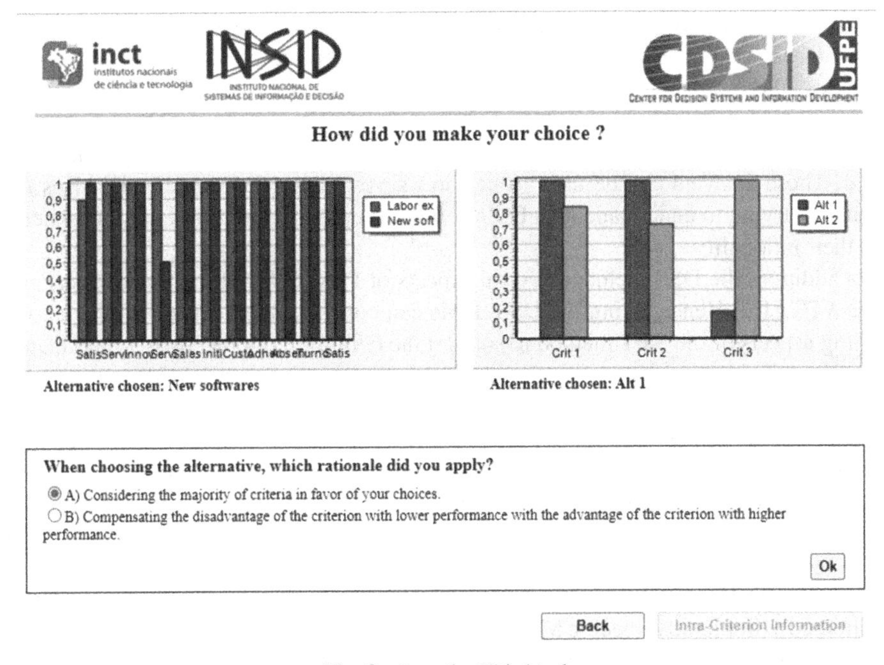

Fig. 8. Step 6 – Third task

At the end of this step, several aspects of DMs preference structure have been considered. The DIS. The Decision and Information System has been developed until this step to implement the framework. In general, steps 7 to 10 have been performed by DSS which implement specific MCDM/A methods, such as the DSS constructed to implement the FITradeoff method [19, 20].

This system aimed to support DMs (users) to solve MCDM/A problems in a structured way. This system does not intend to suggest an MCDM/A method to be used, but to structure the problem and model it, indicating at the final of this phase which MCDM/A method is more appropriate to be used to solve the problem.

4 Conclusion

This study shows the usability of the Decision and Information System constructed to implement the framework for building decision models. The system intends to support building a decision model considering the preferences expressed by DMs according to the steps of the framework. In addition, this system can be used for documentation of important steps of the process, and support DMs to conduct a rational process.

The DSS pretends to provide flexibility for users since they can return to previous steps to review or adjust some points. Also, the DSS is interactive since the users present preferential information in all steps.

Moreover, the system collects preferential information expressed by DMs and investigated it using graphical evaluations. Three tasks are elaborated in order to test how consistent are DMs in declaring their rationality during the preference modeling process, which is one of the most important steps of the framework.

Therefore, an important result obtained with the DSS is the answers provided in step 6. These answers can be used to perform several behavioral studies in terms of rationality, trying to understand how DMs define a rationally, and if they stay consistent with their rationality.

In addition, the DSS included several aspects of PSM [14–17], in special concepts of the VFT [15]. Hence, using the DSS, DMs can construct the decision model incorporating aspects of the VFT method to support the definition of objectives, criteria, and alternatives. This feature improves the new version of the framework [13].

To illustrate the usability of the DIS a decision problem involving a company that wants to perform a study on what are the possible initiatives to optimize the productivity of employees in the home office regime. This company aimed to model the decision problem, defining objectives, criteria, and alternatives. After that, the preference modeling has been done.

At the moment, the DSS presents limitations since does not have all the steps of the framework. However, in general, steps 7 to 10 have been performed using DSS constructed to implement specific MCDM/A methods. In this context, the most important phases to construct a decision model have been already presented in the DIS discussed in this study. Moreover, the implementation of the framework in a DSS represents an important advance of the studies in this field [30].

As future research, the system intends to be continued in order to present other features presented in the process of building decision models. Based on the answers provided by DMs in Step 6, several behavioral studies can be performed. This field can bring several advances to MCDM/A approach since systems can be transformed by the inclusion of behavioral insights [30, 31].

Acknowledgment. This work had partial support from the Brazilian Research Council (CNPq) [grant 308531/2015–9;312695/2020–9] and the Foundation of Support in Science and Technology of the State of Pernambuco (FACEPE) [APQ-0484–3.08/17].

References

1. Fogli, D., Guida, G.: Knowledge-centered design of decision support systems for emergency management. Decis. Support. Syst. **55**(1), 336–347 (2013)
2. Jones, K.: Knowledge management as a foundation for decision support systems. J. Comput. Inf. Syst. **46**(4), 116–124 (2006)
3. Sui, L.: Decision support systems based on knowledge management. In: Proceedings of ICSSSM 2005. 2005 International Conference on Services Systems and Services Management, 2005, vol. 2, pp. 1153–1156. IEEE (2005)
4. Cortés, U., et al.: Knowledge management in environmental decision support systems. AI Commun. **14**(1), 3–12 (2001)

5. Zopounidis, C., Doumpos, M., Matsatsinis, N.F.: On the use of knowledge-based decision support systems in financial management: a survey. Decis. Support. Syst. **20**(3), 259–277 (1997)
6. Keeney, R.L., Raiffa, H.: Decisions With Multiple Objectives: Preferences, and Value Tradeoffs. Wiley, New York (1976)
7. Roy, B.: Multicriteria methodology for decision aiding. Springer, New York (1996)
8. Belton, V., Stewart, T.: Multiple Criteria Decision Analysis. Kluwer Academic Publishers, Dordrecht (2002)
9. de Almeida, A.T., Cavalcante, C., Alencar, M., Ferreira, R., de Almeida-Filho, A.T., Garcez, T.: Multicriteria and multi-objective models for risk, reliability and maintenance decision analysis. In: International Series in Operations Research & Management Science, vol. 231. Springer, New York (2015). https://doi.org/10.1007/978-3-319-17969-8
10. Simon, H.A.: The New Science of Management Decision. Harper & Row Publishers Inc, New York (1960)
11. Polmerol, J.-C., Barba-Romero, S.: Multicriterion Decision in Management: Principles and Practice. Kluwer, Boston (2000)
12. Box, G., Draper, N.: Empirical Model-Building and Response Surfaces. Wiley, New York (1987)
13. Roselli, L.R.P., Frej, E.A., de Almeida, A.T., de Almeida A.T.: A Framework for Building Multicriteria Decision Models (work in progress)
14. Mingers, J., Taylor, S.: The use of soft systems methodology in practice. J. Oper. Res. Soc. **43**, 321–332 (1992). https://doi.org/10.1057/jors.1992.47
15. Keeney, R.L.: Value-Focused Thinking: A Path to Creative Decision-Making. Harvard University Press, London (1992)
16. Eden, C.: Cognitive mapping. Eur. J. Oper. Res. **36**(1), 1–13 (1988)
17. Eden, C., Ackermann, F.: SODA the principles. In: Rosenhead, J., Mingers, J. (eds.) Rational Analysis for A Problematic World Revisited, 2nd edn. Wiley, Chichester (2004)
18. Veloso, M.F.C.S., Ferreira, E.B., Almeida, A.T.D.: Uso do método FITradeoff para priorização de iniciativas que visam aumentar a produtividade de colaboradores no regime de teletrabalho. In: Simpósio Brasileiro De Pesquisa Operacional, 54, 2022, Juiz de Fora, Anais. SBPO (2022)
19. de Almeida, A.T., Almeida, J.A., Costa, A.P.C.S., Almeida-Filho, A.T.: A new method for elicitation of criteria weights in additive models: flexible and interactive tradeoff. Eur. J. Oper. Res. **250**(1), 179–191 (2016)
20. de Almeida, A.T., Frej, E.A., Roselli, L.R.P.: Combining holistic and decomposition paradigms in preference modeling with the flexibility of FITradeoff. CEJOR (2021). https://doi.org/10.1007/s10100-020-00728-z
21. Kang, T.H.A., Júnior, A.M.D.C.S., de Almeida, A.T.: Evaluating electric power generation technologies: a multicriteria analysis based on the FITradeoff method. Energy **165**, 10–20 (2018)
22. Carrillo, P.A.A., Roselli, L.R.P., Frej, E.A., de Almeida, A.T.: Selecting an agricultural technology package based on the flexible and interactive tradeoff method. Ann. Oper. Res., 1–16 (2018)
23. Monte, M.B.S., Morais, D.C.: A decision model for identifying and solving problems in an urban water supply system. Water Res. Manage. **33**(14), 4835–4848 (2019)
24. Fossile, D.K., Frej, E.A., da Costa, S.E.G., de Lima, E.P., de Almeida, A.T.: Selecting the most viable renewable energy source for brazilian ports using the FITradeoff method. J. Clean. Prod. **260**, 121107 (2020)
25. Dell'Ovo, M., Oppio, A., Capolongo, S.: Decision Support System for the Location of Healthcare Facilities SitHealth Evaluation Tool. PoliMI SpringerBriefs. Springer, Cham (2020). https://doi.org/10.1007/978-3-030-50173-0

26. Pergher, I., Frej, E.A., Roselli, L.R.P., de Almeida, A.T.: Integrating simulation and FITradeoff method for scheduling rules selection in job-shop production systems. Int. J. Prod. Econ. **227**, 107669 (2020)
27. Santos, I.M., Ro Frej, E.A., de Almeida, A.T., Costa, A.P.C.S.: Using data visualization for ranking alternatives with partial information and interactive tradeoff elicitation. Oper. Res. Int. J. **2019**, 1–23 (2019)
28. Frej, E.A., Roselli, L.R.P., Araújo de Almeida, J., de Almeida, A.T.: A multicriteria decision model for supplier selection in a food industry based on FITradeoff method. Math. Prob. Eng. 2017 (2017)
29. Barla, S.B.: A case study of supplier selection for lean supply by using a mathematical model. Logist. Inf. Manag. **16**, 451–459 (2003)
30. Korhonen, P., Wallenius, J.: Behavioral issues in MCDM: Neglected research questions. In: Clímaco, J. (ed.) Multicriteria Analysis. Springer, Heidelberg (1997)
31. Wallenius, H., Wallenius, J.: Implications of world mega trends for MCDM research. In: Amor, S.B., de Almeida, A.T., de Miranda, J.L., Aktas, E. (eds.) Advanced Studies in Multi-Criteria Decision Making. Series in Operations Research, 1st ed, pp. 1–10. Chapman and Hall/CRC, New York (2020)
32. de Almeida, A., Rosselli, L., Costa Morais, D., Costa, A.: Neuroscience tools for behavioural studies in group decision and negotiation. In: Kilgour, D.M., Eden, C. (eds.) Handbook of Group Decision and Negotiation, 1st edn., pp. 1–24. Springer International Publishing, Dordrecht, Netherlands (2020)
33. Roselli, L.R.P., de Almeida, A.T.: Use of the alpha-theta diagram as a decision neuroscience tool for analyzing holistic evaluation in decision making. Ann. Oper. Res. **312**, 1197–1219 (2022)
34. Roselli, L.R.P., de Almeida, A.T.: The use of the success-based decision rule to support the holistic evaluation process in FITradeoff. Int. Trans. Oper. Res.Oper. Res. **30**, 1299–1319 (2021)

Graph Representation Learning for Recommendation Systems: A Short Review

Khouloud Ammar[1(✉)], Wissem Inoubli[2], Sami Zghal[1,3], and Engelbert Mephu Nguifo[4]

[1] Faculty of Sciences of Tunis, LIPAH-LR11ES14, University of Tunis El Manar, 2092 Tunis, Tunisia
khouloud.ammar@fst.utm.tn
[2] University of Lorraine, LORIA, Nancy, France
wissem.inoubli@loria.fr
[3] Faculty of Law, Economics and Management Sciences of Jendouba University Campus, University of Jendouba, 8189 Jendouba, Tunisia
sami.zghal@fsjegj.rnu.tn
[4] University Clermont Auvergne, CNRS, LIMOS, Clermont-Ferrand, France
engelbert.mephu-nguifo@uca.fr

Abstract. Since the information explosion, a large number of items are present on the web, making it difficult for users to find the appropriate item from the available set of options. The Recommender System (RS) solves the problem of information overload by suggesting items of interest to the user. It has grown in popularity over the last few decades, and a significant amount of research has been conducted in this field. Among them, Collaborative Filtering (CF) is the most popular and widely used approach for RS, attempting to analyze the user's interest in the target item based on the opinions of other like-minded users. But recent years have witnessed the fast development of the emerging topic of Heterogeneous information networks Recommender Systems. Heterogeneous Information Network (HIN) based recommender systems offer a unified approach to combining various additional information, which can be combined with mainstream recommendation algorithms to effectively improve model performance and interpretability, and have thus been applied in a wide range of recommendation tasks. This paper provides a brief overview of various approaches used for recommender systems, as well as an understanding of the Collaborative Filtering technique. We also discussed HIN-based techniques, and finally, we focus on research challenges that must be addressed.

Keywords: Recommender systems · Collaborative filtering · Heterogeneous information network

1 Introduction

In recent years, recommender systems, one of most prominent applications of Artificial Intelligence (AI), have become popular on the web and widely adopted

I. Saad et al. (Eds.): ICIKS 2023, LNBIP 486, pp. 33–48, 2024.
https://doi.org/10.1007/978-3-031-51664-1_3

to help users, of many widely used web content sharing and e-Commerce platforms to find relevant content, products, or services more easily. Especially, with the internet's rapid development, people are experiencing information overload, with unimportant data. Information overload significantly reduces people's efficiency in obtaining useful information. RS [35] work to address this issue by providing users with a filter of items they might be interested in. In fact, RS have experienced decades of development and have found success in a variety of industries, including e-commerce [39]. On the other hand, companies that sell products or create web content are putting their money on this kind of service because they can pique users' interest and boost their sales as a result. A RS [36] has knowledge, processes, techniques, and challenges that researchers must address in order to enhance the outcomes and user experience.

The recommendation process can be broken down into two parts. The first part involves estimating the ratings or usefulness of items that a user has not yet encountered, which is typically done using the user's or other users' past behavior. The second part involves presenting the items with the highest estimated ratings as recommendations to the user. Various approaches have been proposed in the literature to accomplish this process. Among them, Collaborative Filtering which is a popular recommender system algorithm that identifies user opinion for an item based on the interests of other like-minded people.

CF first appeared in the early 1990s to deal with excessive online messages, and it later developed into an automated collaborative filtering system (GroupLens) [34], during which collaborative filtering was half labor and half machine learning. In the late 1990s, recommended techniques were used in the commercial sector [39], with Amazon's product recommendation system being the most well-known example [1]. Following the footsteps of Amazon, many e-commerce sites and online systems have begun to implement CF recommendation systems.

In the meantime, Graph Learning (GL), or machine learning on graph structure data, is an emerging AI technique that has shown great promise in recent years. Taking advantage of GL's ability to learn relational data, an emerging RS paradigm based on GL, namely Graph Learning based Recommender Systems (GLRS), has been proposed and extensively studied in recent years. In fact, the main motivation for graph learning in RS is that most of the objects in RS are explicitly or implicitly connected with each other. For example, objects here considered include users, items, attributes, etc., and influence each other via various relations [16]. Since heterogeneous information networks [43] are a type of graph structure data with complex networks that consist of multiple types of nodes and edges, modeling such an interaction system with heterogeneous graphs not only naturally preserves the entities and relationships in the recommender system, but also effectively incorporates various auxiliary information.

Using HIN to model various relations in RS is a natural and wise choice because it is one of the most promising machine learning techniques, which will effectively alleviating data sparsity and cold start problems (new user and new item), and to some extent, improving the interpretability of recommender systems. In this paper we will investigates several related works in HIN context and discussed their limitations and some future directions. The main contributions of this paper are:

- We provide an up-to-date overview of the field of recommender systems, highlighting recent advancements in collaborative filtering and graph representation learning in Heterogeneous Information Networks (HIN).
- We explore the potential of graph learning techniques in the field of recommender systems, and highlights the advantages of using graph learning in developing explainable recommender systems.
- We discuss some open research directions of HIN and CF.

The remainder of this paper is organized as follows: Sect. 2 presents some of the traditional CF approaches. Section 3 Provides a summary of recommender systems, focusing on the recent progress in Collaborative Filtering and Graph Representation Learning in HIN. Section 4 highlights future research directions in this field. Finally, Sect. 5 concludes the paper.

2 Traditional Collaborative Filtering

The collaborative filtering approach is a traditional recommendation method that generates recommendations based on shared user preferences and previous interactions [65]. A CF system is made up of m users and n items that form a $m \times n$ rating matrix. Users experience items of their choice and express their opinions about them as rating scores. This rating database is used to determine user similarity. Essentially, CF recommendation systems seeks to identify the top N products that these users have jointly purchased or rated based on previous user evaluations and the recommendee's similarity to other users. Memory-based methods and model-based methods are two subsets of this approach.

2.1 Memory-Based

Memory-based methods [4] memorize user-item rating matrix and use full rating database to determine item/user similarity [6]. The KNN algorithm is the most popular memory-based approach algorithm, this algorithm employs some traditional similarity measures such as Pearson correlation, Spearman, Cosine, Jaccard, etc. [64]. Predictions are typically made in two ways: user-based and item-based. The user-based method assumes that individuals with similar past preferences are probable to have similar future preferences as well. Thus, based on similar users' ratings on given items, we can predict the active user's missing ratings on specific items. This was first used in GroupLens to recommend news articles based on the interests or ratings of other users [34]. This concept was also applied to recommending music albums [41] and video recommendations [13]. In user-based CF, the rating score is calculated using the user-vector for common items. User-based algorithms find the similarity between users by comparing their ratings for common items. On the other hand item-based CF recommendation approach compare the ratings of different users to determine the similarity of two items. The underlying concept of this approach is that users tend to have similar opinions about items that are alike, based on their past experiences [38]. It is explained that the user may want to add items that are similar to the items in the shopping cart [27]. In contrast to user-based CF, item-based CF identifies similarities between the ratings patterns of the target item and other items. Item similarity is measured to determine the user preference for a specific item by analyzing how other people have rated that specific item. For instance, if a significant number of users have assigned similar ratings to multiple items, it suggests that those items are alike. The similarity between items can be determined by comparing the ratings provided by various users.

2.2 Model-Based

Model-based methods, on the contrary hand (or matrix-based decomposition), employ modern machine learning algorithms to discover patterns in training data and build models [36]. These methods construct a model that learns or observes user item interactions using low-dimensional representations (user and item feature vectors). These methods, also known as Latent Factor Models, can be implemented as Matrix Factorization MF [25] and its variants [57,63]. Matrix Factorization attempts to characterize both items and users using feature vectors of low dimension derived from user rating patterns. The most commonly used are. The latent factor model (LFM), which factorizes the user-item rating matrix into two low-rank matrices: the user feature and item feature matrices, is the most competitive and widely used to implement RS. It can reduce data sparsity using dimensionality reduction techniques and typically produces more accurate recommendations than the memory-based CF approach, while drastically reducing memory requirements and computation complexity [18,30].

The memory-based RS is simple to implement, and the algorithms are straightforward. However, when dealing with large numbers of users and items, the memory-based RS is unsuitable for practical applications. Model-based [4] methods were developed to address the space and scalability issues of memory-based approaches.

3 Recommender Systems Based on HINs

Over the years of developing theory on recommender systems, collaborative filtering, has been one of the main generations of this field. Meanwhile, research on this generation of RS is still continued, but new generation of using HIN in RS is growing to be increasingly interesting. In this section we will discuss new CF in heterogeneous information network and the graph learning approaches in HIN. Heterogeneous information networks have received a lot of attention in recent years because of their ability to handle a wide range of data mining activities, particularly in the field of recommendation systems. HINs not only include a variety of objects, such as users, movies, actors, and interest groups in movie recommendations, but they also depict various types of relationships between these items, such as viewing information, social relations, and attribute information. This rich and diverse data set can be used to improve recommendations by incorporating multiple types of data. The Fig. 1 illustrates the significant interactions between various object types in a recommendation system, which is a crucial task in data mining. This task involves a diverse range of object types, including users, movies, actors, and interest groups in movie recommendation. Given the multitude of interactions between these object types, utilizing HINs is an excellent approach to generate enhanced recommendations. In this section, we will highlight the collaborative filtering techniques in Heterogeneous Information Networks as the most widely used method and examine graph learning-based techniques in HIN as a highly promising approach in the field of recommender systems (RS).

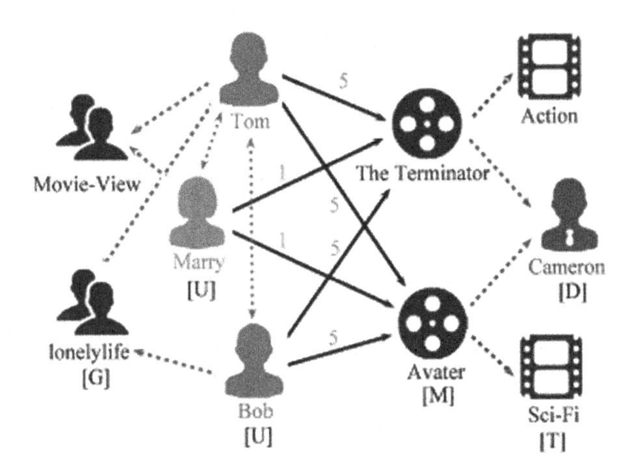

Fig. 1. An example of HIN-based movie recommender system [28].

3.1 Collaborative Filtering in HIN

The collaborative filtering methods calculate the similarity based on the interaction history between the user and the item as discussed in Sect. 2. Initially, similarity CF were defined only for homogeneous information networks. For example, P-PageRank [21] evaluates the probability of moving from the source object to the target object by restarting random walks, and SimRank [20] evaluates object similarity based on the similarity of two objects' neighbors. These algorithms, however, ignore the various types of objects and connections and are therefore unsuitable for recommender systems modeled as heterogeneous information networks. To address this issue, researchers proposed a set of algorithms for measuring the similarity of entities in heterogeneous information networks. We categorized the methods into two main types: memory-based and model-based. Within the memory-based category, there are two subcategories: random walk and metapath-based approaches.

Memory-Based Random Walk. Random Walk is a technique used in various applications such as graph theory, machine learning, and data mining. In the context of graph theory, random walk is a method to traverse a graph by moving from one node to another in a randomly selected manner. Memory-based random walk methods employ the random walk algorithm, which assumes that the relevance of an item to a user can be calculated from the user's random walk probability to the item, with higher probability indicating greater relevance. For instance, RHSN [56] estimates the global importance score for each entity using a random walk model and proposes a pair-wise learning algorithm to learn the weight of each type of relation. Furthermore, as a similarity measurement, ECTD [8] computes the average commute time between any entity pair, which is identified as the symmetrized quantity of the average number of steps required by a random walker to reach the item for the first time. HeteRS [33] represents the heterogeneous information network as a series of relation-based transformation matrices and employs the multivariate Markov chain to obtain the recommendation result for a given query node. To achieve personalized modeling of user preferences, HeteLearn [22] learns link weights based on the random walk and bayesian personalized ranking technology. Methods based on random walks aim to learn relation-level transition probability without explicitly using higher level semantic information.

Memory Based Based Meta-path. In comparison, memory based methods based on meta-path introduce additional prior knowledge by manually introducing meaningful meta-paths. A meta-path is a sequence of nodes in a graph that defines a specific relationship between the nodes (*e.g.,* user-watch-movie). The designed meta-paths satisfy some desirable properties (such as symmetry and self-maximum), which aid in the capture of long-range semantic information. The basic concept behind using the aforementioned meta-paths for collaborative filtering is to generate a large number of candidate similarities using various meta-paths, and then either manually assign or automatically learn the weights of those candidate similarities. Therefore, the methods based on meta-path also have the benefit of strong interpretability because of the readable meta-path and the weight of different meta-paths [15]. HeteRecom [44] determines the weighted meta-path user similarity and then determines the weight using a heuristic weight learning method. Based on the constrained meta-path, MP-PRF [29] reorders the paper objects in the heterogeneous citation network.

Most of the approaches discussed above are the first to use heterogeneous similarity measurement for CF-based recommender systems, which require a large amount of interactive information and make it difficult to provide recommendations to users and items that lack connectivity.

Model Based. To provide recommendations to users and items that lack connectivity, recommender system researchers proposed a matrix factorization model (or model-based RS). The central idea is to decompose the rating matrix to extract the hidden vectors of users and items, and then make recommendations based on the similarity of the hidden vectors. Traditional matrix factorization models use the implicit vector reconstruction co-occurrence matrix as the optimization objective and are unable to take advantage of the rich semantic information found in heterogeneous information networks. Many researchers have proposed matrix factorization methods suitable for heterogeneous information network modeling. The classical matrix factorization model only reconstructs the item's user rating matrix and cannot make use of other types of entities and relations in the heterogeneous information network. To compensate for this shortcoming, many improvements have included heterogeneous regularization terms in the optimization objective. These methods, in particular, calculate the similarity matrix using various similarity measurement methods, and then use the implicit vector to reconstruct the similarity matrix as a separate regularization item, or combine the similarity matrix and the score matrix as a unified matrix to be decomposed. CMF [45], for example, creates a collaborative matrix factorization model by decomposing multiple matrices at the same time and sharing parameters. HeteroMF [19] creates a context-dependent matrix factorization model based on the assumption that different relationships should have different effects.

Some other methods use similarity measures based on meta-paths to generate similarity matrices as regularization items of matrix factorization in order to use the high-level semantic information of meta-paths. For example, Hete-MF [55] calculates item-item similarity based on meta-paths first, and then uses linear regression to weight the similarity scores of different meta-paths. Furthermore, Hete-CF [31] extends Hete-MF [55], employing PathSim [47] to measure the relationships between user-user, item-item, and user-item, and then employing a unified matrix factorization model to integrate the aforementioned three types of heterogeneous information into the social recommendation task.

Model Based Neural Networks. Due to neural networks' exceptional aptitude to adapt to any function, neural matrix factorization [12] and its variants have achieved good results in recommendation tasks with the rise of deep learning. The traditional neural matrix factorization method, on the other hand, only uses one-hot representations of user and item IDs as input features and cannot take advantage of the rich semantic information of heterogeneous information networks. For example, HNAFM [5] calculates the exchange matrix using different meta-paths, which learns user and item representations using multi-layer perceptrons, and then uses a hierarchical attention mechanism to fuse multiple representations from meta-paths. RLDB [11] begins with a meta-path-based random walk to obtain the connection sequence of users and items, then employs a convolutional neural network to learn structural and textual representations, and finally employs a multi-view machine to effectively combine structural and textual representations.

There are several benefits associated with the utilization of matrix factorization methods. These methods generate hidden vectors for users and items, enabling the prediction of scores for any user-item pair. This improves the generalization capability and reduces space complexity. However, due to its linear modeling nature, matrix factorization struggles to effectively capture the intricate preferences of users. It faces challenges in representing high-order interactions between users and items, leading to suboptimal recommendation results. To overcome these limitations, recommendation models have incorporated graph representation learning techniques, which have been increasingly adopted and well-received in the field.

3.2 Graph Representation Learning RS in HINs

Most CF-based recommendation systems personalize recommendations for users by profiling them using unstructured data such as ratings, reviews, or images. Existing approaches are effective, but they struggle to model the explicit relationships between the information we have about users and items. By encoding rich information from multi-type user behaviors and item properties into user/item embeddings, we can improve recommendation performance while preserving the internal structure of the knowledge. Providing structured knowledge to recommender systems also allows the system to generate informed explanations for the recommended items. Personalized explanations for recommendations have been shown by researchers to improve the persuasiveness, efficiency, effectiveness, and transparency of recommender systems [58]. Because of its strong feature cross capability and flexibility in model architecture design, the neural network-based recommendation model has achieved better recommendation results with the development of deep learning. Traditional neural networks, on the other hand, cannot directly model graph structures. Researchers have been exploring ways to incorporate graph representation learning technology into recommendation models in order to leverage the rich structural and semantic information present in graph data. The popularity of graph representation learning has driven the development of these models, which aim to provide more accurate and comprehensive recommendations. We will categorize the methods in this study based on the type of learning they use, which can be either supervised or unsupervised. Supervised learning entails training a model on labeled data, in which the model is given both input and desired output and adjusts its parameters to minimize the difference between predicted and actual output. Unsupervised learning, on the other hand, entails training a model on unlabeled data, with the model attempting to discover patterns or relationships in the data without prior knowledge of the desired output. These two types of learning have distinct applications and characteristics, and our classification will shed light on the advantages and disadvantages of each approach in the context of recommendation systems.

Unsupervised. The goal of unsupervised graph representation learning is to learn low dimensional vector representations of graph (embedding) structure that can be easily stored and applied to various downstream tasks. Unsupervised graph embedding methods are used to generate structural features and feed them into the recommendation model. Unsupervised matrix factorization is, in fact, a bipartite graph embedding model. A bipartite graph is a type of graph that divides vertices into two disjoint sets, connecting vertices only between sets, commonly used to model relationships between two distinct types of entities. The difference between the matrix factorization recommendation model and the non-supervised matrix factorization recommendation model is that the decomposed matrix of non-supervised matrix factorization is based on the similarity matrix obtained by the meta path rather than the rating matrix containing supervisory information. For example, FMG [59] creates a similarity measurement method based on meta-graphs and performs an unsupervised matrix factorization of the user-item similarity matrix obtained from meta-graphs to learn the embedding vectors of users and items. The embedding vector is then input to the factorization machine as a structural feature, thereby modeling the high-order relationship between the features. MoHINRec [60], like FMG [59], proposes a motif-enhanced meta-path that captures the high-order relationship between nodes of the same type and then feeds the embedding representation into the factorization machine for training. However, the cost of decomposing large-scale matrices is typically very high. Researchers have designed many unsupervised heterogeneous graph embedding methods based on random walks, such as metapath2vec [7] and HIN2Vec [9], and applied them to the structural feature generation of recommender systems, inspired by DeepWalk [32], node2vec [10], and other methods. For example, HERec [42] generates object sequences using meta-path based random walks and node2vec [10] to learn object embeddings and then computes embedding similarity for recommendation. Furthermore, HueRec [53] assumes that users or items share semantics across meta-paths and then uses all meta-paths to learn a unified representation of users and items. HetNERec [62] divides a heterogeneous network into multi-subnetworks based on meta-paths and regularizes matrix factorization using the learned embeddings. Furthermore, some methods do not require the meta-path to be specified manually. HRLHG [23], for example, proposes a hierarchical random walk algorithm that implements a two-level random walk guided by two different sets of distributions. The local random walk is based on the definition of relational transformation distribution and is used to model non-task-related local semantic information. The global random walk is used to model task-specific network patterns and is defined based on the usefulness distribution of relation types. To generate entity embedding, NREP [54] optimizes node-content correlation, structure preservation, and edge prediction. Furthermore, ECFKG [2] generates node embedding using a translation-based model [3], whereas HRec [46] uses PTE [49].

Supervised. Graph neural network is a representative neural network that processes graph structures and supports end-to-end training. Early graph neural networks are only suitable for homogeneous graphs, including Graph Convolution Network (GCN) [24], Graph Attention Network (GAT) [50], etc. In order to apply graph neural networks to heterogeneous information networks, researchers have proposed many heterogeneous graph neural networks and applied them to recommender systems. For heterogeneous information networks with rich relation types, researchers usually use relation-aware graph neural networks (*e.g.* RGCN [40]), and apply them to recommender systems. A representative work is IntentGC [61], which transforms complex heterogeneous graphs into bipartite graphs consisting only of users and items based on second-order relations, and performs vector convolution instead of bitwise convolution to avoid unnecessary feature interaction and improve model robustness respectively. For example, DisenHAN [52] projects node attributes into different sub-spaces and uses two-layer attention mechanism for intra- and inter-relation aggregation to learn node representations. IPRec [26] designs an intra- and inter-package attention network for package modeling, and a dual-level attention network for user modeling. In the other hand, GNewsRec [17] uses a sampling-based intra-type aggregation and a linear average combination of inter-type aggregation for both item representation and long-term user preference modeling, and an attention-based LSTM model for short-term user preference modeling.

Unsupervised methods do not incorporate any form of supervision during the graph embedding stage, so the generated graph embedding representation is difficult to be suitable for various recommendation tasks. In contrast, the supervised embedding method can use the supervision information while learning the graph embedding, so as to improve the matching degree of the graph embedding and the recommendation tasks.

Graph Neural Network Meta-path Based. Another type of heterogeneous graph neural network is one that is based on meta-path modeling, for example HAN [51], which can effectively introduce prior knowledge and capture high-level semantic information in heterogeneous graphs. CDPRec [37] extracts a homogeneous graph from a heterogeneous graph using a meta-path based random walker, and then jointly trains multi-head attention and RNN for sequential recommendation. In fact, some potential relationships exist between relation-based methods and meta-path-based methods. Furthermore, some methods proposed prior to the rise of graph neural networks explicitly learn the embedding representation of the meta-path during model training. For example, MCRec [14] learns meta-path embedding representations using convolutional neural networks and proposes a common attention mechanism to mutually improve the meta-path-based context, user and item representations. RKGE [48] employs several RNNs to generate embeddings of different paths connecting the same entity pair and then pools the path embeddings to represent the user's overall preference for an item.

The graph representation learning-based recommendation model can effectively learn the rich structure and semantic information in the graph and can be combined with other deep learning technologies in a variety of ways. Unsupervised graph representation learning, for example, can be used as pre-training to generate feature vectors, which can then be fused with other feature vectors and input to subsequent models. Supervised graph representation learning, on the other hand, can be used as the embedding layer of the recommendation model to achieve end-to-end training. However, the existing graph representation learning technology still faces many challenges, such as scalability and interpretability, on which research is currently focused.

4 Limitations and Future Work

The first few sections introduced some classic HIN-based recommendation methods that are commonly used. It has been discovered that the heterogeneous information network has been used in a variety of recommender system models and recommendation tasks as a modeling method for fusion of auxiliary information, with benefits in improving model performance and interpretability. We will discuss some limitations and several potential future research directions in this section.

4.1 Limitations

Scalability: Is a major challenge, particularly when dealing with large-scale datasets. Traditional CF approaches can become computationally infeasible as the number of users and items grows. Various approaches, such as matrix factorization, neighborhood-based methods, and deep learning models, have been proposed to address this issue, but they still have limitations in terms of scalability and computational efficiency.

Cold Start Problem: The cold start problem refers to the difficulty in CF while making recommendations for new users or items with little or no previous interaction data. Traditional CF approaches frequently fail to address this issue, and additional information, such as demographic or side information, is needed to overcome it.

Sparsity: Another difficulty in CF is the sparsity of interaction data. When users only interact with a small percentage of the available items, making accurate recommendations becomes difficult. This can be addressed by leveraging other sources of information, such as item metadata or user preferences, but this necessitates more data and increases the recommendation system's complexity.

Privacy: Privacy is an issue, particularly when dealing with sensitive information like medical records or financial transactions. Traditional CF approaches can reveal sensitive user information, such as preferences and habits, which can be used to re-identify them. Various privacy-preserving CF approaches, such as differential privacy and federated learning, have been proposed to address this issue, but they still have limitations in terms of accuracy and scalability.

4.2 Future Directions

Scalable Algorithms: Creating scalable algorithms capable of handling large-scale and in real time datasets is an important area of future research in CF and HIN recommender systems. This problem is exacerbated in HIN because graph structure data is typically larger and requires more time and space to process, let alone complex machine learning techniques to generate recommendations. This could entail leveraging advances in parallel computing and distributed systems, or developing more efficient algorithms capable of handling large amounts of data.

Cold Start Problem: Another important area of future work in CF is addressing the cold start problem. This may entail creating more sophisticated algorithms that can make recommendations for new users or items based on additional sources of information, such as social network data or demographic information.

Heterogeneous Information Networks: CF and graph representation learning in HIN are still in their infancy, and there is plenty of room for future research. This could include creating more advanced algorithms for modeling heterogeneous data or investigating new applications of HIN-based methods, such as in healthcare, finance, or social media.

Privacy-Preserving Recommendations: Developing privacy-preserving approaches is an important area of future CF research. This could include leveraging advances in differential privacy and federated learning, as well as developing new privacy-preserving methods for handling sensitive data in a secure and privacy-sensitive manner.

Interpretability: Another critical area of future research is the development of interpretable recommendation systems that can provide users with clear and understandable explanations for why certain recommendations are made. This could include making algorithms more transparent or visualizing the recommendation process to help users understand how recommendations are made.

5 Conclusion

Recommender Systems are extremely useful in providing solutions to information overload. There are several methods available to complete the task of recommendations. Collaborative Filtering is one of the most widely used and successful approaches, and HIN-based RS is a newer trend. The heterogeneous information network modeling method can effectively use the rich structural and semantic information in auxiliary information and apply it to recommender system modeling. In this paper we investigated several related works and discussed some CF and HIN-based challenges. Many researchers have provided solutions to these problems, but there is still a long way to go to achieve satisfactory results.

References

1. Aggarwal, C.C.: An introduction to recommender systems. In: Aggarwal, C.C. (ed.) Recommender Systems, pp. 1–28. Springer, Cham (2016). https://doi.org/10.1007/978-3-319-29659-3_1
2. Ai, Q., Azizi, V., Chen, X., Zhang, Y.: Learning heterogeneous knowledge base embeddings for explainable recommendation. Algorithms **11**(9), 137 (2018)

3. Bordes, A., Usunier, N., Garcia-Duran, A., Weston, J., Yakhnenko, O.: Translating embeddings for modeling multi-relational data. In: Advances in Neural Information Processing Systems, vol. 26 (2013)
4. Breese, J.S., Heckerman, D., Kadie, C.: Empirical analysis of predictive algorithms for collaborative filtering. arXiv preprint arXiv:1301.7363 (2013)
5. Chen, L., Liu, Y., Zheng, Z., Yu, P.: Heterogeneous neural attentive factorization machine for rating prediction. In: Proceedings of the 27th ACM International Conference on Information and Knowledge Management, pp. 833–842 (2018)
6. Delgado, J., Ishii, N.: Memory-based weighted majority prediction. In: SIGIR Workshop Recommender Systems, p. 85. Citeseer (1999)
7. Dong, Y., Chawla, N.V., Swami, A.: metapath2vec: scalable representation learning for heterogeneous networks. In: Proceedings of the 23rd ACM SIGKDD International Conference on Knowledge Discovery and Data Mining, pp. 135–144 (2017)
8. Fouss, F., Pirotte, A., Renders, J.M., Saerens, M.: Random-walk computation of similarities between nodes of a graph with application to collaborative recommendation. IEEE Trans. Knowl. Data Eng. **19**(3), 355–369 (2007)
9. Fu, T., Lee, W.C., Lei, Z.: Hin2vec: explore meta-paths in heterogeneous information networks for representation learning. In: Proceedings of the 2017 ACM on Conference on Information and Knowledge Management, pp. 1797–1806 (2017)
10. Grover, A., Leskovec, J.: node2vec: scalable feature learning for networks (2016)
11. Han, X., Shi, C., Zheng, L., Yu, P.S., Li, J., Lu, Y.: Representation learning with depth and breadth for recommendation using multi-view data. In: Cai, Y., Ishikawa, Y., Xu, J. (eds.) APWeb-WAIM 2018. LNCS, vol. 10987, pp. 181–188. Springer, Cham (2018). https://doi.org/10.1007/978-3-319-96890-2_15
12. He, X., Liao, L., Zhang, H., Nie, L., Hu, X., Chua, T.S.: Neural collaborative filtering. In: Proceedings of the 26th International Conference on World Wide Web, pp. 173–182 (2017)
13. Hill, W., Stead, L., Rosenstein, M., Furnas, G.: Recommending and evaluating choices in a virtual community of use. In: Proceedings of the SIGCHI Conference on Human Factors in Computing Systems, pp. 194–201 (1995)
14. Hu, B., Shi, C., Zhao, W.X., Yu, P.S.: Leveraging meta-path based context for top-n recommendation with a neural co-attention model. In: Proceedings of the 24th ACM SIGKDD International Conference on Knowledge Discovery & Data Mining, pp. 1531–1540 (2018)
15. Hu, J., Zhang, Z., Liu, J., Shi, C., Yu, P.S., Wang, B.: RecExp: a semantic recommender system with explanation based on heterogeneous information network. In: Proceedings of the 10th ACM Conference on Recommender Systems, pp. 401–402 (2016)
16. Hu, L., Cao, J., Xu, G., Cao, L., Gu, Z., Cao, W.: Deep modeling of group preferences for group-based recommendation. In: Proceedings of the AAAI Conference on Artificial Intelligence, vol. 28 (2014)
17. Hu, L., Li, C., Shi, C., Yang, C., Shao, C.: Graph neural news recommendation with long-term and short-term interest modeling. Inf. Process. Manag. **57**(2), 102142 (2020)
18. Huang, S., Ma, J., Cheng, P., Wang, S.: A hybrid multigroup coclustering recommendation framework based on information fusion. ACM Trans. Intell. Syst. Technol. (TIST) **6**(2), 1–22 (2015)
19. Jamali, M., Lakshmanan, L.: HeteroMF: recommendation in heterogeneous information networks using context dependent factor models. In: Proceedings of the 22nd International Conference on World Wide Web, pp. 643–654 (2013)

20. Jeh, G., Widom, J.: SimRank: a measure of structural-context similarity. In: Proceedings of the Eighth ACM SIGKDD International Conference on Knowledge Discovery and Data Mining, pp. 538–543 (2002)
21. Jeh, G., Widom, J.: Scaling personalized web search. In: Proceedings of the 12th International Conference on World Wide Web, pp. 271–279 (2003)
22. Jiang, Z., Liu, H., Fu, B., Wu, Z., Zhang, T.: Recommendation in heterogeneous information networks based on generalized random walk model and Bayesian personalized ranking. In: Proceedings of the Eleventh ACM International Conference on Web Search and Data Mining, pp. 288–296 (2018)
23. Jiang, Z., Yin, Y., Gao, L., Lu, Y., Liu, X.: Cross-language citation recommendation via hierarchical representation learning on heterogeneous graph. In: The 41st International ACM SIGIR Conference on Research & Development in Information Retrieval, pp. 635–644 (2018)
24. Kipf, T.N., Welling, M.: Semi-supervised classification with graph convolutional networks (2017)
25. Koren, Y., Bell, R., Volinsky, C.: Matrix factorization techniques for recommender systems. Computer **42**(8), 30–37 (2009)
26. Li, C., et al.: Package recommendation with intra-and inter-package attention networks. In: Proceedings of the 44th International ACM SIGIR Conference on Research and Development in Information Retrieval, pp. 595–604 (2021)
27. Linden, G., Smith, B., York, J.: Amazon.com recommendations: item-to-item collaborative filtering. IEEE Internet Comput. **7**(1), 76–80 (2003)
28. Liu, J., Shi, C., Yang, C., Lu, Z., Philip, S.Y.: A survey on heterogeneous information network based recommender systems: concepts, methods, applications and resources. AI Open **3**, 40–57 (2022)
29. Liu, X., Yu, Y., Guo, C., Sun, Y.: Meta-path-based ranking with pseudo relevance feedback on heterogeneous graph for citation recommendation. In: Proceedings of the 23rd ACM International Conference on Conference on Information and Knowledge Management, pp. 121–130 (2014)
30. Lü, L., Medo, M., Yeung, C.H., Zhang, Y.C., Zhang, Z.K., Zhou, T.: Recommender systems. Phys. Rep. **519**(1), 1–49 (2012)
31. Luo, C., Pang, W., Wang, Z., Lin, C.: Hete-CF: social-based collaborative filtering recommendation using heterogeneous relations. In: 2014 IEEE International Conference on Data Mining, pp. 917–922. IEEE (2014)
32. Perozzi, B., Al-Rfou, R., Skiena, S.: Deepwalk. In: Proceedings of the 20th ACM SIGKDD International Conference on Knowledge Discovery and Data Mining (2014). https://doi.org/10.1145/2623330.2623732
33. Pham, T.A.N., Li, X., Cong, G., Zhang, Z.: A general recommendation model for heterogeneous networks. IEEE Trans. Knowl. Data Eng. **28**(12), 3140–3153 (2016)
34. Resnick, P., Iacovou, N., Suchak, M., Bergstrom, P., Riedl, J.: GroupLens: an open architecture for collaborative filtering of netnews. In: Proceedings of the 1994 ACM Conference on Computer Supported Cooperative Work, pp. 175–186 (1994)
35. ResnickP, V.: Recommender systems. Commun. ACM **40**(3), 56–58 (1997)
36. Ricci, F., Rokach, L., Shapira, B.: Introduction to recommender systems handbook. In: Ricci, F., Rokach, L., Shapira, B., Kantor, P.B. (eds.) Recommender Systems Handbook, pp. 1–35. Springer, Boston (2011). https://doi.org/10.1007/978-0-387-85820-3_1
37. Sang, L., Xu, M., Qian, S., Martin, M., Li, P., Wu, X.: Context-dependent propagating-based video recommendation in multimodal heterogeneous information networks. IEEE Trans. Multimed. **23**, 2019–2032 (2020)

38. Sarwar, B., Karypis, G., Konstan, J., Riedl, J.: Item-based collaborative filtering recommendation algorithms. In: Proceedings of the 10th International Conference on World Wide Web, pp. 285–295 (2001)
39. Schafer, J.B., Konstan, J.A., Riedl, J.: E-commerce recommendation applications. Data Min. Knowl. Disc. **5**, 115–153 (2001)
40. Schlichtkrull, M., Kipf, T.N., Bloem, P., van den Berg, R., Titov, I., Welling, M.: Modeling relational data with graph convolutional networks. In: Gangemi, A., et al. (eds.) ESWC 2018. LNCS, vol. 10843, pp. 593–607. Springer, Cham (2018). https://doi.org/10.1007/978-3-319-93417-4_38
41. Shardanand, U., Maes, P.: Social information filtering: algorithms for automating "word of mouth". In: Proceedings of the SIGCHI Conference on Human Factors in Computing Systems, pp. 210–217 (1995)
42. Shi, C., Hu, B., Zhao, W.X., Philip, S.Y.: Heterogeneous information network embedding for recommendation. IEEE Trans. Knowl. Data Eng. **31**(2), 357–370 (2018)
43. Shi, C., Li, Y., Zhang, J., Sun, Y., Yu, P.S.: A survey of heterogeneous information network analysis. IEEE Trans. Knowl. Data Eng. **29**(1), 17–37 (2017). https://doi.org/10.1109/TKDE.2016.2598561
44. Shi, C., Zhou, C., Kong, X., Yu, P.S., Liu, G., Wang, B.: HeteRecom: a semantic-based recommendation system in heterogeneous networks. In: Proceedings of the 18th ACM SIGKDD International Conference on Knowledge Discovery and Data Mining, pp. 1552–1555 (2012)
45. Singh, A.P., Gordon, G.J.: Relational learning via collective matrix factorization. In: Proceedings of the 14th ACM SIGKDD International Conference on Knowledge Discovery and Data Mining, pp. 650–658 (2008)
46. Su, Y., et al.: HRec: heterogeneous graph embedding-based personalized point-of-interest recommendation. In: Gedeon, T., Wong, K.W., Lee, M. (eds.) ICONIP 2019. LNCS, vol. 11955, pp. 37–49. Springer, Cham (2019). https://doi.org/10.1007/978-3-030-36718-3_4
47. Sun, Y., Han, J., Yan, X., Yu, P.S., Wu, T.: PathSim: meta path-based top-k similarity search in heterogeneous information networks. Proc. VLDB Endow. **4**(11), 992–1003 (2011)
48. Sun, Z., Yang, J., Zhang, J., Bozzon, A., Huang, L.K., Xu, C.: Recurrent knowledge graph embedding for effective recommendation. In: Proceedings of the 12th ACM Conference on Recommender Systems, pp. 297–305 (2018)
49. Tang, J., Qu, M., Mei, Q.: PTE: predictive text embedding through large-scale heterogeneous text networks. In: Proceedings of the 21th ACM SIGKDD International Conference on Knowledge Discovery and Data Mining, pp. 1165–1174 (2015)
50. Veličković, P., Cucurull, G., Casanova, A., Romero, A., Lio, P., Bengio, Y.: Graph attention networks. arXiv preprint arXiv:1710.10903 (2017)
51. Wang, X., et al.: Heterogeneous graph attention network (2021)
52. Wang, Y., Tang, S., Lei, Y., Song, W., Wang, S., Zhang, M.: DisenHAN: disentangled heterogeneous graph attention network for recommendation. In: Proceedings of the 29th ACM International Conference on Information & Knowledge Management, pp. 1605–1614 (2020)
53. Wang, Z., Liu, H., Du, Y., Wu, Z., Zhang, X.: Unified embedding model over heterogeneous information network for personalized recommendation. In: Proceedings of the 28th International Joint Conference on Artificial Intelligence, pp. 3813–3819 (2019)

54. Yang, L., Zhang, Z., Cai, X., Guo, L.: Citation recommendation as edge prediction in heterogeneous bibliographic network: a network representation approach. IEEE Access **7**, 23232–23239 (2019)
55. Yu, X., Ren, X., Gu, Q., Sun, Y., Han, J.: Collaborative filtering with entity similarity regularization in heterogeneous information networks. IJCAI HINA **27** (2013)
56. Zhang, J., et al.: Recommendation over a heterogeneous social network. In: 2008 The Ninth International Conference on Web-Age Information Management, pp. 309–316. IEEE (2008)
57. Zhang, L., Chen, Z., Zheng, M., He, X.: Robust non-negative matrix factorization. Front. Electr. Electron. Eng. China **6**, 192–200 (2011)
58. Zhang, Y., Chen, X.: Explainable recommendation: a survey and new perspectives. CoRR abs/1804.11192 (2018). http://arxiv.org/abs/1804.11192
59. Zhao, H., Yao, Q., Li, J., Song, Y., Lee, D.L.: Meta-graph based recommendation fusion over heterogeneous information networks. In: Proceedings of the 23rd ACM SIGKDD International Conference on Knowledge Discovery and Data Mining, pp. 635–644 (2017)
60. Zhao, H., Zhou, Y., Song, Y., Lee, D.L.: Motif enhanced recommendation over heterogeneous information network. In: Proceedings of the 28th ACM International Conference on Information and Knowledge Management, pp. 2189–2192 (2019)
61. Zhao, J., et al.: IntentGC: a scalable graph convolution framework fusing heterogeneous information for recommendation. In: Proceedings of the 25th ACM SIGKDD International Conference on Knowledge Discovery & Data Mining, pp. 2347–2357 (2019)
62. Zhao, Z., Zhang, X., Zhou, H., Li, C., Gong, M., Wang, Y.: HetNERec: heterogeneous network embedding based recommendation. Knowl.-Based Syst. **204**, 106218 (2020)
63. Zhou, X., He, J., Huang, G., Zhang, Y.: SVD-based incremental approaches for recommender systems. J. Comput. Syst. Sci. **81**(4), 717–733 (2015)
64. Zhu, B., Hurtado, R., Bobadilla, J., Ortega, F.: An efficient recommender system method based on the numerical relevances and the non-numerical structures of the ratings. IEEE Access **6**, 49935–49954 (2018)
65. Zou, X.: A survey on application of knowledge graph. In: Journal of Physics: Conference Series, vol. 1487, no. 1, p. 012016 (2020). https://doi.org/10.1088/1742-6596/1487/1/012016

FITradeoff Decision Support System Applied to Solve a Supplier Selection Problem

Lucia Reis Peixoto Roselli(✉) and Adiel Teixeira de Almeida

Center for Decision Systems and Information Development (CDSID), Universidade Federal de Pernambuco, Recife, PE, Brazil
{lrpr,almeida}@cdsid.org.br

Abstract. The world's digital transformation changed organizations' routines completely. Decision Support Systems have been developed to support Decision-Makers to capture information intended to permit competitive advances. In this context, regarding Multi-Criteria Decision-Making/Aiding problems, several DSSs have been proposed to support DMs to deal with complex problems. This paper presents the DSS constructed to implement the FITradeoff method. Several steps of the FITradeoff DSS have been described considering a supplier selection process. In addition, the paper presents behavioral studies performed to investigate the FITradeoff method in a synthesized way. The probability of success, which is an important feature proposed by behavioral studies, has been discussed since it can be used by analysts to advise decision-makers during the decision process. This feature uses previous knowledge of holistic evaluation studies to advise decision-makers to use or not use visualizations to perform holistic evaluations in the FITradeoff Decision Support System. Behavioral studies in the FITradeoff DSS provide modulation (transformations) on it, bringing important advances to this area of knowledge.

Keywords: Decision Support System · FITradeoff Method · Elicitation process · Holistic evaluation · Multi-Criteria Decision Making/Aiding (MCDM/A)

1 Introduction

In recent year, organizations' routine has been completely changed by digital transformation. Considering the amount of information captured and treated by organizations, several Decision Support Systems (DSS) have been developed to support Decision-Makers (DMs) to obtain competitive advances for organizations.

In the context of Multi-Criteria Decision Making/Aiding (MCDM/A) problems, several DSS have been constructed to support DMs to solve these problems. The supplier selection problem is a famous example of MCDM/A, which is faced by several organizations [1–4]. Thus, MCDM/A problems have been characterized as problems in which conflicting objectives are presented, and more than one alternative is considered [5–8].

I. Saad et al. (Eds.): ICIKS 2023, LNBIP 486, pp. 49–62, 2024.
https://doi.org/10.1007/978-3-031-51664-1_4

In order to support Decision-Makers (DMs), several MCDM/A methods have been proposed pretending to collect DMs preferences and incorporate them into mathematical models. In general, MCDM/A methods are implemented by DSSs. Thus, in this field, the DSSs aim to interact with DMs to collect their preferences concerning the consequences of those problems.

The Flexible and Interactive Tradeoff (FITradeoff) method [9, 10] is an MCDM/A method in the context of Multi-Attribute Value Theory (MAVT) [5]. This method has been implemented by a DSS which is available for free at www.fitradeoff.org.

The DSS constructed to implement the FITradeoff method [9, 10] is the focus of this paper. The FITradeoff DSS is flexible and interactive. It is flexible since DMs can express preferences using the two types of preference modeling presented in the FITradeoff method – the elicitation by decomposition and the holistic evaluation [10]. Moreover, it is interactive since DMs participate in the whole process.

In this study, this DSS has been illustrated to show its usability to solve a supplier selection problem presented in the literature [2]. Moreover, behavioral analysis has been discussed in order to improve the FITradeoff DSS.

Several studies have been performed to investigate the FITradeoff method. These studies intend to capture behavioral aspects during the decision process when DMs interact with the DSS. Behavioral studies in MCDM/A approach aim to advance this area of knowledge since they can modulate (or transform) those methods by the inclusion of aspects that are not controlled by DMs but as inherent in the decision process [11, 12].

Most MCDM/A methods wish to conduct a rational decision-making process but do not incorporate those aspects with are inherent in the process, such as emotions [13, 14]. Thus, the solutions obtained may not represent the real preference of DMs.

An important modulation performed in the FITradeoff method has been the use of two perspectives of preference modeling – the elicitation by decomposition and the holistic evaluation – in the FITradeoff DSS. At the beginning, the holistic evaluation can be done at the end of the process, to finalize the decision process by selecting the best alternative [9, 10]. Now, DMs can perform holistic evaluations in the middle of the process. The FITradeoff method uses a Linear Programming Problem (LPP) to obtain the solutions [9, 10]. Hence, inequalities are included in the LPP model as constraints to represent DMs preferences expressed during the elicitation by decomposition and holistic evaluations. Several behavioral studies have been done to investigate both paradigms of preference modeling and based on them, several improvements have been done in the FITradeoff DSS [14–25].

The paper describes the FITradeoff DSS considering a supplier selection problem [2], and moreover discuss a behavioral experiment performed to investigate the holistic evaluation process using graphical and tabular visualizations. The paper as divided as follows. Section 2 presents the FITradeoff method. Section 3 describes the FITradeoff DSS. Section 4 discuss the behavioral studies and results. Finally, Sect. 5 remarks on conclusions and future studies.

2 FITradeoff Method

The FITradeoff method [9, 10] is a method used to elicit scaling constants in the context of MAVT [5]. The FITradeoff method considers the axiomatic structure of the Tradeoff procedure [5] but incorporates partial information concepts [9].

This method can be used to solve several decision problems, in Economic, Ambiental, and Social contexts [26–31]. Hence, the method has been constructed to support different types of problematics, such as choice [9], ranking [32], sorting [33], and portfolio problems [34, 35].

In order to permit Decision-Makers (DMs) to use the FITradeoff method, it has been implemented in a DSS. This DSS is flexible and interactive. It is flexible since permits DMs to express preferences for consequences in pairwise comparison during the decomposition process or between alternatives during the holistic evaluation [10].

The combination of these two paradigms of preference modeling is an important transformation made in the FITradeoff method. This feature provides flexibility for DMs since they can use the paradigm that is more adequate with their cognitive style. In other words, in the middle of the decision process, the DMs can alternate between these two perspectives, using those that judge more adequately to express preferences. Thus, DMs can compare consequences during the elicitation by decomposition, or alternatives during the holistic evaluation [10].

Moreover, DMs interact with the DSS throughout the whole process. For each preference expressed by DMs, a new inequality is inserted in the LPP model as a constraint. Thus, scaling constant space (or weight space) can be reduced after each interaction with the DM. During the whole process, the solutions are updated, and DMs can observe partial results. The process stops when a solution has been found, or when the partial result is considered satisfactory (sufficient) for DMs to solve the problem.

The next section describes the FITradeoff DSS applied to solve a supplier selection problem which is already presented in the literature [2].

3 Decision Support System

In this section, the DSS has been described to solve a supplier selection problem which presents five suppliers evaluated against seven criteria. The criteria are price, freight, accuracy, promptness, quality, lead time, and flexibility. The FITradeoff DSS is available for free at www.fitradeoff.org, as illustrated in Fig. 1.

In this problem, the DM wishes to obtain the best alternative, or the subset of potentially optimal alternatives (POAs) [9], thus the DSS for choice problematic has been considered, as illustrated in Fig. 2. After that step, The DM should insert the consequence matrix into the system. For more details about the consequence matrix see [2].

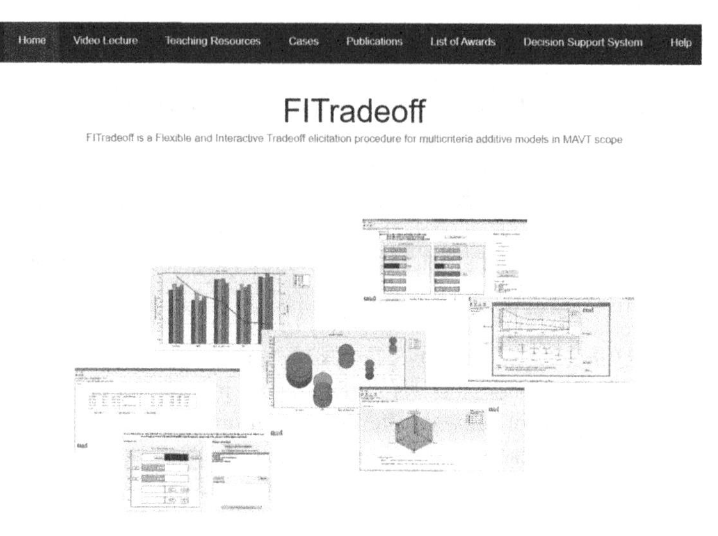

Fig. 1. FITradeoff DSS

Fig. 2. Problematic Selection

After insert the decision matrix into the DSS, the first preferential information presented by DMs is collected. This preferential information regards to the order of scaling constant associated with each criterion. Using the DSS, the DM should select the criterion which had the highest value of scaling constant and clicks on the button "Choose". When the criterion is selected, the bar presents yellow color, and when DM clicks on the button, the bar stays green (Fig. 3). For this problem, the ranking of the scaling constant is represented by Eq. (1).

$$K_{\text{price}} > K_{\text{freight}} > K_{\text{accuracy}} > K_{\text{quality}} > K_{\text{flexibility}} > K_{\text{lead time}} > K_{\text{promptness}} \quad (1)$$

It is worth mentioning that eliciting the scaling constant is the major challenge in methods in the MAVT context [5]. Thus, the FITradeoff method intend to elicit these constants considering using partial information concepts [9, 10]. In addition, to order the scaling constants, DMs can consider a global evaluation, which is important in practice as it reduces the cognitive effort required from DMs, as illustrated in Fig. 3.

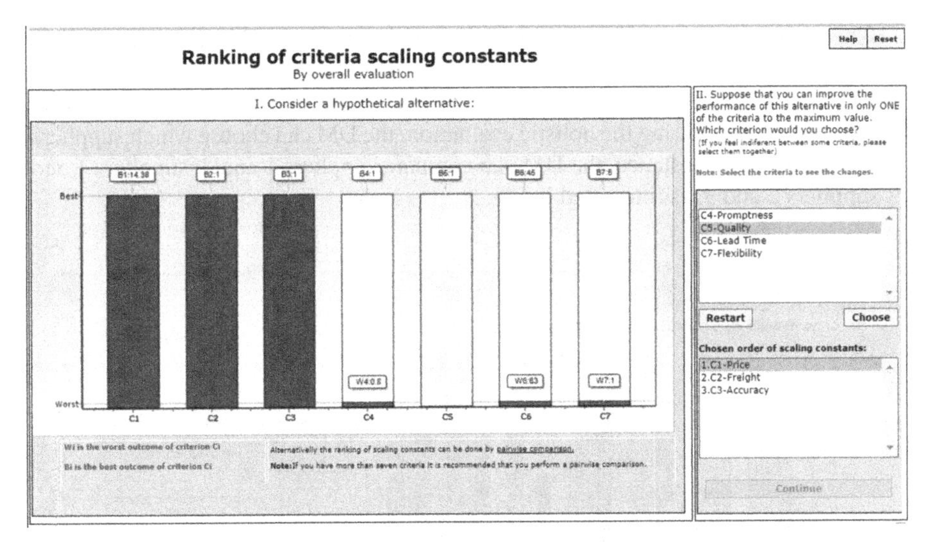

Fig. 3. Ranking of scaling constants

After that, considering only the order of scaling constants, the number of suppliers has been reduced to three. In most cases, after the ranking of criteria weights, the subset of POAs is reduced to up to 5 alternatives at this step of the decision process [36]. In other words, with this preferential information, two suppliers have been eliminated from the problem.

At this moment, the DSS presents the two paradigms of preference modeling – the elicitation by decomposition and the holistic evaluation. Thus, the DM can select which paradigm prefers to express preference relations. For the decision process, the DM prefers the goes to the holistic evaluation step in order to compare the three suppliers (Fig. 4).

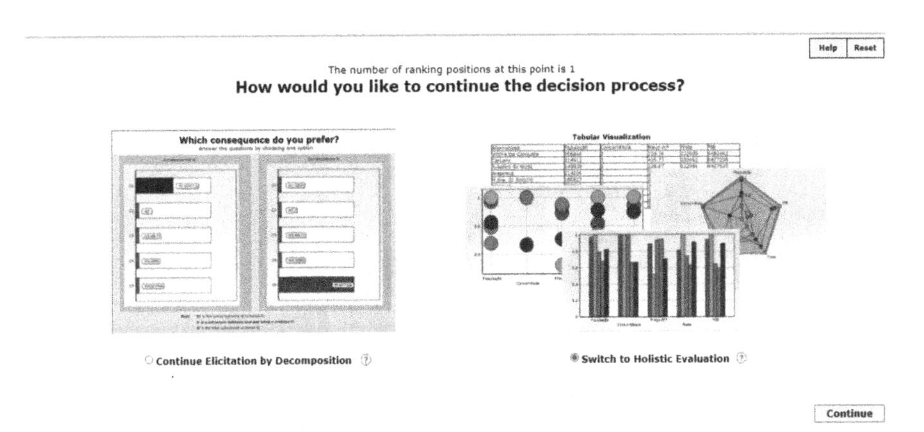

Fig. 4. Two paradigms of preference modelling

Hence, considering the bar graph, the DM can observe the performance of suppliers 1, 2, and 3 against the seven criteria. If the DM wishes, he/she can define a dominance relation between them, selecting the best one in the group or eliminating the worst in the group. Moreover, during the holistic evaluation, the DM can choose which suppliers wish to compare. For instance, the DM can compare suppliers 1 and 2, suppliers 1 and 3 or suppliers 2 and 3, as illustrated in Fig. 5

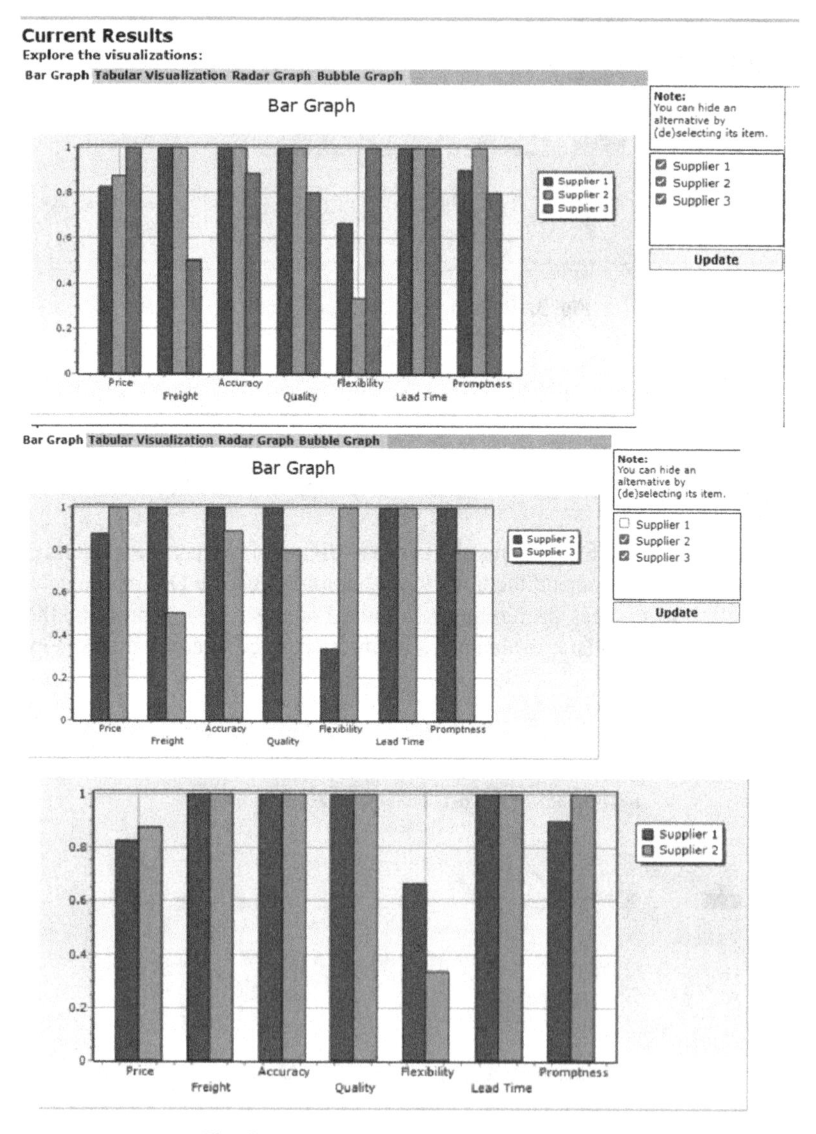

Fig. 5. Bar graphs to compare the suppliers.

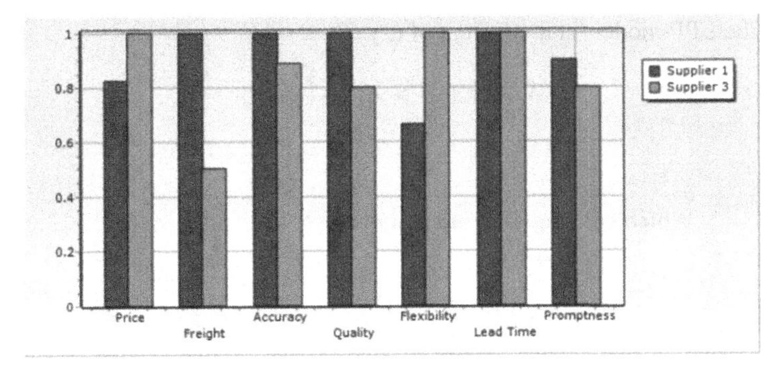

Fig. 5. (*continued*)

In addition, the DM can use another type of graphical or tabular visualization. The DSS presents four types of visualizations: bar graph, spider graph, bubble graph, and tables. For instance, for this problem, the DM uses the table to compare the performance of these three suppliers, as illustrated in Fig. 6.

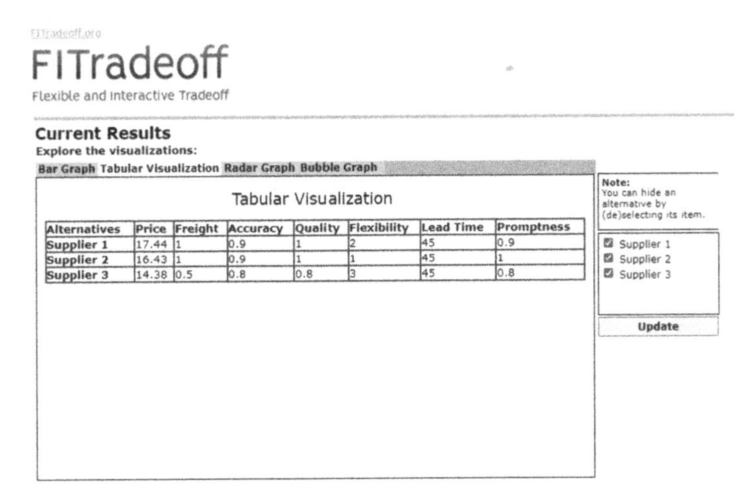

Alternatives	Price	Freight	Accuracy	Quality	Flexibility	Lead Time	Promptness
Supplier 1	17.44	1	0.9	1	2	45	0.9
Supplier 2	16.43	1	0.9	1	1	45	1
Supplier 3	14.38	0.5	0.8	0.8	3	45	0.8

Fig. 6. Table to compare the suppliers.

However, since the suppliers present similar performances in criteria, the DM does not feel comfortable expressing a dominance relation between them during the holistic evaluation. For instance, the alternatives present similar values of price, accuracy, lead time, and promptness. Hence, the DM judges as difficult to define a dominance relation between these alternatives.

In this context, the DM prefers to continue the elicitation by decomposition comparing pairs of consequences. Figure 7 presents a pairwise comparison presented during the decomposition process. Thus, for the comparison of an intermediate consequence in the criterion Price (Consequence A) versus the best consequence in the criterion promptness (Consequence B), the DM prefers Consequence A. In this context, another inequality as inserted in the LPP model, in the format of (2).

$$K_{price} v_i(x_i) > K_{\text{promptness}} \tag{2}$$

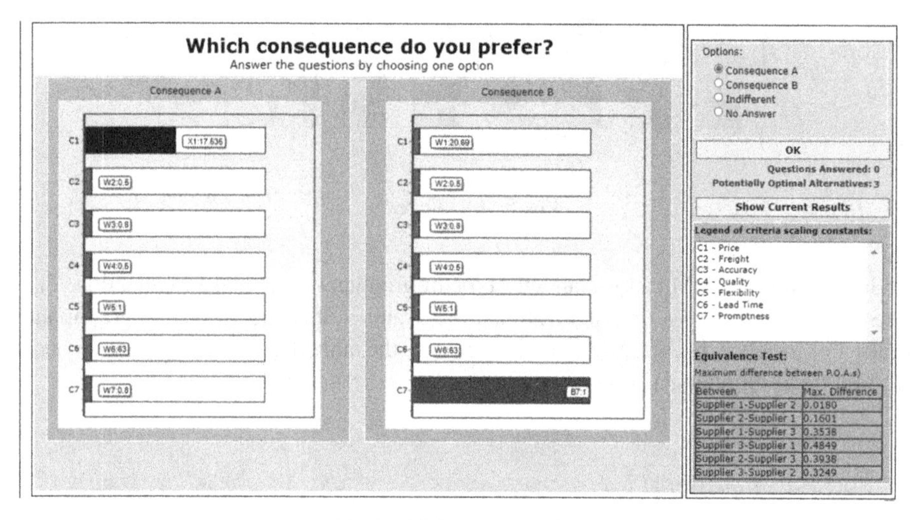

Fig. 7. Elicitation question

The DM continue in the decomposition paradigm comparing consequences, and after one more question the solution has been found. Supplier 2 has been selected as the best one, as illustrated in Fig. 8.

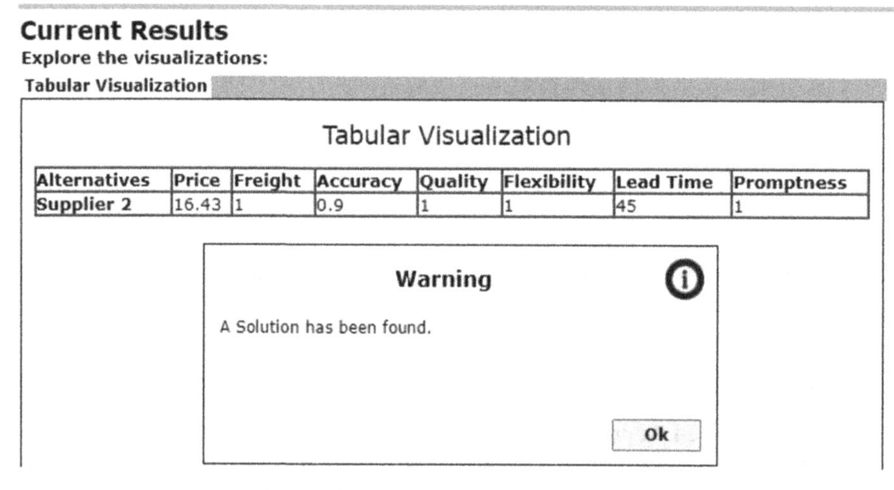

Alternatives	Price	Freight	Accuracy	Quality	Flexibility	Lead Time	Promptness
Supplier 2	16.43	1	0.9	1	1	45	1

Fig. 8. A solution has been found.

4 Behavioral Studies to investigate the Holistic Evaluation

The previous section describes the use of the DSS to solve a supplier selection process [2]. This section presents an overview of behavioral studies performed to investigate how DMs use visualizations to perform holistic evaluations in the FITradeoff DSS.

Several experiments have been done to compare different types of visualizations in order to understand how DMs use these visualizations to express dominance relations between alternatives [9, 10]. In order words, these studies have been done to investigate how users select or eliminate alternatives during the holistic evaluation of the FITradeoff DSS.

Most of the studies have been performed in the NSID (Laboratory for Neuroscience and Behavioral Studies in Decision) laboratory which is located at the Federal University of Pernambuco. To perform the experiments two neuroscience tools have been used: the Electroencephalograph (EEG) with 14 channels by Emotiv and the X120 Eye-Tracking by Tobii Studio [14–25]. The experiments follow a similar scheme, as described below:

- Phase 1: The experiments have been scheduled with all the participants.
- Phase 2: Participant arrives in the NSID lab, signs documents and receives instructions.
- Phase 3: Equipment calibration (EEG, Eye-Tracking and, Computer Screen)
- Phase 4: Experiment Starts
- Phase 5: When the experiment is finished participants answered a preferential survey.

The experiments have been done individually. The sample is composed of graduate and postgraduate students of the Management Engineering course at the Federal University of Pernambuco (UFPE). These students attended MCDM/A Classes and the experiments were an extra activity The total sample is composed of approximately one hundred participants, most of them are post-graduate students.

Moreover, the experiments have been scheduled after a specific class in which the students learn about the MAVT theory [5]. Thus, the students present similar knowledge to perform the experiments. In addition, they received the same information before the beginning of the experiments and had similar conditions to do the required tasks. The experiments have been approved by the Ethical Committee of the university.

The experiments were composed of different types of graphical and tabular visualizations which were shown in a random sequence. For each visualization presented on the screen, the participant should select the best or eliminate the worst alternative, depending on the task required. This is a mandatory task; the participants cannot skip the question. Thus, they had to select the best alternative if asked for it or eliminate the worst alternative, depending on the question presented on the screen. The experiments do not have a pre-defined time. Thus, the participant can perform the holistic evaluation without pressure of time.

The graphical and tabular visualizations used have been constructed to illustrate MCDM/A problems without context. Hence, the alternatives are named by letter (A, B, C) and the criteria by numbers (Crit1, Crit2, Crit3). Problems without context have been used to permit general conclusions.

The aim of the experiments has been to investigate how DMs perform mathematical compensation between alternative performances in order to select the best and the worst alternative of the group. Compensations between consequences should be done since the FITradeoff method follows the MAVT context [5].

Concerning the criteria weights, some visualizations had the same values for criteria weights. Other present different values for criteria weights. In this case, the weights had been obtained following an arithmetic progression [20, 21].

Figure 9 illustrates a bar graph, and a table constructed to represent a problem that has three alternatives evaluated against five criteria. For this problem, the criteria had different values for the weights. Moreover, Crit 1 is those which had the upper value, and Crit 5 is those which had the lower value of weights.

The bar graphic and the table represent the same problem, but the order of alternatives has been changed (alternative A stays in the same position in both visualizations, but alternatives B and C had order changed). Acronyms have been used to define visualizations. For instance, Sel 3A5C BD represents a bar graph (B) which had three alternatives (3A) evaluated against five criteria (5C) which had different values for the weights (D). Also, for this bar graph, the participants had to select the best alternative (Sel).

It is worst to mention that two decision processes have been considered on experiments: the selection of the best alternative and the elimination of the worst alternative. All participants performed both decision processes. Thus, they had to select the best alternative or eliminate the worst alternative in different times.

In order to compare the visualizations presented in the experiments, an important feature has been proposed in a previous study - the probability of success [16]. The probability of success is estimated using the Hit Rate and following a Bernoulli distribution. In this context, three important issues are investigated using the probabilities of success:

- Participants performed better during the selection process or during the elimination process.
- Participants presented higher probabilities of success using which type of visualization (bar graph, spider graph, bubble graph, or table).
- Participants performed better in visualizations that had the same values for the criteria weights or in visualizations that had different values for the weights.

Thus, by computing the probabilities of successes it is possible to observe which configuration participants performed better. The probability of success can be used by DMs and Analysts to support holistic evaluations. In order words, DMs can consult previous probabilities of success which had been already computed for different types of visualizations to proceed with the holistic evaluation or to go to the decomposition process [16].

For instance, if a specific visualization presents a lower probability of success, the DM may not use this visualization to express the dominance relation and can return to the decomposition process in order to express preferences for consequences, as illustrated in Fig. 7.

The probability of success is an important feature developed from behavioral studies to support DMs during the process. In special this feature should be used by analysts to advise DMs on how to proceed with the decision process, performing holistic evaluations or decomposition process in the FITradeoff DSS.

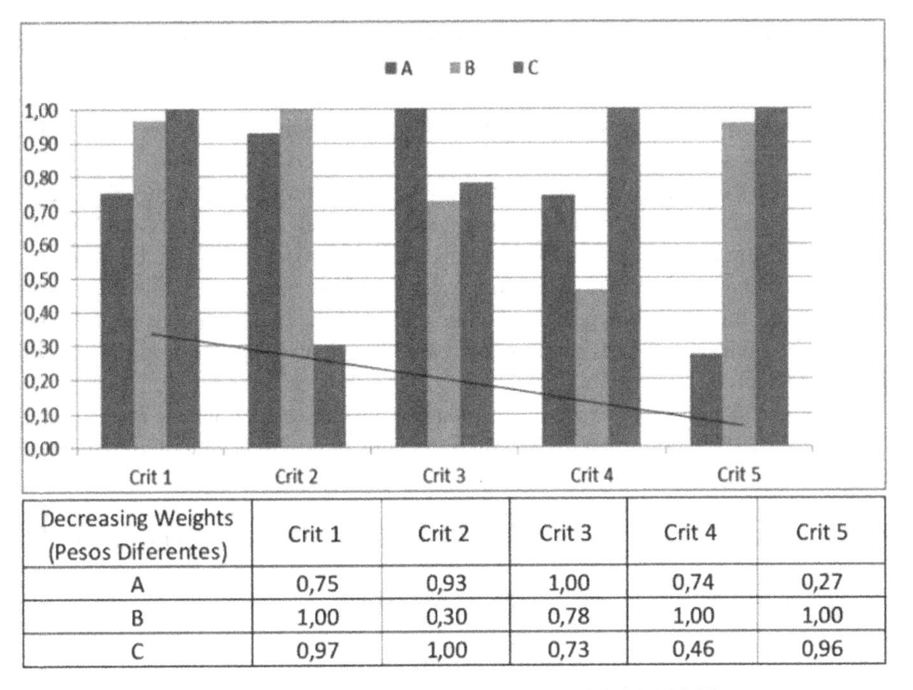

Decreasing Weights (Pesos Diferentes)	Crit 1	Crit 2	Crit 3	Crit 4	Crit 5
A	0,75	0,93	1,00	0,74	0,27
B	1,00	0,30	0,78	1,00	1,00
C	0,97	1,00	0,73	0,46	0,96

Fig. 9. Selection process Sel 3A5C BD and Sel 3A5C TD

5 Conclusion

Decision Support Systems present an important role in organizations' routines. They can be used to support DMs to capture and treat information. DSS permits knowledge management by the access to specific knowledge and dissemination of knowledge between the time.

Concerning to MCDM/A problems, several DSSs have been proposed to support DMs to deal with problems like that, which are very common in organizations routine, such as the supplier selection problem. For instance, the FITradeoff method is proposed to solve problems like that, in the context of MAVT [5].

In this context, this paper describes the DSS constructed to implement the FITradeoff method considering a supplier selection process which is already been published in the literature [2]. Several steps of the DSS have been illustrated and described in order to present the DSS.

Moreover, the paper summarizes several behavioral studies which have been performed to investigate the FITradeoff method. These studies have been performed to transform the FITradeoff DSS including behavioral suggestions captured during neuroscience experiments [14–25].

An important feature proposed with these studies is the probability of success. The probability of success can be used to improve the holistic evaluation process of the FITradeoff method.

In other words, the probability of success is a feature that can be used to advise DMs concerning the use of graphical and tabular visualizations to perform the holistic evaluation.

The probability of success represents an important improvement of the FITradeoff decision process since permits the use of previous knowledge to support DMs during the decision process. Also, it represents an important improvement in the FITradeoff DSS since it can be included in the DSS to provide knowledge for analysts and DMs. In special, this feature can support analysts in the advising process performed with DMs.

Based on behavioral studies several improvements have been done in the FITradeoff DSS. The most important one permits the integration of the two perspectives of preference modeling during the decision process of the FITradeoff. This integration has been described in Sect. 3.

For future research, more experiments intend to be done to continue the investigation of the use of the DSS. For instance, the two paradigms of preference modeling intend to be investigated conjointly.

Acknowledgment. This work had partial support from the Brazilian Research Council (CNPq) [grant 308531/2015-9; 312695/2020-9] and the Foundation of Support in Science and Technology of the State of Pernambuco (FACEPE) [APQ-0484-3.08/17].

References

1. Santos, I.M., Roselli, L.R.P., da Silva, A.L.G., Alencar, L.H.: A supplier selection model for a wholesaler and retailer company based on FITradeoff Multicriteria method. Math. Probl. Eng. **2020** (2020)
2. Frej, E.A., Roselli, L.R.P., Araújo de Almeida, J., de Almeida, A.T.: A multicriteria decision model for supplier selection in a food industry based on FITradeoff method. Math. Probl. Eng. **2017** (2017)
3. Chai, J., Liu, J., Ngai, E.: Application of decision-making techniques in supplier selection: a systematic review of literature. Expert Syst. Appl. **40**(10), 3872–3885 (2013)
4. Barla, S.B.: A case study of supplier selection for lean supply by using a mathematical model. Logist. Inf. Manag. **16**, 451–459 (2003)
5. Keeney, R.L., Raiffa, H.: Decisions with Multiple Objectives: Preferences, and Value Tradeoffs. Wiley, New York (1976)
6. Belton, V., Stewart, T.: Multiple Criteria Decision Analysis. Kluwer Academic Publishers, Dordrecht (2002)
7. Figueira, J., Greco, S., Ehrgott, M. (eds.): Multiple Criteria Decision Analysis: State of the Art Surveys. Springer, Berlin (2005)
8. de Almeida, A.T., Cavalcante, C., Alencar, M., Ferreira, R., de Almeida-Filho, A.T., Garcez, T.: Multicriteria and Multi-objective Models for Risk, Reliability and Maintenance Decision Analysis. ISOR, vol. 231. Springer, New York (2015)
9. de Almeida, A.T., Almeida, J.A., Costa, A.P.C.S., Almeida-Filho, A.T.: A new method for elicitation of criteria weights in additive models: flexible and interactive tradeoff. Eur. J. Oper. Res. **250**(1), 179–191 (2016)
10. de Almeida, A.T., Frej, E.A., Roselli, L.R.P.: Combining holistic and decomposition paradigms in preference modeling with the flexibility of FITradeoff. Central Eur. J. Oper. Res. (2021). https://doi.org/10.1007/s10100-020-00728-z

11. Korhonen, P., Wallenius, J.: Behavioral issues in MCDM: neglected research questions. In: Clímaco, J. (ed.) Multicriteria analysis, pp. 412–422. Springer, Heidelberg (1997). https://doi.org/10.1007/978-3-642-60667-0_39
12. Wallenius, H., Wallenius, J.: Implications of world mega trends for MCDM research. In: Ben Amor, S., De Almeida, A., De Miranda, J., Aktas, E. (eds.) Advanced Studies in Multi-Criteria Decision Making. Series in Operations Research, 1st edn., pp. 1–10. Chapman and Hall/CRC, New York (2020)
13. Eagleman, D.: The Brain: A Story of You. Pantheon Books, New York (2015)
14. de Almeida, A., Rosselli, L., Costa Morais, D., Costa, A.: Neuroscience tools for behavioural studies in group decision and negotiation. In: Kilgour, D.M., Eden, C. (eds.) Handbook of Group Decision and Negotiation, 1st edn., pp. 1–24. Springer, Dordrecht (2020). https://doi.org/10.1007/978-3-030-12051-1_53-1
15. Roselli, L.R.P., de Almeida, A.T.: Use of the Alpha-Theta Diagram as a decision neuroscience tool for analyzing holistic evaluation in decision making. Ann. Oper. Res. (2022)
16. Roselli, L.R.P., de Almeida, A.T.: The use of the success-based decision rule to support the holistic evaluation process in FITradeoff. Int. Trans. Oper. Res. (2021)
17. Reis Peixoto Roselli, L., de Almeida, A.T.: Analysis of graphical visualizations for multi-criteria decision making in FITradeoff method using a decision neuroscience experiment. In: Moreno-Jiménez, J.M., Linden, I., Dargam, F., Jayawickrama, U. (eds.) ICDSST 2020. LNBIP, vol. 384, pp. 30–42. Springer, Cham (2020). https://doi.org/10.1007/978-3-030-46224-6_3
18. Roselli, L.R.P., de Almeida, A.T.: Improvements in the FITradeoff decision support system for ranking order problematic based in a behavioral study with NeuroIS tools. In: Davis, F.D., Riedl, R., vom Brocke, J., Léger, P.-M., Randolph, A.B., Fischer, T. (eds.) NeuroIS 2020. LNISO, vol. 43, pp. 121–132. Springer, Cham (2020). https://doi.org/10.1007/978-3-030-60073-0_14
19. Roselli, L.R.P., De Sousa Pereira, L., De Almeida, A.T., Morais, D.C., Costa, A.P.C.S.: Neuroscience experiment applied to investigate decision-maker behavior in the tradeoff elicitation procedure. Ann. Oper. Res. 1–18 (2019)
20. Roselli, L.R.P., De Almeida, A.T., Frej, E.A.: Decision neuroscience for improving data visualization of decision support in the FITradeoff method. Oper. Res. Int. J. **19**, 1–21 (2019)
21. Roselli, L.R.P., Frej, E.A., de Almeida, A.T.: Neuroscience experiment for graphical visualization in the FITradeoff decision support system. In: Chen, Ye., Kersten, G., Vetschera, R., Xu, H. (eds.) GDN 2018. LNBIP, vol. 315, pp. 56–69. Springer, Cham (2018). https://doi.org/10.1007/978-3-319-92874-6_5
22. Pessoa, M.E.B.T., Roselli, L.R.P., de Almeida, A.T.: A neuroscience experiment to investigate the selection decision process versus the elimination decision process in the FITradeoff method. In: EWG-DSS 7th International Conference on Decision Support System Technology, Loughborough, UK (2021)
23. da Silva, A.L.C.D.L., Cabral Seixas Costa, A.P., de Almeida, A.T.: Analysis of the cognitive aspects of the preference elicitation process in the compensatory context: a neuroscience experiment with FITradeoff. Int. Trans. Oper. Res. **31** (2022). https://doi.org/10.1111/itor.13210
24. da Silva, A.L.C.D.L., Costa, A.P.C.S., de Almeida, A.T.: Exploring cognitive aspects of FITradeoff method using neuroscience tools. Ann. Oper. Res. 1–23 (2021)
25. da Silva, A.L.C.L., Costa, A.P.C.S.: FITradeoff decision support system: an exploratory study with neuroscience tools. In: NeuroIS Retreat 2019, Viena. NeuroIS Retreat (2019)
26. Kang, T.H.A., Júnior, A.M.D.C.S., de Almeida, A.T.: Evaluating electric power generation technologies: a multicriteria analysis based on the FITradeoff method. Energy **165**, 10–20 (2018)

27. Carrillo, P.A.A., Roselli, L.R.P., Frej, E.A., de Almeida, A.T.: Selecting an agricultural technology package based on the flexible and interactive tradeoff method. Ann. Oper. Res. 1–16 (2018)
28. Monte, M.B.S., Morais, D.C.: A decision model for identifying and solving problems in an urban water supply system. Water Resour. Manag. **33**(14), 4835–4848 (2019)
29. Fossile, D.K., Frej, E.A., da Costa, S.E.G., de Lima, E.P., de Almeida, A.T.: Selecting the most viable renewable energy source for Brazilian ports using the FITradeoff method. J. Cleaner Prod. 121107 (2020)
30. Dell'Ovo, M., Oppio, A., Capolongo, S.: Decision Support System for the Location of Healthcare Facilities: SitHealth Evaluation Tool. BRIEFSPOLIMI, Springer, Heidelberg (2020). https://doi.org/10.1007/978-3-030-50173-0
31. Pergher, I., Frej, E.A., Roselli, L.R.P., de Almeida, A.T.: Integrating simulation and FITradeoff method for scheduling rules selection in job-shop production systems. Int. J. Prod. Econ. **227**, 107669 (2020)
32. Frej, E.A., de Almeida, A.T., Costa, A.P.C.S.: Using data visualization for ranking alternatives with partial information and interactive tradeoff elicitation. Oper. Res. Int. J. **2019**, 1–23 (2019)
33. Kang, T.H.A., Frej, E.A., de Almeida, A.T.: Flexible and interactive tradeoff elicitation for multicriteria sorting problems. Asia Pac. J. Oper. Res. **37**, 2050020 (2020)
34. Frej, E.A., Ekel, P., de Almeida, A.T.: A benefit-to-cost ratio based approach for portfolio selection under multiple criteria with incomplete preference information. Inf. Sci. **545**, 487–498 (2021)
35. Marques, A.C., Frej, E.A., de Almeida, A.T.: Multicriteria decision support for project portfolio selection with the FITradeoff method. Omega-Int. J. Manag. Sci. **111**, 102661 (2022)
36. MendeS, J.A.J., Frej, E.A., De Almeida, A.T., De Almeida, J.A.: Evaluation of flexible and interactive tradeoff method based on numerical simulation experiments. Pesquisa Operacional **40**, e231191, 1–25 (2020). https://doi.org/10.1590/0101-7438.2020.040.00231191

A Step-By-Step Decision Process for Application Migration to Cloud-Native Architecture

Daniel Olabanji(✉), Tineke Fitch, and Olumuyiwa Matthew

University of Portsmouth, Portsmouth, UK
daniel.olabanji@port.ac.uk

Abstract. Cloud-native architecture is changing the cloud computing environment and introducing new dimensions to cloud computing adoption through the advancement in virtualisation, scalability and automation. However, many organisations found it challenging to adopt cloud-native architectures because of some issues, mainly the decision-making around migration because of the need for the adoption decision process. However, there are many decision-support systems or processes for cloud computing. Nevertheless, the additional characteristic and complexity of cloud-native architecture made many organisations find it challenging to adopt cloud-native architecture solutions. The need for a new decision-support process for the migration of either a native cloud application or legacy application to cloud-native architecture is highly needed to help the developers to be able to select the choice of model and whether to migrate or not. Therefore, this paper provides a step-by-step decision process for an organisation to migrate their application to cloud-native architecture to fill the gap. The decision process depends on the step-by-step cross-analysis of the steps and the tasks to help identify and resolve critical constraints that can affect migration into cloud-native architecture. An expert evaluation method was used to evaluate the decision process and steps. The decision process was helpful as they informed the decision makers about the needful steps and tasks to consider before migrating to the cloud-native architecture.

Keywords: Cloud-native architecture · Decision Process · Portability · Migration

1 Introduction

Cloud-native architecture is changing the cloud computing environment and introducing new dimensions to cloud computing adoption through the advancement in virtualisation, scalability and automation. Cloud-native architecture is not stand-alone technology but part of the cloud computing paradigm. The term has been used in the early period of the development of cloud computing technology to mean applications developed solely for cloud computing [1]. The progression in cloud computing through the development of several parts of the technology, cloud-native architecture, evolved to mean something different from the early definition. The cloud-native architecture, as

I. Saad et al. (Eds.): ICIKS 2023, LNBIP 486, pp. 63–77, 2024.
https://doi.org/10.1007/978-3-031-51664-1_5

defined by the Cloud-Native Computing Foundation (CNCF), is the sole convener of the technology. Cloud-native is a set of technologies that empower organisations to build and run scalable applications in modern, dynamic environments such as public, private, and hybrid clouds, and the approach is exemplified by "containers, service meshes, microservices, immutable infrastructure, and declarative APIs" [2]. Furthermore, CNCF stated that these techniques enable loosely coupled systems that are resilient, manageable, and observable with robust automation that allows engineers to make high-impact changes frequently and predictably with minimal work [2]. Cloud-native technologies are used to develop applications built with services packaged in containers, deployed as microservices, and managed on elastic infrastructures through agile DevOps processes and continuous delivery workflows [3].

Cloud-native architecture is a methodology that complies with deploying and developing applications built with services packaged in containers [4], deployed as microservices [5], and managed on elastic infrastructure through agile DevOps processes and continuous delivery workflows. Such an application must follow the design methodology before it can be referred to as a cloud-native application. In agreement with the above, [6] discussed that cloud-native deployment or development of applications should agree with the cloud-native's principles, properties, methods and architecture. The cloud-native architecture consists of four critical pillars: containers, microservices, DevOps and Continuous Integration/Continuous Delivery, which was indicated in all the definitions. Moreover, a clearer view of how the architecture worked and was implemented was expatriated, as seen in Fig. 1.

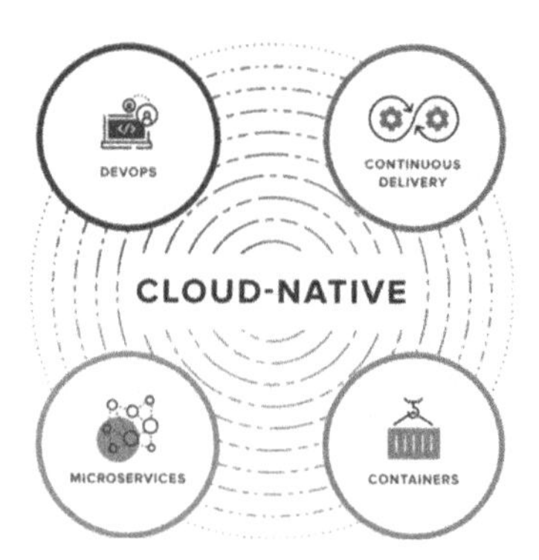

Fig. 1. Cloud-native application component adapted from [3]

Cloud-native architecture has reduced IT expenses to the barest minimum and made scaling both horizontally and vertically easier and more achievable in several organisations that adopt the architecture; with several benefits derived from using the cloud-native architecture, many other businesses are interested in migrating into this architecture. However, several challenges such as security, technicality, multitenancy and others affect the migration. These challenges bring the need for guidelines and decision support tools for enterprises that are considering migrating. To fill this gap, this paper presents a step-by-step process aimed at supporting organisations and application developers in making informed migration decisions into cloud-native architecture. The process relies on several pieces of research work that have been done based on cloud-native architecture criteria concerning critical characteristics of organisations. This decision process enables the stakeholders to cross-analysed the steps and tasks to identify and resolve critical constraints that can affect a smooth migration of data or applications into cloud-native architecture.

The rest of the paper is organised as follows. Section 2 details the decision process, describing its main activities flow. Also, it will consist of the phases of the decision process. Section 3 illustrates the use of expert review to evaluate the decision process by experts in cloud computing. Finally, section 4 offers some concluding remarks and outlines our future work.

2 Decision Process for Cloud-Native Migration

2.1 Process Overview

The proposed decision support process is designed for enterprises already using cloud-native architecture or considering the adoption of cloud-native architecture in their businesses. Organisations can use the framework to check their business considerations, needs, benefits and the thoughts put in place before and after porting. The need for a framework based on the decision-making process for an organisation is essential. It is a challenging task due to the wide and ever-expanding variety of IaaS service offerings, with the need for more knowledge about the selection criteria in a diverse market that is likely to confuse the customer [7]. Decision-making processes include a series of steps, structure the decision problem, access the possible impact of alternatives, determine the decision-makers preferences, and evaluate and compare alternatives [8].

The decision process can be used to develop a framework whose core function is to act as an assessment tool for crucial stakeholders when selecting cloud services and to guide decision-makers [8]. A regulatory framework for decision-making is one of the guidelines to support effective decision-making processes for the enterprise [9].

2.2 The Decision Process and Steps

The proposed step-by-step decision process was designed to support organisations in their quests to move data and applications to the cloud or from one cloud service or deployment model to another without significantly redesigning or reconfiguring data or applications. The proposed decision process was divided into mini steps and tasks for easy understanding and to give a clear path that can provide organisations with the support needed to implement and use the framework.

As seen in Fig. 2, The decision process in the framework was divided into four (4) key phases, which are:

- Phase 1 Intention and planning
- Phase 2 Model selection and evaluation
- Phase 3 Contract and Execution implementation
- Phase 4 Testing and Management

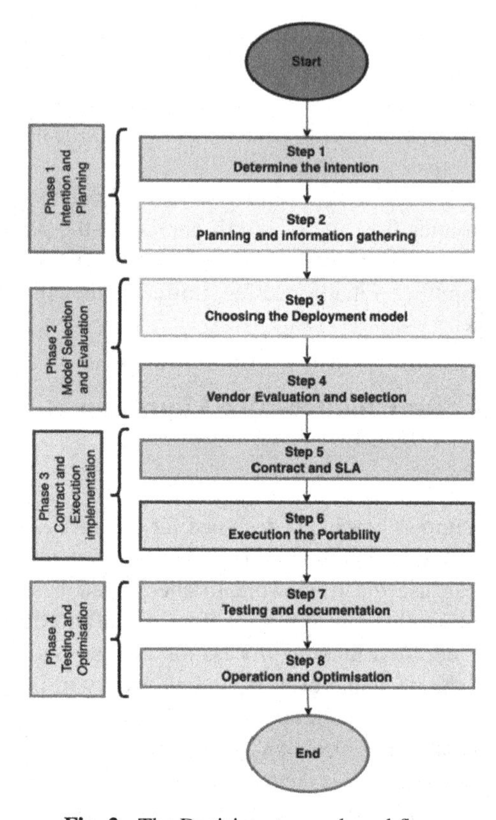

Fig. 2. The Decision step and workflow

The phases have eight (8) steps within them, and each step has tasks and considerations, which are discussed in the following subsection.

1) **Phase 1 Intention and Planning**

In this phase, the intention to adopt and use cloud-native architecture, the stakeholder and the organisation should be considered. The phase also includes strategic planning borrowed from project management, including tasks such as pricing, risk assessment and budgeting concerning decision-making. Such considerations are rare in most cloud computing decision-support frameworks. After literature reviews and several considerations, the content of the phase is critical and needed by organisations migrating or porting into cloud computing through cloud-native architecture, especially in lower-middle-income countries. The phase was further divided into two steps which are:

A) *Step 1: Determine the Intention.*

Determine the intention to help the business establish the migration's purpose to cloud-native architecture and answer the Why question from different perspectives. The following questions can be used to determine the intention of migrating or adopting cloud-native architecture.

- Who was the initiator of the migration?
- How easy is the usage of the technology around cloud native architecture?
- What is the organisation's attitude to new technology, especially cloud-native architecture?
- What is the technical experience level of the organisation's personnel?
- What advantage will the migration have on the organisation?

Answering this question and doing the task in Step 1 of the decision process allows the organisation to have explicit knowledge of the reason behind the migration and adoption of cloud-native architecture. The task can be seen in Fig. 3.

B) *Step 2: Planning and Information Gathering*

Step 2 focuses on the initial planning and information gathering needed after the intention is more precise and understood. The planning involves all stakeholders, including project managers, engineers, and others, because budgeting, requirement analysis and budgeting are considered at this stage. As shown in Fig. 3, this step laid the foundation for the next stage of the framework in which the evaluation, decision-making, and implementation occur. The following question can be used to assess the planning and readiness for the migration and to gather more information about the organisation before migrating.

- What is the organisation or financial constraints (including debt)?
- How does this migration impact the daily activities of the organisation?
- What are the factors influencing the stakeholder decision for migration?
- What is the organisation's IT budget, and how much is going into the migration?
- What are the threats and risks that migration can cause, and how can the risk be mitigated?

Furthermore, combining a precise intention and comprehensive planning before the porting or migration will produce a successful adoption.

2) **Phase 2 Model Selection and Evaluation**

The decision-making part of the decision process or framework starts from this phase, which is the phase of model selection and evaluation phase. This phase is essential because it introduces a new cloud decision support system research perspective. The choice of cloud deployment model is paramount to deciding data or application migration into the cloud-native architecture. It is noteworthy that most cloud computing adoption and decision-support systems focus mainly on the public cloud. However, several kinds of research work have been done about the difference between cloud computing deployment models. Considering the difference and how cloud-native architecture interacts with the deployment model. This addition gives the organisation using the decision process to consider options for the model that they will want to migrate or use. The phase consists of two steps, shown in Fig. 4, and more details about the step will be discussed in the following section.

A) *Step 3: Choosing the Deployment Model*

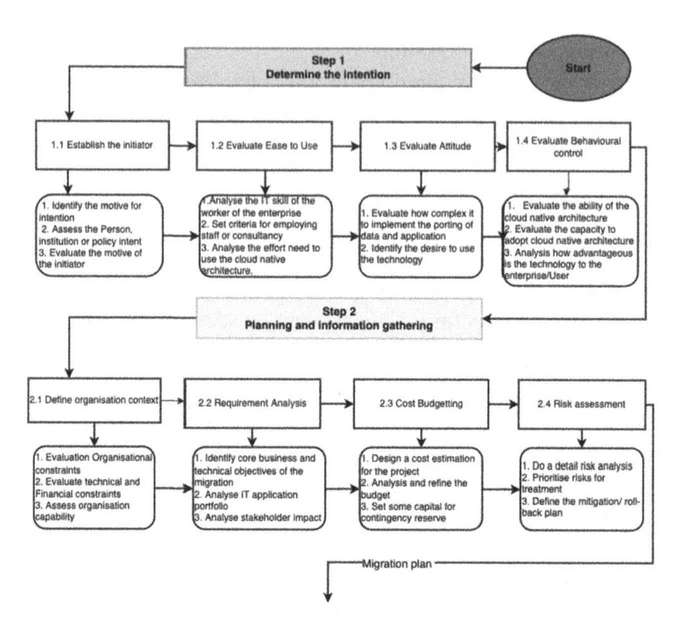

Fig. 3. Phase 1 of the decision support process

Step 3 of the decision process focuses on the choice of cloud deployment model required for application or data. At this stage, the user or organisation is allowed to make a potential decision before going deep into the implementation of the porting or migration into cloud-native architecture. The steps dealt with factors responsible such as the policy, Capital expenditures (CAPEX) or Operation expenses (OPEX) cost, security and growth consideration before making such decisions as considered in the sub-step and tasks. The stakeholder needs to answer the following question:

- What are the policies or governance standards that determine how and where the application data are stored?
- Are there any security considerations, and if there are, what are they?
- What is the cloud Capex or Opex for the application or data storage?
- Is there any forecast metric for the usage and growth of the application and data?

Furthermore, many following steps depend on the choice of cloud deployment model. After the organisation's choice is made on the deployment model, other steps can start first through the consideration of Step 4.

B) *Step 4: Vendor Evaluation and Selection*

In this study, vendor evaluation and selection should be understood through the perceptive of both private and public cloud deployment models. In the case of public cloud computing, the vendors are cloud service providers, which could be any of Amazon Web Services (AWS), Microsoft Azure, Google Cloud, IBM, Oracle and others. A step-by-step examination of the service provider is needed before selecting the porting and migrating plan in the public cloud. However, in the private cloud, the vendors are those

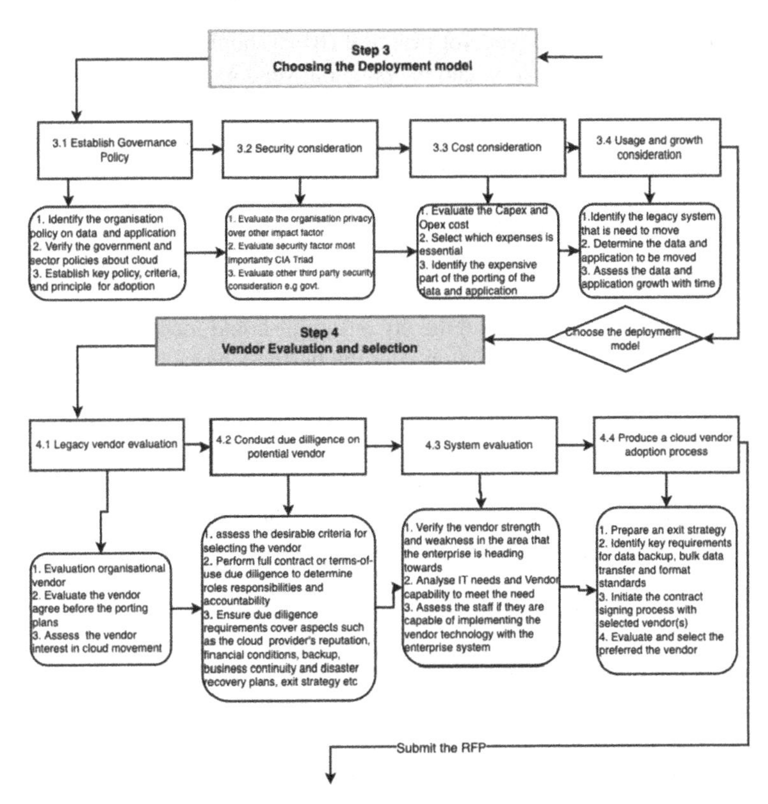

Fig. 4. Phase 2 of the decision support process

companies that rent or sell hardware, software, and service with an economic relationship between seller and buyer. Businesses such as Dell, Cisco, Hewlett-Packard (HP), Microsoft and others fall into this qualification. In both the public and private cloud, the vendors could be examined with the tasks in Step 4. This step involves sub-steps such as checking the legacy vendor for continuity and perhaps news solution that the organisation can still benefit from, conducting due diligence on the choice of vendors before choosing, evaluating the system, and producing a cloud vendor process. In order to achieve the vendor evaluation and selection, the following questions needed to be answered, and tasks in the steps needed to be done.

- What are the learning points derived from the evaluation of the legacy system or past cloud computing?
- Is the new vendor evaluated based on the contract, and what is covered and what is not covered?
- Does the organisation understand the need for the technology and how the vendor(s) can cater for the needs?
- Is there a new vendor request to proposal processed, signed and evaluated by the requirement of the application migration to the cloud-native architecture?

At the end of Phase 2, a request for proposal (RFP) should be developed. RFP is a document between the organisation and the potential vendors that should be developed and sent, leading to the next phase of the decision-support process. It should be noted that the choice of deployment model will determine how the organisation or user will deal with the vendor selection. The reason is that if the user selects either public, private, or hybrid, some details will need to be more focused on than others, which may be insignificant to the best deployment model for the organisation.

3) **Phase 3 Contract and Execution**

The next thing the developers and the organisation should focus on is developing a contract and implementing the migration after the business has agreed on the contract. This phase involves several parts that the contract covers and the execution of the migration or porting, which can be achieved through detailed consideration of the Service Level Agreement (SLA). The contract and SLA part of the phase covered the chosen deployment model and the contract's design and acceptance. As seen in Fig. 5, this phase is divided into two main steps, which are discussed in the section below.

A) *Step 5 Contract and SLA*

Step 5 is another important part of the decision process to create a contract between the organisation and service provider for the cloud deployment model. The public cloud adopter can use the step to make a contract and SLA between the cloud service provider and the organisation. At the same time, for the private deployment model, the relationship between the organisation and IT vendors can be the step to prepare for executing the porting. Likewise, the commitment between the service provider or supplier and the customer to the quality and availability of the service using the service-level agreement (SLA). It is noteworthy that service providers or vendor organisations are designed to make a profit, which may involve advertising and presenting solutions or products that the organisation may or may not need for the business. However, such solutions or products still need to be fully understood and analysed and agreed upon through a clear contract to guide the organisation about their responsibility in the contract. The following question helped track and evaluate if the contract and SLA were put together positively impacting the organisation.

- Does the organisation have a negotiation team?
- Is any negotiation process going on with the vendor(s) in the organisation?
- At what point is the agreement between the organisation and vendor made?
- Did the final SLA mirror all the needs of the organisation?
- When and who is signing and agreeing to the SLA?

Some of the questions may look basic, but for a business, it is needed, and the combination of all these sub-steps is designed to foster the relationship between the organisation and service provider based on contact and agreement.

B) *Step 6 Execution of the Porting or migration*

Several considerations are put in this step because this is where the implantation starts. Developers and organisations need to understand that when applications or services are being moved from one position to the other in cloud-native architecture, the

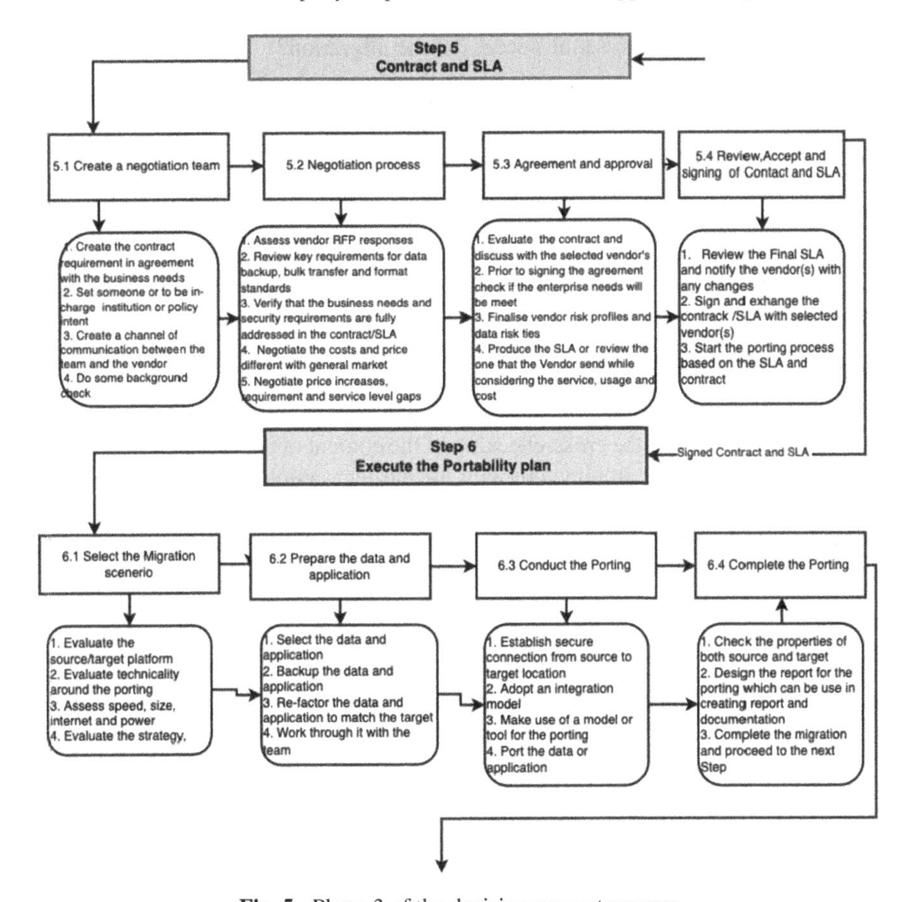

Fig. 5. Phase 3 of the decision support process

implementation is different from cloud computing or Service oriented application architecture. In this step, success of Steps 1 to 5 brings about the success of this step. Also, the migration or porting should have been successful at the end of the step. The step includes using technical skills and cloud-native architecture knowledge, including the configuration of the systems, orchestrator, pod and namespace, networking, database design and development and others. However, the different approaches to the deployment model and the porting or migration to the cloud-native architecture should be seriously considered and implemented. To achieve success in this step and to be able to move to the next phase, the following question needs to be answered concerning the task presented in Fig. 5.

- What scenario and technicality are the application or data migration following or needs?
- Are the application and data well prepared from migration with the backup plans?
- What is the technical detail (Memory size, Hard drive size, CPU speed needed) of the application or data to be migrated?
- How secure is the migration process?

- How can we rate and analyse the success of the migration?

The step comprises checking both the source and destination by ensuring that all parts of the data, application or service are duly and fully ported into the necessary cloud deployment model that fits the organisation's needs. Furthermore, the step provides a detailed view of the porting process step by step, following the consideration of the sub-step by the organisation. This is done first to provide adequate reporting of the porting process, which may include the success story and vice versa and documentation purposes needed in the next phase. After completing the step, the organisation can move to the next phase of the framework.

4) **Phase 4 Testing and Management**

Testing and managing the ported service, data or application is essential in the decision process. In this phase, the cross-checking of the content of the application or service migrated into cloud-native architecture with the earlier version. Also, this phase helps to check if the SLA is followed subsequently with the agreement in the SLA document. For this phase to be achieved, the phase is also divided into two steps which are Step 7, Testing and Documentation and Step 8, Operation and Optimisation, respectively. Details about the step can are in the subsection below.

A) *Step 7 Testing and Documentation*

After a detailed evaluation of the porting and the report generated in step 6, the next is details testing and documentation of the porting activities, which is the whole reason for step 7, and the first thing the organisation must do at this stage is to build the testing team. However, the organisation needed skillful team members while the remaining parts of the step were achieved. Following the team building, these are the questions that step 7 provided and answering the questions answered the testing and documentation part of the decision process and framework.

- Does the organisation have a team of testers and testing procedures to follow?
- Does the testing cover both the functional and non-functional parts of the application?
- Who will produce and assess the document from the documentation from the migration?
- How many errors will be encountered in the implementation, and who and when will document them?

B) *Step 8 Operation and Optimisation*

Operating and optimising a working system is an added task. The step provides a tested approach to the daily operational task of the organisation, as seen in Fig. 6. The phase consists of the last part of the framework, and at this stage of the framework, the organisation can at the end of the phase; the framework does not mean that the system cannot be rollback or terminated when necessary. Notably, all steps of the framework are related and influence each other to get to the final part of the decision process. This task includes creating an exit strategy, managing the relationship with the vendor, and managing the configuration documentation and strategy for continuing and discontinuing

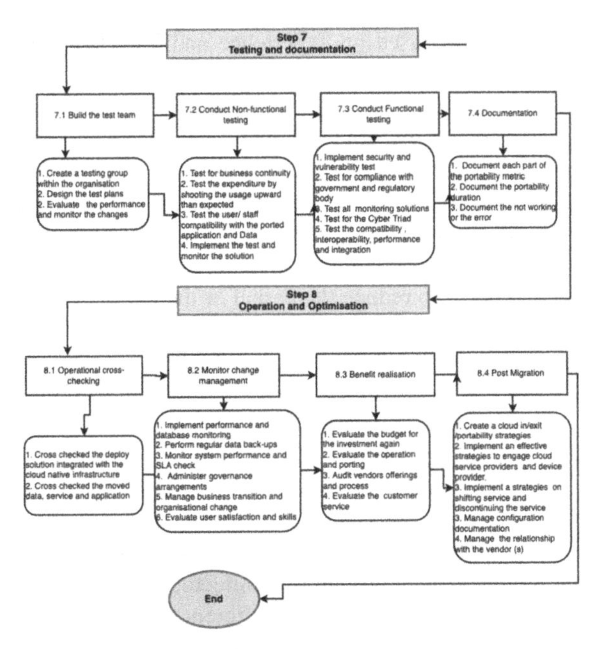

Fig. 6. Phase 4 of the decision support process

the service. The following questions are to be asked by the stakeholders or developer in order to achieve Step 8, which is operation and optimisation.

- Does the migrated application able to achieve the set operation aim of the application?
- Does the user get satisfaction with using the migrated application or data?
- How do we manage the success and the benefit realised?
- How do we keep track of any other changes that the application or data migrations caused?

These systematically arranged steps and tasks provide a straightforward way to monitor all parts of the system through a systematic transition or transformation of the organisation's technologies and information system. This could be done by looking closely at the changes happening in the business through migration and strategically providing business continuity through continuous monitoring and change management.

3 Evaluation of the Proposed Eight-Step Decision Process and Framework for Cloud-Native Architecture Migration

This section presents the empirical evaluation of the usability and effectiveness of the novel 8-step decision framework through the expert evaluation method. The evaluation is done through the expert evaluation method to evaluate the framework's usability with the help of experts specialised in the subject matter. The expert evaluation method is a method that is used to evaluate parameters of complex objects effectively and make decisions based on them. Experts that know how to receive information, digest it and provide enhanced information based on experience and knowledge are the right and frequently set of people to be contacted for such evaluation [10]. The expert evaluation method has been used in several pieces of research. [11]. In evaluating the decision-making framework for cloud computing [12], policy research discussed expert evaluation as a highly considered internal reliability and result-based legitimacy [13] and several other researchers. Determining the levels of verbal scales used to evaluate quantitative and qualitative parameters in expert evaluation is a serious issue when using expert assessment for decision-making evaluation [14]. However, to mitigate this issue, the five-point Likert scale was adopted to reduce the verbal or the use of technical terminology that may impact the evaluation and a sample of the question from [12] is followed in the designing of the question.

Recruiting participants for expert evaluation was selected using a non-probability sampling method to sample experts on the subject area that are willing to provide their feedback or evaluation of the framework. Up to 50 participants are contacted for the evaluation. The target population is the potential participants for this kind of evaluation because they are more interested in the subject matter and are experts and ready to give such an evaluation [15]. The participant needed to understand the migration or portability in cloud computing, focusing on cloud-native architecture. Twenty participants in total participated in the evaluation. They come from different education levels, starting from the Nation Diploma to PhD and from different range of career levels; all this help to balance the bias that may come from the selection.

3.1 The Appropriateness of the Steps

The expert review on the acceptability and suitability of the steps shows that many of the reviewers understand the step-by-step arrangement of the decision process. Table 1 below shows the data analysis of the expert view on how appropriate the steps are. Many of them agreed that all the steps are very appropriate, and the location of the steps is also appropriate.

Table 1. The appropriate of the steps and important steps can be seen in

	Variance	Mean	Standard Deviation
Step 1 – Determine the intention	0.63	4.15	0.79
Step 2 – Planning and information gathering	0.35	4.42	0.59
Step 3 – Choosing the deployment model	0.7	4	0.84
Step 4 – Vendor evaluation and selection	0.53	4.15	0.73
Step 5 – Contract and SLA	0.46	4.2	0.68
Step 6 - Execute the portability plan	0.44	4.37	0.67
Step 7 – Testing and documentation	0.74	4.4	0.86
Step 8 – Management and Optimisation	0.63	4.35	0.79

Table 1 provides a clear view of the expert consideration and how the data is well spread through the mean and the standard deviation. Comparing the steps based on the descriptive statistics computation shows that the difference between average mean values for the steps is less than the least standard deviation (i.e., 0.59). This means that the most significant standard deviation from the sample mean (i.e., 0.86) is less than twice the smallest (2*0.59 = 1.18), illustrating that the survey data is reasonably standard. After detail considering the output of the analysed data, it can be inferred that the steps are appropriate, and the evaluation results of the decision process can be considered reasonable.

3.2 The Importance of the Steps?

The expert evaluation method was used to evaluate the importance of the individual steps and their position in the decision process. The expert response to the question in the questionnaire set to them reflected this, and a descriptive statistical method was used to generate the data in Table 2. Data in the table agreed with the participant's consideration about how impactive the arrangement and arrangement of the steps and the individual location of the step.

Table 2. Important is the decision process Steps.

Steps	Variance	Mean	Standard Deviation
Step 1 – Determine the intention	0.53	4.15	0.73
Step 2 – Planning and information gathering	0.43	4.32	0.65
Step 3 – Choosing the deployment model	0.6	4	0.77
Step 4 – Vendor evaluation and selection	0.53	4.25	0.73
Step 5 – Contract and SLA	0.21	4.3	0.46
Step 6 - Execute the portability plan	0.33	4.35	0.57
Step 7 – Testing and documentation	0.25	4.55	0.5
Step 8 – Management and Optimisation	0.59	4.25	0.77

Using the descriptive statistics computation to show the difference between average mean values for the steps is less than the least standard deviation (i.e., 0.50) for the step. This means that the most significant standard deviation from the sample mean (i.e., 0.77) is less than twice the smallest (2*0.50 = 1), which further illustrates that the survey data is reasonably expected. The output of the analysed data shows that the steps are essential, and the evaluation results of the novel decision framework can be considered important.

3.3 Effectiveness of the Decision Process or Framework

This study further questioned the expert's view on the effectiveness of the decision process based on their understanding of the subject area and the decision process, and their responses were documented in Table 3. As seen in Table 3, most of the reviewers agreed that the decision process was effective, and some agreed that it was Extremely effective. Based on the feedback, the decision process is very effective and is needed now.

The result from reviewers depicts the decision process to be effective. As seen in Fig. 7, Forty-five per cent (45%) of the respondents consider the framework an "extremely effective framework", and another 45% also consider this framework a "quite effective" one. From the result shown in the graph and table and the response by IT professionals and cloud experts, one can conclude the effectiveness of each task as opposed to its non-effectiveness. None of the participants indicated that the decision process could be more effective. One is moderate, and the other is slightly effective (see Fig. 7).

Table 3. Response of reviewers over the effectiveness of the Decision process

Pointer	Response	
Extremely effective (1)	45%	9
Quite effective (2)	45%	9
Moderately effective (3)	5%	1
Slightly effective (4)	5%	1
Not at all effective (5)	0%	0
Total	**100%**	**20**

Fig. 7. A figure carpeting the respondents consider the decision process.

4 Conclusions and Future Work

In conclusion, the decision process or the decision framework for cloud native architecture migration or porting is a very needed process for organisations and stakeholders that are considering the adoption and migration to the architectures. This study also shows the difference between the decision-supporting process for cloud-native architecture and cloud computing. These differences require a decision process and steps tailored to the cloud-native architecture. The decision process provided the stakeholders with a different perspective on the decision by adding the decision about the deployment model as one to be considered through the steps and the tasks in the framework. Furthermore, the expert evaluation method used in this study also points out the need for the process.

However, there are areas that this study can improve the process by implementing the process on an application, and the knowledge base tool, such as the expert system. This can be developed to facilitate the framework through the designing of a system that can automate the expert views. The future work for this research is the provision of a case study where the framework has been utilised and the evaluation of such implementation will be provided in a future research. In addition, the used of machine learning and artificial intelligence to test the framework in future publication.

References

1. Andrikopoulos, V., Binz, T., Leymann, F., Strauch, S.: How to adapt applications for the Cloud environment: challenges and solutions in migrating applications to the Cloud. Computing **95**(6) 493–535 (2013). https://doi.org/10.1007/s00607-012-0248-2
2. CNCF. Cloud-native Definition v1.0 (2019). https://github.com/cncf/toc/blob/master/DEFINITION.md. Accessed 11 Dec 2021
3. Toffetti, G., Brunner, S., Blöchlinger, M., Spillner, J., Bohnert, T.M.: Self-managing cloud-native applications: design, implementation, and experience. Future Gener. Comput. Syst. **72**, 165–179 (2017). https://doi.org/10.1016/j.future.2016.09.002
4. Pahl, C., Brogi, A., Soldani, J., Jamshidi, P.: Cloud container technologies: a state-of-the-art review. IEEE Trans. Cloud Comput. **7**(3), Art. no. 3 (2019). https://doi.org/10.1109/TCC.2017.2702586
5. Soldani, J., Tamburri, D.A., Van Den Heuvel, W.-J.: The pains and gains of microservices: a systematic grey literature review. J. Syst. Softw. **146**, 215–232 (2018). https://doi.org/10.1016/j.jss.2018.09.082
6. Kratzke, N., Quint, P.-C.: Understanding cloud-native applications after 10 years of cloud computing-a systematic mapping study. J. Syst. Softw. **126**, 1–16 (2017)
7. Samreen, F., Blair, G.S., Elkhatib, Y.: Transferable knowledge for low-cost decision making in cloud environments. IEEE Trans. Cloud Comput. **10**(3), Art. no. 3 (2020)
8. Opara-Martins, J.: A decision framework to mitigate vendor lock-in risks in cloud (SaaS category) migration (2017)
9. Cunningham, P., Cunningham, M., Eds.: 2021 IST-Africa Conference: 10-14 May 2021, virtual conference. Dublin, Ireland: IIMC International Information Management Corporation (2021)
10. Poleshchuk, M.: Creation of linguistic scales for expert evaluation of parameters of complex objects based on semantic scopes. In: 2018 International Russian Automation Conference (RusAutoCon), pp. 1–6 (2018)
11. Schömig, N., et al.: Checklist for expert evaluation of HMIs of automated vehicles—discussions on its value and adaptions of the method within an expert workshop. Information **11**(4), 233 (2020)
12. Opara-Martins, J., Sahandi, M., Tian, F.: A holistic decision framework to avoid vendor lock-in for cloud saas migration. Comput. Inf. Sci. **10**(3) (2017)
13. Sager, F., Mavrot, C.: Participatory vs expert evaluation styles. In: The Routledge Handbook of Policy Styles, pp. 395–407. Routledge, Milton Park (2021)
14. Litvak, B.: Expert assessments and decision making. M patent, p. 298 (1996)
15. Zikmund, W.G., Babin, B.J., Carr, J.C., Griffin, M.: Business Research Methods. 6th edn. Fort Worth, Texas (2000)

Information Systems and Machine Learning

ACTIVE SMOTE for Imbalanced Medical Data Classification

Raul Sena and Sana Ben Hamida(✉)

Paris Dauphine University, PSL Research University, CNRS, UMR[7243], LAMSADE, Paris, France
raul.sena-rojas@dauphine.com, sana.mrabet@dauphine.psl.eu

Abstract. Classifying imbalanced data is a big challenge for machine learning techniques, especially for medical data. To deal with this challenge, many solutions have been proposed. The most famous methods are based on the Synthetic Minority Over-sampling Technique (SMOTE), which creates new synthetic instances in the minority class. In this paper, we study the efficiency of the SMOTE-based methods on some imbalanced data sets. We then propose extending these techniques with Active Learning to control the evolution of the minority class better. Active Learning uses uncertainty and diversity sampling to choose wisely the data points from which the synthetic samples will be generated. To evaluate our approach, we make comprehensive experimental studies on two medical data sets for diabetes diagnosis and breast cancer diagnosis.

Keywords: Imbalanced medical data · Machine Learning · SMOTE · Active Learning · Diversity Sampling · Uncertainty Sampling · Diabetes Diagnosis · Breast Cancer Detection

1 Introduction

The main objective of machine learning applications in the medical field is to propose efficient diagnostic tools with null or very low error probability. Nowadays, the availability of data is no more a difficulty. The amount of information available allows to train different types of learners. However, medical data is often extremely imbalance, where, minority class (known as the "positive" class) are far less than majority classes (known as the "negative" class).

The main difficulty with imbalanced data is that the classification algorithm could be biased towards the majority classes. This Bias induces a higher misclassification rate in the minority class [3]. This problem takes on other dimensions with medical data because classification is applied to generate models for diagnosing some diseases such as cancer, diabetes, etc. In this case, bias in the diagnostic models is not tolerated. Indeed, in medical diagnostics, mislabeling a patient as a healthy individual is expensive and often can lead to deadly consequences. Thus, addressing the class imbalance for medical data is crucial for machine learning tasks [5,17,20].

I. Saad et al. (Eds.): ICIKS 2023, LNBIP 486, pp. 81–97, 2024.
https://doi.org/10.1007/978-3-031-51664-1_6

Many solutions have been proposed to deal with the class imbalance problem in Machine Learning, that could be classified into three categories: Cost-sensitive methods, algorithmic modification methods, and data pre-processing. The last method uses either the under or over sampling techniques, that eliminate or replicate instances until the classes are balanced. Data pre-processing includes also the Synthetic Minority Over-sampling Technique (SMOTE) and its variation, which achieves the same purpose by creating new synthetic instances from the minority class [7]. The efficiency of each approach depends on the context. For medical diagnostics, using the under/oversampling could induce a loss of information in the training sample. Moreover, with a high imbalance ratio, the synthetic samples generated by SMOTE in the positive class could overcome the original samples relative to patients diagnostic as positive cases.

In this paper, we propose a novel scheme to solve the imbalanced data problem for medical diagnostic. This scheme combines SMOTE with Active Learning, and we call it Active SMOTE. It proposes a twofold contribution. First, instead of choosing a sample at random from the training set to use as the pivot point to generate the synthetic samples (as classical SMOTE does), Active SMOTE chooses the points intelligently with uncertainty and diversity sampling, which are two techniques of Active Learning. The training sample is balanced progressively in incremental steps, that we call training epochs. Thus, at each step, the current synthetic samples could be used with the original samples to generate new synthetic samples in the minority class.

This paper is organized as follows. In Sect. 2, we summarize the previous solutions to solve the class imbalance problem in Machine Learning (ML). Section 3 introduces the Active SMOTE method and details how SMOTE is combined with Active Learning. Section 4 explains the methodology used to compare our proposed algorithm with other sampling techniques. In Sect. 5, we evaluate with some graphs and figures the performance of our proposed algorithm. Finally, Sect. 6 makes general conclusions and gives some research ideas to continue forward.

2 Handling Class Imbalance for Classification - Some Related Works

2.1 Cost Sensitive Learning

Cost-sensitive learning is an aspect of algorithm-level modifications for class imbalance. This modification refers to a specific set of algorithms sensitive to different costs associated with certain characteristics of considered problems. These costs can be learned during the classifier training phase or be provided by a domain expert. There exist two different views on cost-sensitive learning in the literature. These are the following:

1. **Cost associated with classes:** This technique considers that making errors on instances coming from a particular class is associated with a higher cost [18]. There are two views for this approach: A financial perspective (e.g.,

giving credit to a person with a bad credit score will potentially cause higher losses to a bank than declining credit to a person with a good score) or the other scenario priority/health/ethical issues (e.g., sending a cancer patient home is much more costly than assigning additional tests to a healthy person). In general, the misclassification cost of the minority examples must be higher than that of the majority examples [2].

2. **Cost associated with features:** This method supposes that obtaining a particular feature is connected to a given cost, also known as test cost. We can view this from a monetary perspective (e.g., a feature is more expensive to obtain as it requires more resources) or other inconveniences (e.g., the measurement procedure is unpleasant, puts a person at risk, or is difficult to obtain). In other words, this approach aims at creating a classifier that obtains the best possible predictive performance while utilizing features that can be obtained at the lowest cost possible.

2.2 Data Level Preprocessing Methods

Data preprocessing methods consist of procedures to modify the imbalanced dataset to a more adequate or balanced data distribution [9]. This is helpful for many classifiers because rebalancing the dataset significantly improves their performance. This subsection will review the undersampling and oversampling techniques such as SMOTE. These techniques are simple and easy to implement. However, no clear rule tells us which technique works best. Resampling techniques can be categorized into three groups:

1. **Oversampling methods:** This method replicates some instances or creates new instances from existing ones, thus creating a superset of the original dataset.
2. **Undersampling methods:** Create a subset of the original dataset by eliminating instances (usually negative class instances).
3. **Hybrid methods:** A combination of Oversampling and Undersampling techniques.

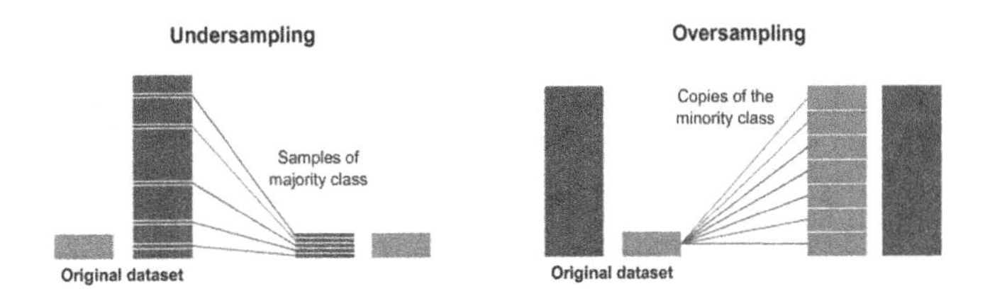

Fig. 1. Random Under and Over Sampling

Random Under and over Sampling. There are many ways to implement the previous techniques, where the simplest preprocessing are non-heuristic methods like random undersampling and random oversampling, as shown in Fig. 1. Nevertheless, these techniques have some drawbacks. In the case of undersampling, the major problem is that it can discard potentially valuable data that could be used in the training process, reducing our dataset's variability. On the other hand, our classifier can occur overfitting with random oversampling because it makes exact copies of existing instances.

To tackle the previous problems, more sophisticated methods have been proposed. The "Synthetic Minority Oversampling Technique" (SMOTE) has gained popularity among them. In short, its main idea is to overcome overfitting posed by random oversampling with the generation of new instances with the help of interpolating between the positive instances closer to each other. However, SMOTE could generate noise samples, boundary samples and overlapping samples [19]. Thus, many variants have been proposed, such as Borderline SMOTE and ADASYN, that we present below since they are used in the experimental study.

SMOTE: Synthetic Minority Oversampling Technique. As stated before, the problem of random oversampling is that because it replicates the exact copies of existing instances, no new information is added to the model's training; therefore, there is a high risk of overfitting. Here is where SMOTE comes in handy because instead of applying a simple replication of the minority class, the central idea of SMOTE is to generate new synthetic samples. This procedure focuses on the "feature space" rather than the "data space" since these new examples are created by interpolating several minority class instances closer. The process to generate new instances with SMOTE is shown in Fig. 2.

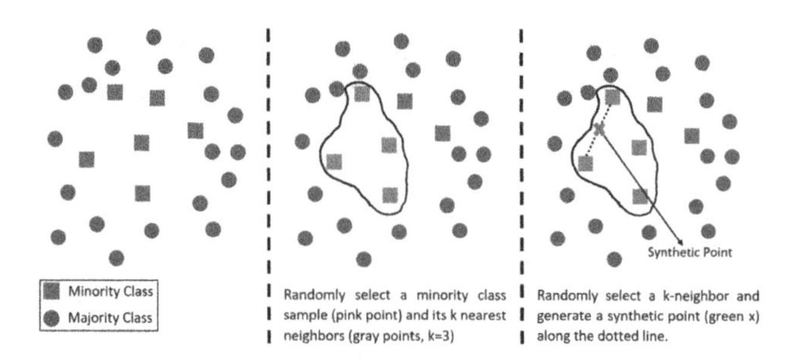

Fig. 2. SMOTE algorithm

Borderline SMOTE. Many Machine Learning algorithms like Logistic Regression and Support Vector Machines use the concept of decision boundary to decide whether an example belongs to one class or another. This decision boundary tries to learn the limits of each class as accurately as possible in the training process. Then when the decision boundary is set, if an example lies far away from it, there is a small probability that it will belong to the opposite class. It is as if the decision boundary divides the space into regions where each region belongs to one class.

Based on the previous statement, the algorithms state that examples away from the borders may contribute little to classification. With this in mind, a new method of oversampling minority class examples was proposed, Borderline-SMOTE [12], in which only the limited examples of the minority class will be over sampled (the ones close to the decision boundary). It is important to clarify that the points close to the borderline are more important, but there is greater uncertainty about which class they belong to, so it is riskier to create synthetic points there. This method differs from the existing ones of oversampling, in which all minority examples or a random subset of the minority class are oversampled [15].

Borderline SMOTE knows which points are closer to the borderline because it classifies them into three categories, as shown in Fig. 3. Then it only uses the danger points to generate synthetic samples.

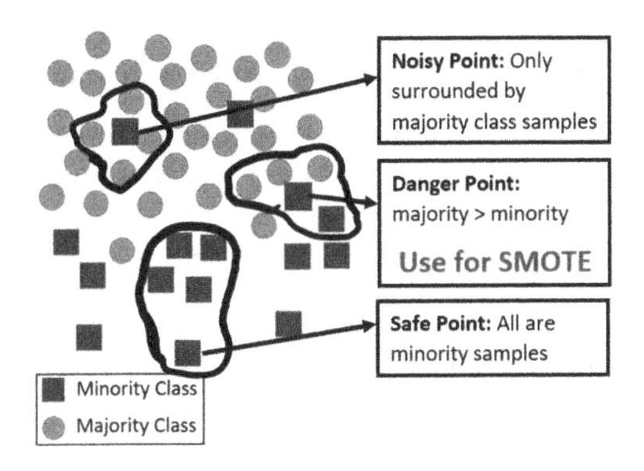

Fig. 3. Borderline Smote

ADASYN. Adaptive synthetic sampling (ADASYN) is an attempt to enhance the SMOTE performance by modifying minority instance selection. It adaptively changes the number of artificial minority examples according to the density of the majority instances around the original minority ones [13]. It reduces the bias introduced by the class imbalance and adaptively shift the classification decision boundary toward the difficult instances.

Other SMOTE Based Methods. SMOTE is the method that has received the most interest from researchers in addressing the imbalance problem. Several techniques were then proposed based on SMOTE with very close objectives. The main purpose of these variants is to avoid noise in the generated training sample. These methods use either techniques to study the disparity and density of the data, or clustering methods to identify safe samples. They can also be extended with Ensemble learning approaches. For example, in RSMOTE (Robust SMOTE) [4], relative density has been introduced to measure the local density of every minority sample, and the non-noisy minority samples are divided into the borderline samples and safe samples adaptively basing their distinguishing characteristics of relative density.

Ma and Fan [20] proposed CURE-SMOTE (Clustering Using Representatives-based Synthetic Minority Over-sampling TEchnique) that uses clustering to sample the training data. Then, SMOTE is performed on the revealed samples. Similarly, Xu et al. [16] proposed KNSMOTE combining k-means clustering with SMOTE in imbalanced medical data. KNSMOTE uses a k-means to cluster the instances and find so-called "safe samples" and remove noise. Then KNSMOTE creates synthetic samples based on founded "safe samples". Many other variants could be found in the literature such as DBSMOTE that uses density-based approach, MWMOTE that analyzes the most difficult minority examples, SMOTECSELM (Synthetic Minority Over-Sampling TEchnique based on Class-Specific Extreme Learning Machine) and many others. An extended review with a comprehensive analysis could be found in [10].

2.3 Algorithm Level Processing with Active Learning

Active Learning is another aspect of algorithm-level modifications. Active learning methods are used to select the instances to be considered in order to control the learning cost [1,11], mainly in the context of massive data, or to select the most informative instances in order to improve the quality of the obtained classifier. The active selection system is integrated in the iterative engine of the learner.

In the context of imbalanced data, active learning can be used to balance the training sample by selecting the most representative instances from the majority class [8], eliminating noisy examples from the minority class [21], and reducing the overall imbalance ratio. Active learning has been tested with iterative learning algorithms, essentially SVCs, particularly with not fully labeled data sets, and Genetic Programming as a scaling solution for classifying large imbalanced data [14].

Although active learning does not directly change the learning procedure, it is considered an algorithm-level solution. It is built into the learning process, unlike preprocessing approaches that are executed before learning begins [11].

The main purpose of Active Learning is to apply a dynamic data sampling to evolve the training data along the training process. The main question is how do we choose samples for a training set? What sample will increase the algorithm performance? At first, this problem may sound disconnected from the imbalanced class problem. However, in our case, the question is: What are the points from the minority class that we need to choose first to generate the synthetic samples, so we can finally have good model performance?

For this purpose, we have selected two sampling approaches: uncertainty sampling and diversity sampling.

2.4 Uncertainty Sampling

The general idea of active learning is to iteratively provide the algorithm with new data, allowing it to improve the performance of the generated classification models. Intuitively, the instance selection method must be driven by the quality of the obtained classifiers, that can be measured with the uncertainty. The more uncertain a prediction is, the more useful the selected instances will be for the learner. Uncertainty sampling is the set of techniques for identifying the least confident samples with the highest uncertainty near a decision boundary, to be inserted in the new training sample. It uses uncertainty measures for a classified item. There are many ways of measuring uncertainty, like least, margin or ratio of confidence, and entropy.

- Margin of confidence sampling: It computes the difference between the two most confident predictions.
- Least confidence sampling: It is defined by the difference between the most confident prediction and the maximum confidence (100%).
- Ratio of confidence sampling: Ratio between the two most confident predictions.
- Entropy-based sampling: Difference between all predictions, as defined by information theory. In our example, entropy-based sampling would capture how much every confidence differed from every other. In the binary classification problem, entropy base sampling is the same as the margin of confidence.

2.5 Diversity Sampling

This type of sampling tackles the problem of identifying where the model might be confident but wrong due to undersampled or non-representative data. It is based on various data sampling approaches helpful in identifying gaps in the model's knowledge, such as clustering, representative sampling, and methods that identify and reduce real-world bias in the models. Collectively, these techniques are known as diversity sampling.

In the Figs. 4 and 5, we can see how uncertainty sampling chooses items closer to the borderline for different ML algorithms like Decision Trees, SVC, Random Forest, and Naive Bayes. In contrast, diversity sampling selects samples that differ from one another or, in other words, are more different.

3 Active SMOTE: Combining SMOTE with Active Learning - Proposed Algorithm

Active SMOTE aims to combine SMOTE with Active Learning. In other words, instead of choosing a point at random from the training set to use as the pivot point to generate the synthetic samples, we will choose the points intelligently with uncertainty and diversity sampling. Practically, we can say there are two main phases of the new proposed algorithm. An uncertainty sampling phase and a diversity sampling phase. These phases are shown in the following images:

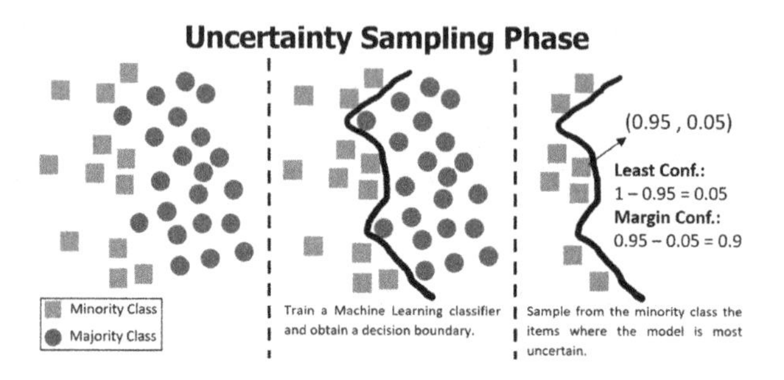

Fig. 4. Uncertainty Sampling Phase of SMOTE with Active Learning

The uncertainty sampling phase shown in Fig. 4 serves to select the items that are close to the decision boundary. First, we train a machine learning model with all the data, then calculate the probability of belonging to the minority class of all minority class samples. Finally, we compute the uncertainty of the model base in an uncertainty measure and select a percentage of the most uncertain minority class samples. The percentage of the items that we are going to select is a hyperparameter. After that, we proceed to the diversity sampling phase shown in Fig. 5.

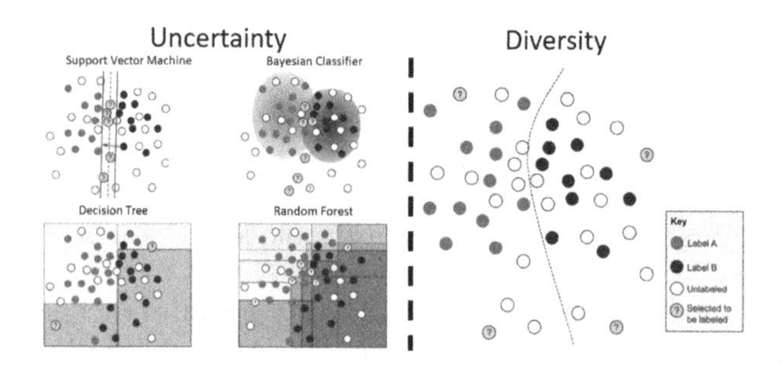

Fig. 5. Diversity Sampling Phase of SMOTE with Active Learning

The diversity sampling phase aims to make a diverse sample of the selected most uncertain items. First, we divide the most uncertain items in k clusters using K-means, and then we make a stratified sample from every cluster. The number of k clusters is a hyperparameter that needs to be tuned, as well as the number of items we will select from each cluster. We selected K-means as the clustering method because it is the most used clustering strategy. However, a good line of research could be to compare other clustering methods.

The different steps of Active SMOTE method are illustrated in Algorithm 1. First, at each epoch, Active SMOTE apply the uncertainty sampling on the minority class sample $Xmin$ (lines 2 to 6). The confidence measure, margin or least, is a required parameter of the algorithm ($uncertain\_measure$). After retaining the appropriate portion ($perc_{uncertain}$) from the resulting sample $X_{uncertain}$ (line 8), this last one is clustered with K-means, and some points are selected from each cluster to obtain the final sample $X_{diversity}$ (lines 9 and 10). The third step consists on generating one synthetic sample from every diversity cluster in $X_{diversity}$ with the original SMOTE using $k = 5$ as number of neighbors to generate a new sample, and save it on the X_{smote} set (line 11). Finally, the ML algorithm is trained on the over sampled set gathering X_{maj}, X_{min} and X_{smote}.

Algorithm 1. Active SMOTE Algorithm

Parameters:
X_{min}: Minority class sample
X_{maj}: Majority class sample
epochs: Number of iterations
$perc_{uncertain}$: % of most uncertain samples from T to retain (uncertainty sampling)
$perc_{diversity}$: % of samples from uncertainty sampling that are going to pass to diversity sampling
$uncertain\_measure$: Uncertainty measure to use for confidence (margin or least confidence)
k: Number of clusters for k-means in diversity sampling
N: Number of samples points

1: **for** $i = 1$ to *epochs* **do**
2:   **if** $uncertain\_measure$="margin confidence" **then**
3:     $X_{uncertain} \leftarrow$ compute model's margin confidence with X_{min}
4:   **else**
5:     $X_{uncertain} \leftarrow$ compute model's least confidence with X_{min}
6:   **end if**
7:   $X_{uncertain} \leftarrow$ sort $X_{uncertain}$ by descending order
8:   $X_{uncertain} \leftarrow$ retain $perc_{uncertain}$ samples from $X_{uncertain}$
    #Second do Diversity Sampling
9:   $X_{diversity}$ = Cluster $X_{uncertain}$ with K-means obtaining k clusters.
10:   $X_{diversity}$ = Do (cluster sampling of $X_{diversity}$) until size ($X_{diversity} * perc_{diversity}$) is reach
11:   $X_{smote} = SMOTE(X_{diversity}, N = 100, k = 5)$
12:   $X_{train} \leftarrow (X_{min} + X_{smote} + X_{maj})$
13:   model = retrain the model with the new training set X_{train}
14: **end for**
15: Return model

Table 1. ML methods' parameter setting

ML technique	Parameters
Logistic Regression	$0.001 \leq C \leq 2$
SVC	kernel: 'rbf'
	gamma (adjustment degree) $\in \{0.001, 0.01, 0.1, 1\}$
	C (regulation) $\in \{1, 10, 50, 100, 200, 300, 1000\}$
Gradient Boosting	$n\_estimator \in \{10, 20, 50\}$
	learning rate $\in \{0.075, 0.01, 0.005\}$
	max depth $\in \{1, 2, 3, 4, 5\}$

4 Experimentation

4.1 ML Models Evaluated

To evaluate the performance of Active SMOTE, we trained the following three ML classifiers: Logistic Regression (LR), Support Vector Machines (SVCs) and Gradient Boost (GB). The training procedure involves performing a random search with ten-fold cross-validation using the hyperparameter space outlined in the Table 1.

4.2 Sampling Techniques Evaluated

The training data was generated from the following two methods:

Classical Sampling Methods. The classical methods are SMOTE variants, simple random undersampling, and oversampling. All these techniques were used to make a balanced dataset, except with UNBAL. Otherwise, all oversampling techniques generate synthetic samples in the minority class with the specified strategy. A Python implementation of these techniques is available in the library "imbalanced-learn"[1] The abbreviations are the following:

- UNDER: Random undersampling of the majority class.
- OVER: Random oversampling of the minority class.
- SMOTE: Create synthetic samples with the original SMOTE.
- SVM: Generate synthetic samples with SVM SMOTE.
- BORDER: Create synthetic samples with Borderline SMOTE.
- ADASYN: Generate synthetic samples with Adasyn SMOTE.
- UNBAL: Do not change the original training dataset. That is, keeping the classes imbalanced.

[1] https://imbalanced-learn.org/stable/index.html.

New Proposed Methods. The newly proposed methods are a combination of different Active Learning techniques with SMOTE, random (simple SMOTE) is used to evaluate if randomly choosing the pivot points has the same or better results as Active Learning with SMOTE. A more precise description of the newly proposed methods is the following:

- Random (simple SMOTE): At each epoch, randomly generate N synthetic samples with the original SMOTE.
- Diversity & Uncertainty - margin: At each iteration, generate N synthetic samples with the technique described in Algorithm 1. In uncertainty sampling, margin confidence is used as an uncertainty measure.
- Diversity & Uncertainty - least: At each epoch, generate N synthetic samples with the technique described in Algorithm 1. In uncertainty sampling, the least confidence is used as an uncertainty measure.
- Uncertainty - margin: At each epoch, generate N synthetic samples with the technique described in Algorithm 1, only doing uncertainty sampling with margin confidence.

4.3 Data Sets Evaluated

Two medical data sets (obtained from the public UCI data repository[2]) are used to assess the performance of Active SMOTE. The first one is the PIMA data set, which the target is to predict whether or not a patient will have diabetes based on specific diagnostic measurements included in the dataset. It consists of 768 females aged 21 or above, with 500 negative and 268 positive instances.

The second data set is "Breast Cancer Wisconsin" used in [6]. Features are computed from a digitized image of a fine needle aspirate (FNA) of a breast mass. They describe the characteristics of the cell nuclei present in the image. The target is to identify if a tumor is benign or malignant.

..The class Imbalance Ratios (IR) of PIMA and Breast Cancer data sets are respectively 1.6 and 1.8.

4.4 Performance Measures

For the Active SMOTE based models, by the end of each epoch, the classification model is evaluated on the test data set. Results are recorded in a confusion matrix from which accuracy, $F1 - weigth$, recall, sensitivity, and AUC are calculated. The same measures are computed for the other configurations based on the classic SMOTE and its variants by the end of each learning process. All the results corresponding to one machine learning method (LR, SVC or GB) are illustrated by a single figure in Sect. 5

[2] http://archive.ics.uci.edu/ml.

5 Results and Discussion

Figures 6 and 7 illustrate respectively the results obtained on PIMA and Breast Cancer data sets. The different graphs show on the left line plots the newly proposed methods (that's why they have epochs in the x-axis), while the bar plots from the right are the classical sampling techniques previously described in the Subsect. 4.2. All the experiments below generate ten new synthetic samples in every epoch with every method. In epoch 18th, the dataset is balanced.

5.1 Results on PIMA Data Set

The first plot in Fig. 6 demonstrates clearly how the methods of Active SMOTE, when used with Logistic regression, achieve more outstanding performance in recall than the classical ones and are better than simple iterative SMOTE. The better performance of the minority class comes at the expense of the majority class. That is why there is a slight decrease in Precision, F1-Weighted and Balanced Accuracy. Furthermore, Active SMOTE achieved a similar performance in recall as the classical methods at around iteration 12. That means that with only 120 synthetic samples, Active SMOTE achieved equivalent performance as the classical methods. Since the 13th epoch, Active SMOTE with uncertainty margin reaches higher performances of recall and accuracy, largely better than the other variants of balancing.

The second plot in Fig. 6 illustrates the results with SVC on PIMA dataset. There is considerable variability with all the methods when using SVC. In fact, there is a decrease in recall as epochs pass. Our intuition is that we need to augment the diversity sampling cluster size to make a more diverse sampling because the model generates the synthetic samples in a single region. In this case, the new methods like the classical ones are not performing well. The undersampling technique achieves the best results in all the metrics. Our intuition is that the hyperspace where the new synthetic samples are generated is wrong. In this example, we can see one of the advantages of Active Learning with SMOTE. If we see that the model is doing a lousy performance as epochs pass, we can ask an expert to verify if our synthetic samples are ok.

The best results on PIMA data are given by the Ensemble method Gradient Boost (Fig. 6, plot 3). The behavior of these graphs is quite similar to Logistic Regression. There is a considerable increase in recall and not a big decrement in precision with Active SMOTE. As with Logistic Regression, Active SMOTE can achieve similar results as classical methods at around epoch 12. ROC - AUC does not change much as epochs pass.

5.2 Results on Breast Cancer Data Set

From the first plot in Fig. 7, we can remark that when the confidence measure margin is used, there is a decrease in precision with the active learning strategy. Furthermore, there is a slight variation in the metrics as we generate new synthetic samples, which shows that a large margin already separates the classes.

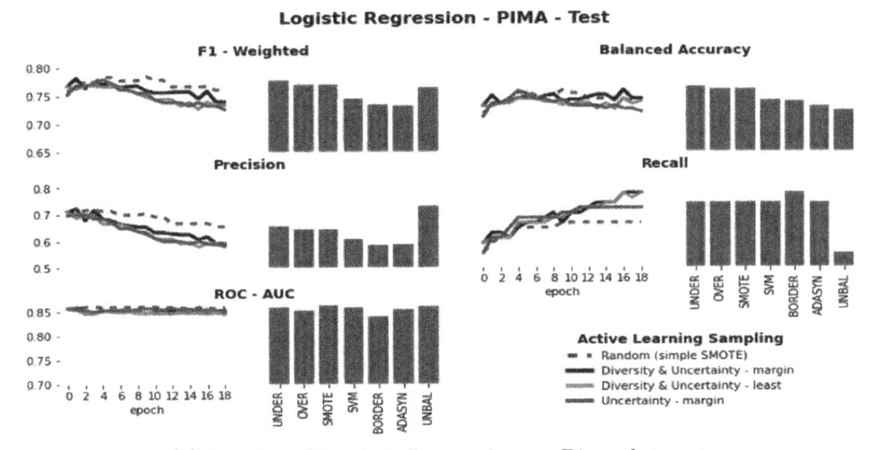

(1) Results of Logistic Regression on Pima data set

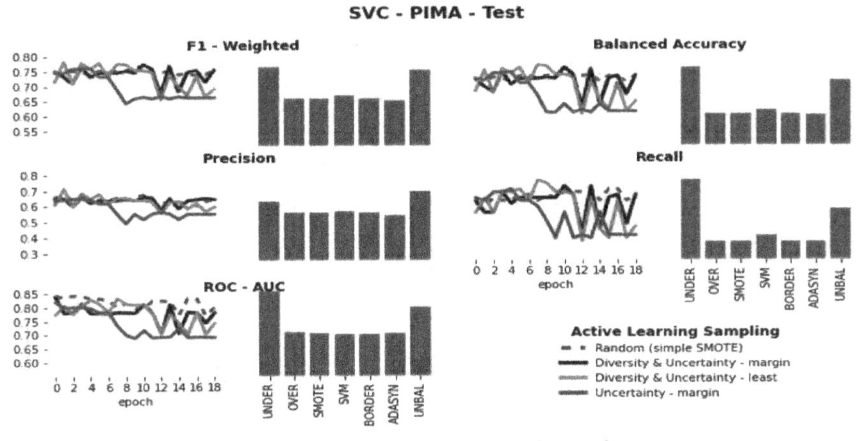

(2) Support Vector Classifier on Pima data set

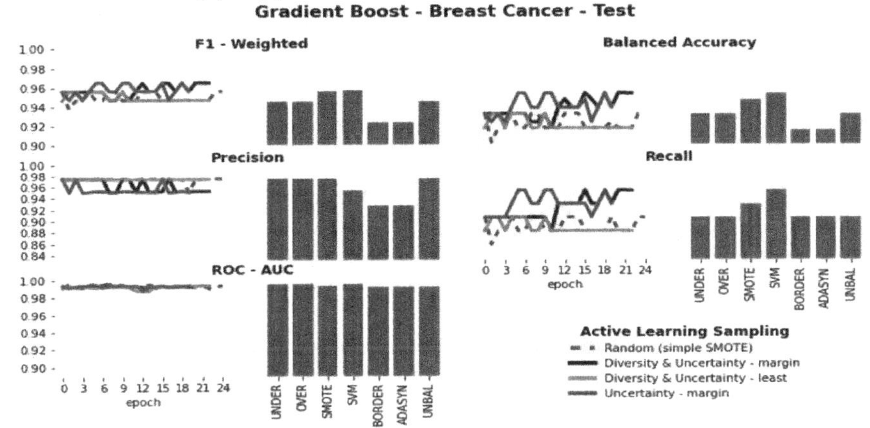

(3) Results of Gradient Boost on Pima data set

Fig. 6. Active SMOTE on Pima data set

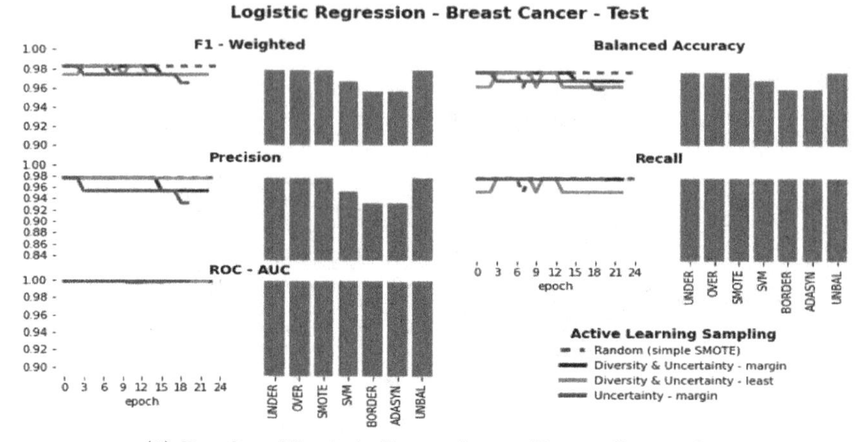

(1) Results of Logistic Regression on Breast Cancer data

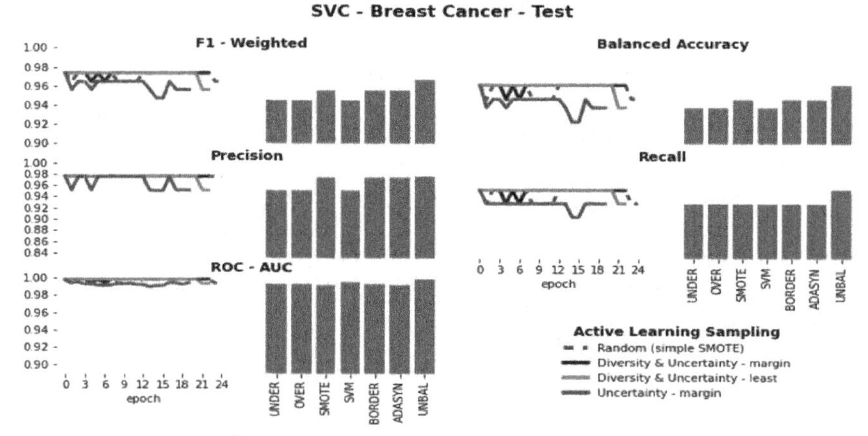

Results Support Vector Classifier on Breast Cancer data

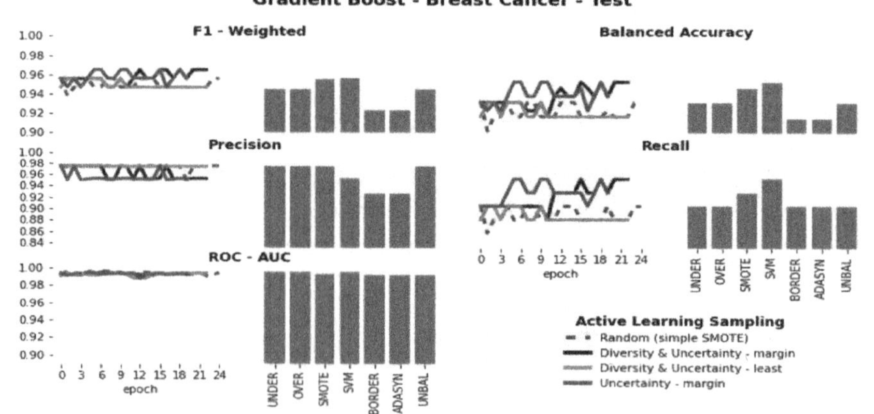

Results Gradient of Boosting on Breast Cancer data set

Fig. 7. Active SMOTE: Results on Breast Cancer data set

Similarly, the second plot, illustrating the results with SVC, shows that the worst performance metric with Active Learning is done with margin of confidence as an uncertainty measure. However, the proposed new methods perform better than classical methods in recall. It is interesting to note that SVC is able to deal with the imbalance of this data set and provide satisfactory results with a balancing strategy (case of UNBAL).

The Logistic Regression and SVC algorithms did not show much variance because they are simpler than ensemble techniques. That is why we will see a more significant variance in the results obtained with Gradient Boost. In other words, this is a case of the Bias and variance trade-off. Otherwise, there is a substantial increase in recall from iterations 9 to 10. Indeed, Active SMOTE with uncertainty margin, with or without diversity sampling, gives the best scores in terms of recall, balanced accuracy (score higher than 0.96) and F1-weighted (score higher than 0.98). Those newly generated samples were of great utility for the model. Precision does not decrease when recall increases.

Further Discussion. The main conclusion that can be deduced from this first experimental study is that it is not necessary to completely balance the data to obtain good learning results. Indeed, Active SMOTE has proven that the learning algorithm can reach an optimal performance with the first synthetic instances generated during the first epochs. This can avoid, in the case of medical data, filling the training set with synthetic positive cases, affecting the quality of the original data. Thus, it is important to adjust the technique of selecting pivot points for SMOTE to the nature of the available data. For example, diversity sampling gave the best performance with the Diabetes database.

Indeed, many datasets are biased toward a specific gender, race, and socioeconomic background. That bias generally occurs in favor of the most privileged demographic: persons from the wealthiest nations, problems from the wealthiest economies, and other biases resulting from a power imbalance. We believe it is vital to research later if doing diversity sampling with SMOTE shows evidence of increasing the diversity of the people who can benefit from models built from data. Active Learning with SMOTE could also be used to know where to do medical studies to maximize model performance. For example, knowing that the model generates synthetic samples in a particular age group could enforce the decision to study that age group in real life.

6 Conclusion

We introduced in this paper Active SMOTE, a new dynamic approach to deal with imbalanced medical data. Active SMOTE combines SMOTE with Active Learning using specific sampling techniques based on diversity and/or uncertainty in the positive class.

Combined with three machine learning techniques, Logistic Regression, Support Vector classifier and Gradient Boost, and applied to two medical data sets for diabetes diagnosis and breast cancer diagnosis, Active SMOTE has proven its ability to improve the performance of the obtained classifiers in several cases, mainly when applied with an Ensemble Learner. It also demonstrated, thanks to the active learning strategy, that it is not necessary to completely balance the training set to reach high satisfactory results, which is very important in the medical context, since it could reduce the synthetic data corresponding to fictive patients.

The results presented in this paper are obtained with a preliminary experimental study. Further experiments will be done on other data sets with higher imbalance rate and for further medical purpose.

References

1. Aggarwal, C.C., Kong, X., Gu, Q., Han, J., Philip, S.Y.: Active learning: a survey. In: Data Classification, pp. 599–634. Chapman and Hall (2014)
2. Bach, F.R., Heckerman, D., Horvitz, E.: Considering cost asymmetry in learning classifiers. J. Mach. Learn. Res. **7**, 1713–1741 (2006)
3. Chawla, N.V., Japkowicz, N., Kotcz, A.: Special issue on learning from imbalanced data sets. ACM SIGKDD Explor. Newsl. **6**(1), 1–6 (2004)
4. Chen, B., Xia, S., Chen, Z., Wang, B., Wang, G.: RSMOTE: a self-adaptive robust smote for imbalanced problems with label noise. Inf. Sci. **553**, 397–428 (2021). https://doi.org/10.1016/j.ins.2020.10.013
5. Devarriya, D., Gulati, C., Mansharamani, V., Sakalle, A., Bhardwaj, A.: Unbalanced breast cancer data classification using novel fitness functions in genetic programming. **140**, 112866. https://doi.org/10.1016/j.eswa.2019.112866
6. Dua, D., Graff, C.: UCI machine learning repository (2017). http://archive.ics.uci.edu/ml
7. Elreedy, D., Atiya, A.F.: A comprehensive analysis of synthetic minority oversampling technique (SMOTE) for handling class imbalance. Inf. Sci. **505**, 32–64 (2019)
8. Ertekin, S., Huang, J., Giles, C.L.: Active learning for class imbalance problem. In: Proceedings of the 30th Annual International ACM SIGIR Conference on Research and Development in Information Retrieval, SIGIR 2007, pp. 823–824. ACM (2007). https://doi.org/10.1145/1277741.1277927
9. Fernández, A., García, S., Galar, M., Prati, R.C., Krawczyk, B., Herrera, F.: Learning from Imbalanced Data Sets, vol. 10. Springer, Cham (2018)
10. Fernandez, A., Garcia, S., Herrera, F., Chawla, N.V.: SMOTE for learning from imbalanced data: progress and challenges, marking the 15-year anniversary. **61**, 863–905 (2018). https://doi.org/10.1613/jair.1.11192
11. Ben Hamida, S., Benjelloun, G., Hmida, H.: Trends of evolutionary machine learning to address big data mining. In: Saad, I., Rosenthal-Sabroux, C., Gargouri, F., Arduin, P.-E. (eds.) ICIKS 2021. LNBIP, vol. 425, pp. 85–99. Springer, Cham (2021). https://doi.org/10.1007/978-3-030-85977-0_7
12. Han, H., Wang, W.-Y., Mao, B.-H.: Borderline-SMOTE: a new over-sampling method in imbalanced data sets learning. In: Huang, D.-S., Zhang, X.-P., Huang, G.-B. (eds.) ICIC 2005. LNCS, vol. 3644, pp. 878–887. Springer, Heidelberg (2005). https://doi.org/10.1007/11538059_91
13. He, H., Bai, Y., Garcia, E.A., Li, S.: ADASYN: adaptive synthetic sampling approach for imbalanced learning. In: 2008 IEEE International Joint Conference on Neural Networks (IEEE World Congress on Computational Intelligence), pp. 1322–1328 (2008)

14. Hmida, H., Hamida, S.B., Borgi, A., Rukoz, M.: Sampling methods in genetic programming learners from large datasets: a comparative study. In: Angelov, P., Manolopoulos, Y., Iliadis, L., Roy, A., Vellasco, M. (eds.) INNS 2016. AISC, vol. 529, pp. 50–60. Springer, Cham (2017). https://doi.org/10.1007/978-3-319-47898-2_6
15. Le, T., Vo, M.T., Vo, B., Lee, M.Y., Baik, S.W.: A hybrid approach using oversampling technique and cost-sensitive learning for bankruptcy prediction. Complexity **2019** (2019)
16. Li, J., et al.: SMOTE-NaN-DE: addressing the noisy and borderline examples problem in imbalanced classification by natural neighbors and differential evolution. Knowl.-Based Syst. **223**, 107056 (2021)
17. Oh, S., Lee, M.S., Zhang, B.T.: Ensemble learning with active example selection for imbalanced biomedical data classification. IEEE/ACM Trans. Comput. Biol. Bioinform. **8**(2), 316–325 (2010). https://doi.org/10.1109/TCBB.2010.96
18. Pazzani, M., Merz, C., Murphy, P., Ali, K., Hume, T., Brunk, C.: Reducing misclassification costs. In: Machine Learning Proceedings, pp. 217–225. Elsevier (1994)
19. Saez, J.A., Luengo, J., Stefanowski, J., Herrera, F.: SMOTE-IPF: addressing the noisy and borderline examples problem in imbalanced classification by a resampling method with filtering. Inf. Sci. **291**, 184–203 (2015). https://doi.org/10.1016/j.ins.2014.08.051
20. Xu, Z., Shen, D., Nie, T., Kou, Y., Yin, N., Han, X.: A cluster-based oversampling algorithm combining smote and k-means for imbalanced medical data. Inf. Sci. **572**, 574–589 (2021)
21. Zhang, J., Wu, X., Shengs, V.S.: Active learning with imbalanced multiple noisy labeling. IEEE Trans. Cybern. **45**(5), 1095–1107 (2015). https://doi.org/10.1109/TCYB.2014.2344674

Evolutionary Graph-Clustering vs Evolutionary Cluster-Detection Approaches for Community Identification in PPI Networks

Marwa Ben M'Barek[1,2](✉), Sana Ben Hmida[2], Amel Borgi[1], and Marta Rukoz[2]

[1] LIPAH, Faculté des Sciences de Tunis, Université de Tunis El Manar, 2092 Tunis, Tunisia
bnembarekmarwa@gmail.com, amel.borgi@insat.rnu.tn

[2] Paris Dauphine University, PSL Research University, CNRS, UMR[7243], LAMSADE, Paris, France
marta.rukoz@lamsade.dauphine.fr, raul.sena-rojas@dauphine.com

Abstract. Community detection in protein-protein interaction networks (PPIs) is an active area of research, and many studies have applied Genetic Algorithms (GAs) to this problem. This paper summarizes the different GA based approaches for community detection in PPIs and provides a taxonomy of these methods. Detailed comparative studies are then provided comparing an evolutionary graph-clustering approach (EGCPI) based on the partitioning paradigm and an evolutionary cluster-detection approach based on an evolutive and incremental search for potential communities in the graphs (GA-PPI-Net). The communities obtained by the two algorithms on Collins PPI network are compared according to the average similarity and interaction between genes, and also according to the recovery rate of known communities in some biological pathways. Experiments tests verify the effectiveness of the GA-PPI-Net approach compared with EGCPI approach.

Keywords: Community detection · Biological networks · PPI networks · Genetic Algorithm · GA-PPI-Net · EGCPI

1 Introduction

Networks can represent many systems such as biology, computer science, linguistics, etc. A network has a set of nodes and a set of edges, it is represented by a graph. The nodes represent the basic components of the system, and the edges represent the links between nodes according to a defined relationship [20]. For example, a network of interactions between proteins is generally represented as an interaction graph, where nodes represent proteins and edges represent pairwise interactions. When a system is modeled by a network, it helps to understand the system easily and identify hidden information. Lots of studies have done around networks and how to use them.

I. Saad et al. (Eds.): ICIKS 2023, LNBIP 486, pp. 98–113, 2024.
https://doi.org/10.1007/978-3-031-51664-1_7

Networks have some common features. The most known feature is the existence of parts, or sub-graphs, more densely connected than others. These parts, which are a set of nodes and edges, are called communities. So, the community can be defined as a group of nodes much more strongly connected to each other than other nodes [24]. These groups, or communities, are usually thought to correspond to groups of nodes that play similar roles or have similar functions within the network.

Communities detection (CD) in networks has become a new field of research [2,10,20,31]. The goal of CD is to identify these groups in a way that is both meaningful and useful for understanding the structure and function of the network. Applications of CD include social network analysis, biology, and computer science, among others. In biological networks, for example, CD has been used to identify groups of genes that are involved in similar biological processes, or groups of proteins that interact to form functional complexes. In social networks, CD has been used to identify groups of individuals with similar interests or social roles.

The purpose of CD in protein-protein interaction networks (PPIs) is to better understand the biological mechanisms and pathways that underlie cellular processes. By identifying groups of proteins that are more likely to interact with each other, CD can help to identify functional modules or pathways within the network, and can provide a starting point for further experimental investigation. CD can also be used for network-based drug discovery and personalized medicine. By identifying communities of proteins that are involved in specific diseases, it may be possible to identify new drug targets or develop personalized treatments based on an individual's unique network structure. This work is multidisciplinary, as it brings the field of biology and computer science in the broad sense.

The community detection can be handled either by heuristic based methods or optimization based methods. Heuristic based algorithms use essentially the edge-betweenness [20]: its value is higher for inter-communities than for intra-communities. Then, communities are identified by sequentially removing edges that do not increase the edge-betweenness for inter-communities. The Optimization based methods maximize an objective function often computed from structural information in the network, such as the modularity (cf Sect. 2). However, computing such function for all possible partitions of the network is a NP-hard problem. In this context, Genetic Algorithms (GAs) have been considered as suitable approaches to obtain an optimal result [8]. Most of the proposed algorithms use the greedy or the partitioning paradigms. EGCPI (Evolutionary Game-based Community detection algorithm for Protein-protein Interaction networks) [11] is a known method in this category. However, recently, a new approach has been proposed by the GA-PPI-Net method [4,5]. It consists of evolving groups of nodes by optimizing their structural and qualitative scores in order to obtain some potential communities in the graph. This approach is known as an incremental approach, since the communities are constructed iteratively along the evolution.

In this study, we perform a systematic quantitative and qualitative evaluation of the capability of a clustering evolutionary method (EGCPI) and an incremental evolutionary approach (GA-PPI-Net) for inferring communities from PPI networks. EGCPI was proposed for detecting communities in PPI networks using evolutionary game theory. The method aims to find the optimal partition of nodes into communities by modeling the interactions between nodes and using a genetic algorithm to get for the best solution [11]. The GA-PPI-Net is a Genetic Algorithm that allows to find one or several communities on a PPIs based on a search strategy. It uses the similarity score as well as the interaction score between proteins or genes and tries to find the best community by maximizing the concept of community measure [5]. To compare the two algorithms, we use the collins dataset[1]. This dataset concerns the genes of the yeast species Saccharomyces Cerevisiae (yeast).

Algorithms like EGCPI, identify protein community in PPI networks based only on network topologies. However, the GA-PPI-Net is not only based on graphical topology but also on semantic similarity between nodes. To our knowledge, GA-PPI-Net is the only communities' detection method that uses both semantic and topological measures. The main contribution of this paper is to evaluate and to compare two genetics based approaches in order to prove that GA-PPI-Net approach is suitable for the CD problem. The comparison is based on Three specific evaluation measures: i) the recovered percentage of each identified communities in an existing networks by using DAVID Tools [12]; ii) the semantic similarity measure and iii) the interaction score.

The rest of the paper is organized as follows. The next section presents an overview of the existing community detection algorithms. Section 3 provides a description of the two evolutionary approaches used in this study. In Sect. 4, experimental results on real data set are presented and analyzed. Finally, Sect. 5 reports the conclusion.

2 Background

The task for network community detection is to divide the whole network into small parts or groups, which are also called communities. In the literature, there is no uniform definition for community, but in academic domain, a community (also called a cluster or a module) is defined as a group of nodes that are connected densely inside the group but connected sparely with the rest of the network. Radicchi et al. [24] propose two definitions of community. These definitions are based on the degree of a node (or valency)[2]. In the first definition, a community is a subgraph in a strong sense: each node has more connections within the community than the rest of the graph. In the second definition, a community is a subgraph in a weak sense: the sum of all incident edges in a node is greater than the sum of the out edges.

[1] https://thebiogrid.org/.

[2] The degree of a node is the number of edges incident to the node.

2.1 Overview of Community Detection Methods

Many community detection methods in networks have been developed over the years. Each method has advantages and disadvantages (simplicity, efficiency, run time etc.). The literature survey of community detection methods in graphs is divided into two categories: methods based on analytical or computational approaches and those based on evolutionary approaches [2,21,31]. The Analytical approaches use well defines rules that systematically explore the search space. Most of them are based on the hierarchical clustering techniques. These methods group nodes of a graph by minimizing the number of links between the different groups, so that the nodes of the same group or cluster are as similar as possible, while the nodes of different communities are as different as possible. As examples of clustering-based methods using graph modularity and density, we list Markov Clustering (MCL [30]), Restricted Neighborhood Search Clustering (RNSC [15]) and ClusterOne [19]. These methods have been applied to biological networks (PPI) to identify protein communities in PPI networks [31]. A recent comparative study of these methods with an evolutionary approach is published in [6] Other computational techniques have been proposed using different approaches, such as Random walk [23] or spectral clustering [32].

Evolutionary methods apply a global search algorithms that implicitly sample the search space and try to zoom on interesting regions based on the quality of the sampled solutions. Similarly, most of Evolutionary techniques are based on graph clustering strategy, except for GA-PPI-Net that uses an iterative an incremental strategy to identify clusters in the graph [5,11]. As the main objective of this paper is to evaluate and to compare two GA based approaches, namely GA-PPI-Net vs EGCPI, the Evolutionary Approaches for community detection are presented in more details in next section.

A complete review of community detection methods in complex networks could be found in [2].

Most of clustering methods, either computational or evolutionary, are based on the modularity metric (Q). Modularity is a metric to measure the quality of partitioning a graph into communities. It is mainly used in social network analysis. It is introduced by M. E. J. Newman [20] and is defined in Eq. 1. It is described as the proportion of edges incident on a given class minus the value that this same proportion would have been if the edges were randomly arranged between the nodes of the graph.

$$Q = \frac{1}{2m} \sum_{i,j} \left[A_{ij} - \frac{k_i k_j}{2m} \right] \delta(C_i, C_j) \tag{1}$$

where:

- m is the total number of edges in the network
- A_{ij} is the ij^{th} element of the adjacency matrix of the network (it indicates whether the pair of nodes i and j are adjacent or not in the graph).
- k_i and k_j are the degrees of nodes i and j, respectively

- C_i and C_j are the communities to which nodes i and j belong
- $\delta(C_i, C_j)$ is the Kronecker delta function, which is 1 if $C_i = C_j$ and 0 otherwise

The modularity has been used as an optimization function for several community detection tasks in graphs [17,20,28]. The goal is to find, among all possible partitions, the one with the best modularity. Another quality function has also been used in [22]. It consists in the determination of a global measure of the quality of a division within communities called *community score.*

2.2 Evolutionary Approaches for Community Detection

The Evolutionary Algorithm's (EA) ability to solve various problems has brought considerable popularity for them in solving optimization problems. These methods start from random individuals, and through keeping and combining the fittest and eliminating the weak solutions, narrow the search space to desired solutions [8]. This seemingly simple logic has shown to be able to find remarkable results for complicated problems. In recent years, several EA-based methods have been proposed to solve CD problems. A majority of evolutionary-based community detection methods are based on Genetic Algorithms (GA). GAs can be applied to community detection in networks by treating the problem as an optimization problem, where the objective is to find a partition of the network's nodes into communities that maximize a quality function. The quality function is typically chosen such as nodes within a community should be densely connected to each other, while nodes in different communities should be sparsely connected. A synthetic taxonomy of the different category of evolutionary methods is given in the following Fig. 1.

The basic idea is to define a representation for each potential partition of the network and to evolve new candidate solutions by two mechanisms:

- the production of new solutions by the genetic operators (crossover and mutation)
- the selection of new solutions that replace earlier solutions in order to improve the overall quality of the population and preserve its diversity.

The fitness of each candidate solution is evaluated based on the quality function, and the fittest solutions are selected for further evolution. The modularity is the commonly used objective function in GA-based approaches. However, other quality functions have been used with GAs such as the conductance that measures the degree to which a community is internally connected and externally disconnected, or the edge betweenness that measures the importance of an edge in the network, based on the number of shortest paths that pass through the edge. The edge betweenness objective function can be used to identify edges that act as bottlenecks or bridges between communities, and can be used in combination with other objective functions such as modularity or conductance.

To design the individuals (solutions) of the population to be evolved, two main strategies were used by the GAs. In the first one, a graph partitioning solution is defined by the whole population. Thus, each individual represents

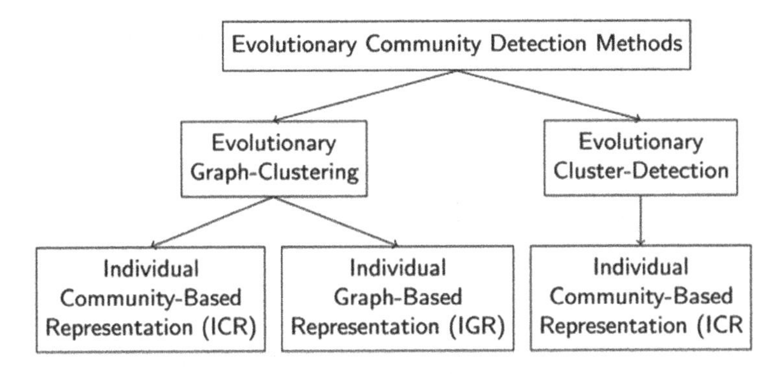

Fig. 1. Synthetic taxonomy of the different community detection methods.

a part of the graph (a cluster). Any modification of a cluster naturally modifies the other clusters. We denote this strategy in the taxonomy (Fig. 1) as the Individual-Community based representation (ICR). In the second strategy, each individual is a full clustering solution. So an individual represents the full clustered graph. We denote this strategy in the taxonomy (Fig. 1) as the Individual Graph-based Representation (IGR). This representation is expensive and difficult to scale to very large graphs. The simplest and least expensive representation that only evolves potential communities in the graph is proposed by GA-PPI-Net presented in the next section. Below some methods in the first two categories.

To the best of our knowledge, Bingol et al. made the first attempt to use EAs to detect the communities of a network [28,29]. They have implemented an algorithm where individuals are graph partitions (IGR) designed with arrays of integer values. Each value indicates which community a particular node belongs to, and the modularity is the fitness function. Liu et al. introduced, in [17], an algorithm in which the optimal partition with the best modularity is obtained through successive bipartitions of the graph. A bipartition given by the GA is accepted only if it increases the total modularity of the graph. In [22], Pizzuti proposed a genetic algorithm named GA-Net to discover communities in social networks. GA-net is one of the famous community detection algorithms based on evolutionary methods, that doesn't use modularity as its fitness function. Two new measures are introduced, the community score and the notion of safe individual. The first one measures the quality of the partitioning of the network into communities and define the objective function for the algorithm.

The application of GA to PPI networks clustering started with MCODE-Evo. This method is an extension of the MCODE algorithm [3] for PPIs Clustering. Based on ICR, it applies a GA-based search to optimize the density and size of candidate communities, and then applies a refinement step using a local search strategy to iteratively remove or add nodes to the communities, while maintaining or increasing their fitness [16]. Later, Genetic Algorithm for Modularity Maximization (GAMMA) [7] is introduced. GAMMA is an ICR-based approach that uses modularity as the objective function for community detection in PPI networks. The algorithm starts with a random partition of the PPI network into communities, and then iteratively applies genetic operators to generate new partitions with higher modularity.

Other GA-based approaches have been proposed for CD in PPIs this last decade, but they still using modularity as objective function, such as HGAC (Hierarchical Genetic Algorithm for Community Detection) [13] or PPI-GA (Clustering Genetic Algorithm to Identify Protein Complexes within PPI) [27]. However, despite the extensive use of modularity, it suffers from scalability problems, which indicates that modularity can't detect communities smaller than a specific scale [9]. Therefore, efforts to propose another measure continue, such with GA-PPI-Net presented in the following section.

3 GA-PPI-Net vs EGCPI

The main objective of this paper is to compare two Evolutionary approaches to identify communities in PPI networks belonging to the two different evolutionary categories in the DC taxonomy described in Sect. 2. The first one, ECGPI [11], is an Evolutionary Graph-Clustering EA using Individual-Graph based representation (IGR). The second one is GA-PPI-Net [5] that is an Evolutionary Cluster-Detection GA using Individual-Community based representation (ICR). Both methods are detailed below.

3.1 EGCPI

EGCPI (Evolutionary Game-based Community detection algorithm for Protein-protein Interaction networks) is a recently proposed method for detecting communities in PPI networks using evolutionary game theory. The method aims to find the optimal partition of nodes into communities by modeling the interactions between nodes as a non-cooperative game, and using a genetic algorithm to search for the best solution [11].

EGCPI performs the task of community protein identification in several steps.

- An Attributed PPI network Graph (APPIG) is constructed based on the PPI network and the attribute information that can be obtained from the Gene Ontology database.
- A weighted Attributed PPI Graph (wAPPIG) is then constructed. Given an APPIG, EGCPI determines a weight to each edge in the graph based on the degree of topological similarity. It measures the weight of each pair of interacting proteins according to how much they are connected.
- An evolutionary algorithm is applied to identify dense graph clusters by maximizing the overall degree of topological similarity in each cluster.

EGCPI evolves an optimal graph clustering arrangement in several steps. To begin, a number of individuals is first initialized. Each individual in the population is encoded as a vector of integers, where each integer represents the community to which the corresponding node belongs. A population of random solutions is generated. Then, it starts with the definition of a fitness function, which quantifies the quality of a partition of nodes into communities. The evolutionary game is then played by iteratively applying the following steps:

1. Reproduction: The fittest individuals in the population are selected to reproduce, and their offspring are generated through crossover and mutation.
2. Fitness evaluation: The fitness of each offspring is evaluated using the fitness function.
3. Game interaction: Each offspring competes with a randomly chosen individual from the previous generation, and the fitter individual is selected to survive to the next generation.

The algorithm terminates when a stopping criterion is met, such as a maximum number of generations or a threshold fitness value. The EGCPI method uses two parameters to control the algorithm's behavior: population size and mutation rate.

EGCPI applies an evolutionary graph clustering. The graph partitioning evolves by optimizing a topological measure called Independence of Cluster (IoC). Thus, in GA semantic similarity is not taken into account during evolution.

$$IoC_{c_i} = \frac{\sum\limits_{v_j, v_k \in C_i} W_{j_k}}{\sum\limits_{v_j \in C_i} W_{j_k}} \tag{2}$$

$$IoC_{wAPPIG} = \frac{\sum\limits_{i=1}^{S} n_{v_i}}{n_v} IoC_{c_i} \tag{3}$$

With:

- c_i: the ith cluster,
- w_{j_k}: the weight assigned to the interaction e_{jk} between the gene j and the gene k
- n_{v_i}: the total number of nodes in the cluster i.
- n_v: the total number of nodes in the graph wAPPIG.
- S: the number of clusters.
- IoC_{c_i}: represents the total weight of the intra-interactions compared to all the interactions linking the genes of the c_i cluster
- IoC_{wAPPIG}: measures the independence between clusters c_i with respect to the other clusters of the population. The latter allows to decrease the interdependence between two clusters.

After obtaining a set of clusters from the wAPPIG graph, EGCPI performs an additional step to identify gene communities. These communities are detected by calculating the degree of homogeneity between each pair of genes in the c_i cluster.

3.2 GA-PPI-NET

GA-PPI-Net is an evolutionary algorithm, that aim at identifying communities of proteins [5]. It allows finding communities having different sizes. It uses the similarity measures as well as the interaction measure between proteins and tries to find the best protein community by maximizing the concept of community measure. Thus, it allows combining two level of information:

1. Semantic level: information contained in biological ontologies such as Gene Ontology (GO) [1] and information obtained by the use of a similarity measure such as GO-based similarity of gene sets (GS2) [25]. It assesses the semantic similarity between proteins or genes.
2. Functional level: information contained in public databases describing the interactions of proteins or genes, such as Search Tool for Recurring Instances of Neighboring Gene (STRING) database [18].

GA-PPI-Net uses an Individual-Community based representation (ICR) and it is based on a specific structure for representing a community. Thus, a solution S is a community, and its quality is assessed thanks to the community measure. This one is used as the fitness function. The community measure, denoted F of a solution, S is computed using Eqs. 4.

$$F(S) = W_1 \ AVGSim(S) + W_2 \ AVGInteraction(S) \tag{4}$$

where W_1 and W_2 are weights $\in [0, 1]$

$$AVGSim(S) = \textstyle\sum_{i,j \ \in \ [1,n], \ i \neq j} SIM_{GS2}(G_i, G_j)/n \tag{5}$$

where: i) G_i and G_j are two different genes in the community S; ii) n: the size of the community S; iii) $SIM_{GS2}(G_i, G_j)$: the similarity value between two genes (G_i, G_j) in S, it is calculated using the semantic similarity measure GS2 [25];

$$AVGInteraction(S) = \sum_{i,j \ \in \ [1,n], \ i \neq j} InteractionValue(G_i, G_j)/n \tag{6}$$

where: $InteractionValue(G_i, G_j)$ is the value of an interaction between two genes (G_i, G_j) in S extracted from STRING Database [18].

This concept provides a solution of communities that are semantically similar and interacting. Moreover, it is based on a new genetic operation that is a specific mutation operator. The algorithm outputs the final community by selectively exploring the search space [5].

4 Comparative Study

4.1 Experimental Protocol

In this section, we propose a comparative study of the algorithm *GA-PPI-Net* with the EGCPI method using the Collins dataset. The latter was downloaded from the BioGRID database platform[3]. The Collins dataset concerns the genes of the yeast species Saccharomyces cerevisiae (yeast). It is composed of 1620 genes and 9064 interactions.

The results obtained by EGCPI on Collins dataset are available online at[4]. In order to apply GA-PPI-NET, a preprocessing step on the Collins dataset is needed in order to determine the semantic similarity using the $GS2$ method and the interaction values from the String database.

The experimental protocol used is as follows:

- We execute *GA-PPI-Net* 20 times using the Collins dataset, and we retain the best detected community each time.
- To perform a fair comparison, the value of common parameters were considered the same for all methods. GA-PPI-Net and EGCPI is run using the same parameters described in [11] namely population size = 100, $PC = 0.6$ and number of generations = 30.
- We retrieve from the data available online all the clusters obtained by EGCPI. Then, we filter the identified communities in order to obtain communities with a size greater than five genes.
- Randomly select 20 communities from online available data of EGCPI having the same size as the communities detected by GA-PPI-Net.

4.2 Performance Measures

We propose to use different metrics to report the comparison results. These metrics inspect the performance of the EGCPI and the GA-PPI-Net with respect to structural and functional quality as well as biological relevance of the identified communities. First, we check if the identified communities exist in real biological pathway databases such as KEGG Reactome, etc. This checking is achieved by the DAVID tools (Database for Annotation Visualization and Integrated Discovery), which compares this community with others in different databases and gives the percentage of proteins that belong to the existing communities in those databases. DAVID bioinformatics resources consist of an integrated biological knowledge-base and analytic tools that aim at systematically extracting biological meaning from large gene/protein lists. It is the most popular functional annotation programs used by biologists [26]. It takes as input a list of proteins and exploits the functional annotations available on these genes in a public database in order to find common functions that are sufficiently specific to these genes.

[3] https://thebiogrid.org/.

[4] https://github.com/he-tiantian/EGCPI.

Moreover, to evaluate these two approaches, we assessed the functional and structural quality of the obtained communities by calculating the semantic similarity and interaction score for each pair of proteins within predicted communities. We then compared the distribution of these values across all communities being studied. The semantic similarity between two proteins was measured by their respective Gene Ontology (GO) annotation terms using the GS2 method [5]. GS2 quantifies the similarity of the Gene Ontology annotations among a set of proteins by averaging the contribution of each all gene's Gene Ontology terms and their ancestor terms with respect to the Gene Ontology vocabulary graph [25]. The semantic similarity of a community was summarized by the average of the semantic similarity ($AVGSim$) of the protein pairs in the community, as defined in (Eq. 5). To determine the structural quality of an identified community, we used the string database. For each community, we calculated the average interaction score ($AVGInteraction$) of the protein pairs within that community, as defined in (Eq. 6).

4.3 Results and Discussions

In this section, we comprehensively evaluate and compare the performance of the GA-PPI-Net [5] with the clustering evolutionary method EGCPI on widely used Collins dataset of real PPI network data. Since these two approaches depend on EA parameters, we have used the same parameters proposed in the study of Tiantian [11].

As said before, to inspect the performance of these two approaches with respect to functional and structural quality of the resulting communities, we first, determined the semantic similarity and the interaction score in each predicted community (i.e. clusters). The different computed scores are summarized in Table 1.

The results are summarized with a box plot in Fig. 2.

The comparison of the results of the two approaches can be made only by comparing the general quality of the obtained communities by the proposed measure of performance based on the similarity and the interaction scores. It should be noted that the similarity is not used by EGCPI but is calculated later on the obtained clusters to identify the communities. The similarity score' values varied between 0.55 and 0.97 for GA-PPI-NET and between 0.55 and 0.92 for EGCPI. Nevertheless, the GA-PPI-Net exhibited the largest average semantic similarity with respect to the EGCPI approach, for example where the size of the identified community is equal to 128, 57 and 30. As for the interaction score, which is used by both methods, it varies between 0.66 and 0.87 for GA-PPINET and between 0.48 and 0.82 for EGCPI. This result reflects that GA-PPI-NET performs slightly better according to this score than the other method. This superiority is confirmed by Fig. 2 illustrating the median value for both cases. Thus, the quality of the communities given by the two methods has a close competition, with a slight superiority of GA-PPI-NET.

Table 1. Semantic similarity and interaction values of the obtained communities with GA-PPI-Net and EGCPI

Method	community size	AVG Similarity	AVG Interaction
GA-PPI-Net	9	0.88	0.71
	30	0.9	0.86
	7	0.67	0.61
	14	0.73	0.7
	8	0.59	0.8
	19	0.84	0.81
	128	**0.97**	**0.85**
	57	**0.91**	**0.83**
	8	0,72	0,66
	10	0,74	0,73
	38	0,67	0,7
	12	0,55	0,76
	7	0,77	0,84
	10	0,91	0,8
	25	**0,91**	**0,87**
	22	0,89	0,8
	12	0,67	0,78
	18	0,84	0,76
	30	0,9	0,79
	16	0,77	0,83
EGCPI: Step 1	9	0,78	0,48
	31	0,76	0,61
	7	0,7	0,6
	14	0,55	0,79
	8	0,69	0,72
	19	0,74	0,71
	123	0,92	0,59
	53	0,86	0,73
	8	0,71	0,74
	10	0,83	0,7
	39	0,85	0,72
	11	0,61	0,76
	7	0,73	0,82
	10	0,81	0,79
	25	0,87	0,84
	22	0,81	0,76
	12	0,65	0,72
	19	0,82	0,72
	29	0,82	0,8
	15	0,79	0,74
EGCPI: Step 2	10	0,92	0,81
	17	0,73	0,92

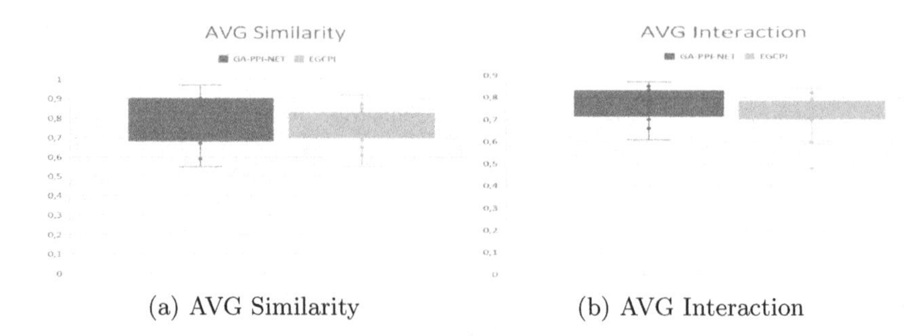

(a) AVG Similarity (b) AVG Interaction

Fig. 2. Synthesis of AVG Similarity and AVG Interaction measures recorded with EGCPI and GA-PPI-Net.

Otherwise, the performance of both GA-PPI-Net and EGCPI approaches with respect to biological relevance of the obtained communities are evaluated by checking whether they exist in multi-species biological pathways such as KEGG [14], Reactome and Ec Number. This assessment is performed using the DAVID tool [12]. Each new identified community is presented to the DAVID tools, which compares this community with others in different biological databases and gives the percentage of proteins that belong to the existing communities in those biological databases. The Table 2 describes the minimum and maximum percent recovery rate of each approach in different biological databases.

Table 2. Min and Max recovery rate of the obtained communities (Collins dataset).

Methods	biologicals databases	% Min	% Max
GA-PPI-Net	Ec Number	2.4%	25.0%
	KEGG	12.5%	**100%**
	Reactome	5.3%	**100%**
EGCPI (Step 1: clustering)	Ec Number	2.5%	28.6%
	KEGG	5.1%	**100%**
	Reactome	6.9%	**100%**
EGCPI (Step 2: Community detection)	Ec Number	–	–
	KEGG	–	100%
	Reactome	30%	100%

The results presented in Table 2 show that the identified communities correspond to some "parts" of real communities existing in other biological pathway databases, and in some cases to a complete network (percentage 100%). Therefore, the two methods were able to efficiently rebuilt communities existing in real biological pathway databases. GA-PPI-Net approach, as well the EGCPI, achieve the highest percentage 100% in two pathway databases: KEGG and Reactome.

The worst percentage value is of 2.4% which corresponds to some "parts" of the real communities. Nevertheless, GA-PPI-Net exhibited the largest Percentage Min in the KEGG database. These tests should be supplemented on a larger scale with other datasets and different communities.

To conclude, the results show the capability of the GA-PPI-Net approach to effectively deal with community detection in PPI networks. Further extensions will be proposed to detect networks with larger size and identify new networks not yet known in the public biological databases.

5 Conclusion

With the continuous growth of the complexity and size of the available networks to be explored, the use of evolutionary algorithms as a powerful meta-heuristic for graph clustering is expanding, and several solutions have been proposed for community detection. This paper classifies the different community detection approaches based on GA in a general taxonomy, with a synthetic overview of the evolutionary methods. It presents in details two evolutionary methods based on different solution encoding and using different fitness function, EGCPI and GA-PPI-Net. A comparative study is then performed on the Collins PPI network. The biological relevance of the obtained communities by the two methods is similar. The qualities of the detected communities by the two methods, in terms of the average of the similarities and the average of the interactions between the genes, are very close, with a slight superiority of the GA-PPI-Net method. The advantage of the latter lies essentially in the simplicity of the representation of the solutions (communities) that makes it able to scale up for very large networks.

Future works aim to extend this comparative study to some analytical approaches for DC, such as Markov Clustering [30], Restricted Neighborhood Search Clustering [15] and ClusterOne [19].

References

1. Ashburner, M., et al.: Gene ontology: tool for the unification of biology. The gene ontology consortium. Nat. Genet. **25**(1), 25–29 (2000). https://doi.org/10.1038/75556
2. Attea, B.A., et al.: A review of heuristics and metaheuristics for community detection in complex networks: current usage, emerging development and future directions. Swarm Evol. Comput. **63**, 100885 (2021). https://doi.org/10.1016/j.swevo.2021.100885. https://www.sciencedirect.com/science/article/pii/S2210650221000468
3. Bader, G.D., Hogue, C.W.: An automated method for finding molecular complexes in large protein interaction networks. BMC Bioinform. **4**(1), 1–27 (2003)
4. Ben M'barek, M., Borgi, A., Bedhiafi, W., Ben Hmida, S.: Genetic algorithm for community detection in biological networks. Proc. Comput. Sci. **126**, 195–204 (2018). https://doi.org/10.1016/j.procs.2018.07.233. Knowledge-Based and Intelligent Information & Engineering Systems: Proceedings of the 22nd International Conference, KES-2018, Belgrade, Serbia

5. Ben M'barek, M., Borgi, A., Ben Hmida, S., Rukoz, M.: Genetic algorithm to detect different sizes' communities from protein-protein interaction networks. In: Proceedings of the 14th International Conference on Software Technologies - Volume 1: ICSOFT, pp. 359–370. SciTePress (2019)
6. Ben M'barek, M., Ben Hmida, S., Borgi, A., Rukoz, M.: GA-PPI-Net approach vs analytical approaches for community detection in PPI networks. Procedia Comput. Sci. 903–912 (2021). https://doi.org/10.1016/j.procs.2021.08.093
7. Bilal, S., Abdelouahab, M.: Evolutionary algorithm and modularity for detecting communities in networks. Phys. A **473**, 89–96 (2017)
8. Cai, Q., Ma, L., Gong, M., Tian, D.: A survey on network community detection based on evolutionary computation. Int. J. Bio-inspired Comput. **8**(2), 84–98 (2016). https://doi.org/10.1504/IJBIC.2016.076329
9. Fortunato, S., Hric, D.: Community detection in networks: a user guide. Phys. Rep. **659**, 1–44 (2016). https://doi.org/10.1016/j.physrep.2016.09.002. arXiv:1608.00163
10. Girvan, M., Newman, M.E.J.: Community structure in social and biological networks. Proc. Natl. Acad. Sci. U.S.A. **99**(12), 7821–7826 (2002). https://doi.org/10.1073/pnas.122653799
11. He, T., Chan, K.C.C.: Evolutionary graph clustering for protein complex identification. IEEE/ACM Trans. Comput. Biol. Bioinform. **15**(3), 892–904 (2018). https://doi.org/10.1109/TCBB.2016.2642107
12. Jiao, X., et al.: DAVID-WS: a stateful web service to facilitate gene/protein list analysis. Bioinformatics **28**(13), 1805–1806 (2012). https://doi.org/10.1093/bioinformatics/bts251
13. Jin, D., Wang, T., Cao, L., Zhang, Y.: HGAC: a hierarchical genetic algorithm for overlapping community detection in social networks. Inf. Sci. **258**, 26–42 (2014)
14. Kanehisa, M., Goto, S.: KEGG: Kyoto encyclopedia of genes and genomes. Nucleic Acids Res. **28**(1), 27–30 (2000)
15. King, A.D., Pržulj, N., Jurisica, I.: Protein complex prediction via cost-based clustering. Bioinformatics **20**(17), 3013–3020 (2004)
16. Leiserson, M.D., et al.: Pan-cancer network analysis identifies combinations of rare somatic mutations across pathways and protein complexes. Nat. Genet. **47**(2), 106–114 (2015)
17. Liu, X., Li, D., Wang, S., Tao, Z.: Effective algorithm for detecting community structure in complex networks based on GA and clustering. In: Shi, Y., van Albada, G.D., Dongarra, J., Sloot, P.M.A. (eds.) ICCS 2007. LNCS, vol. 4488, pp. 657–664. Springer, Heidelberg (2007). https://doi.org/10.1007/978-3-540-72586-2_95
18. von Mering, C., Huynen, M., Jaeggi, D., Schmidt, S., Bork, P., Snel, B.: STRING: a database of predicted functional associations between proteins. Nucl. Acids Res. **31**(1), 258–261 (2003). https://doi.org/10.1093/nar/gkg034
19. Nepusz, T., Yu, H., Paccanaro, A.: Detecting overlapping protein complexes in protein-protein interaction networks. Nat. Methods **9**(5), 471 (2012)
20. Newman, M.E.J., Girvan, M.: Finding and evaluating community structure in networks. Phys. Rev. E **69**(2) (2004). https://doi.org/10.1103/PhysRevE.69.026113. arXiv:cond-mat/0308217
21. Pizzuti, C.: Evolutionary computation for community detection in networks: a review. IEEE Trans. Evol. Comput. **22**(3), 464–483 (2018). https://doi.org/10.1109/TEVC.2017.2737600
22. Pizzuti, C.: GA-Net: a genetic algorithm for community detection in social networks. In: Rudolph, G., Jansen, T., Beume, N., Lucas, S., Poloni, C. (eds.) PPSN

2008. LNCS, vol. 5199, pp. 1081–1090. Springer, Heidelberg (2008). https://doi.org/10.1007/978-3-540-87700-4_107
23. Pons, P., Latapy, M.: Computing communities in large networks using random walks. J. Graph Algorithms Appl. **10**(2), 191–218 (2006)
24. Radicchi, F., Castellano, C., Cecconi, F., Loreto, V., Parisi, D.: Defining and identifying communities in networks. PNAS **101**(9), 2658–2663 (2004). https://doi.org/10.1073/pnas.0400054101
25. Ruths, T., Ruths, D., Nakhleh, L.: GS2: an efficiently computable measure of GO-based similarity of gene sets. Bioinformatics **25**(9), 1178–1184 (2009). https://doi.org/10.1093/bioinformatics/btp128
26. Sherman, B.T., et al.: DAVID knowledgebase: a gene-centered database integrating heterogeneous gene annotation resources to facilitate high-throughput gene functional analysis. BMC Bioinform. **8**, 426 (2007). https://doi.org/10.1186/1471-2105-8-426
27. Shirmohammady, N., Izadkhah, H., Isazadeh, A.: PPI-GA: a novel clustering algorithm to identify protein complexes within protein-protein interaction networks using genetic algorithm. Complex. **2021**, 2132516:1–2132516:14 (2021). https://doi.org/10.1155/2021/2132516
28. Tasgin, M., Bingol, H.: Community detection in complex networks using genetic algorithm. arXiv:cond-mat/0604419 (2006)
29. Tasgin, M., Herdagdelen, A., Bingol, H.: Community detection in complex networks using genetic algorithms. arXiv:0711.0491 [physics] (2007)
30. Van Dongen, S.M.: Graph clustering by flow simulation. Ph.D. thesis, Utrecht University Repository (2000)
31. Wu, Z., Liao, Q., Liu, B.: A comprehensive review and evaluation of computational methods for identifying protein complexes from protein-protein interaction networks. Briefings Bioinform. **21**(5), 1531–1548 (2020)
32. Zhang, Y., Levina, E., Zhu, J.: Detecting overlapping communities in networks using spectral methods. SIAM J. Math. Data Sci. **2**(2), 265–283 (2020)

Predictive Monitoring of Business Process Execution Delays

Walid Ben Fradj(✉) and Mohamed Turki

MIRACL Laboratory, ISIMS, University of Sfax, P.O. Box 242, 3021 Sfax, Tunisia
walid.benfradj@isimsf.u-sfax.tn, mohamed.turki@isims.usf.tn

Abstract. Nowadays, organizations are becoming increasingly aware of the importance of better utilizing their knowledge assets and adopting a quality management model based on a process approach. This can be achieved by adopting a multidisciplinary approach that combines the fields of Knowledge Management, Business Process Management and Process Mining. Therefore, to improve their performance and increase their responsiveness, organizations must identify, manage and monitor all the business processes (BP) that are likely to mobilize crucial knowledge. In fact, implementing an IT system that automates business processes is necessary to achieve these goals. In this context, we propose a new method for predicting the execution times of business processes baptized BPETPM. This method is based on the CRISP-DM approach. We used a Process Mining techniques particularly the machine learning to exploit the execution data of a workflow engine. In order to prove the applicability of this method we have developed an intelligent system for predicting execution time of BP named iBPMS4PET. The applicative framework of this research work is the incoming mail management process as part of a group health insurance.

Keywords: Business Process Management · Process Mining · Knowledge Management · Machine Learning · CRISP-DM

1 Introduction

Currently, businesses are becoming increasingly aware of the importance of optimizing the use of their knowledge and implementing a quality management method based on a process approach. This can be achieved by adopting an interdisciplinary method that combines KM and BPM. To improve their results and increase their flexibility, companies must identify and manage all key processes that can mobilize tacit or explicit knowledge, which is a source of wealth to be valued. To achieve these goals, it is crucial to implement an IT system to automate business processes. According to [1], there are various information systems that are sensitive to processes, including mainly ERP (Enterprise Resource Planning), WFM (Workflow Management), CRM (Customer Relationship Management), SCM (Supply Chain Management), PDM (Product Data Management), and BPM (Business Process Management). Most of these systems use execution engines (or workflow engines). These engines serve, on the one hand, to deploy

I. Saad et al. (Eds.): ICIKS 2023, LNBIP 486, pp. 114–128, 2024.
https://doi.org/10.1007/978-3-031-51664-1_8

processes, and on the other hand, they allow the recording of business information in the form of a database and technical information related to the execution of processes in the form of event logs. These can be exploited by Process Mining techniques for extracting new knowledge that can be useful for process optimization and decision making within the organization.

According to the godfather of Process Mining Van der Aalst [2], there are various types of Process Mining, the most commonly used are: Process Discovery, Compliance Checking, Performance Analysis, Comparative Process Mining, Predictive Process Mining, and Action-Oriented Process Mining. Note that Process Discovery is a crucial step in all Process Mining work. According to [3], the different process discovery algorithms go through a step of filtering event logs by removing infrequent activities in order to have only the dominant behavior. The result obtained is a graphical model, namely, DFG (Directly-Follows Graphs), PN (Petri Net), PT (Process Tree), and BPMN (Business Process Model and Notation). This raises the following question: Can we extract the different transitions, regardless of their frequency, and enrich a database with all the paths followed by process instances? If so, another question arises: Can we predict the execution times of instances based on transitions? On the other hand, following a literature review, we do not find a standard scientific approach to follow for a Process Mining project. This leads to the idea of following the well-known CRISP-DM (Cross Industry Standard Process for Data Mining) procedure in Data Science.

In the context of predictive process monitoring, we present in this article a method that combines process discovery and predictive process mining. Without going through the extraction of a graph model which requires the elimination of less frequent activities, this method allows the extraction of all paths followed by process instances in a database. Then it allows to predict, based on these paths, whether a process instance completes execution on time or late. In this context, we follow the CRISP-DM approach, to exploit event logs that save data from a workflow engine execution and apply process mining techniques to create an intelligent system based on Machine Learning.

In the second section of this document, we present the main concepts related to the field of Knowledge Management (KM) and Business Process Management (BPM). We also discuss current trends in these areas, which involve the integration of artificial intelligence. In the third section, we provide an overview of techniques for predictive monitoring of business processes. In the fourth section, we detail our method for predicting the execution times of business processes, following the CRISP-DM method. In the fifth section, we present the implementation and evaluation of the computer solution developed, by applying it to a business process of managing incoming mail extracted from the health insurance management activity of the I-WAY company in Tunisia. Finally, we conclude this work with a conclusion and research perspectives that we envision.

2 Knowledge Management and Business Process Management: New Trends and Innovation

2.1 Knowledge Management (KM)

According to Grundstein M. (2000) in [4], "Capitalizing on the knowledge of the company is to identify its crucial (essential) knowledge, preserve and make it sustainable while making sure it is shared and used by the greatest number to the benefit of increasing the wealth of the company". Explicit knowledge refers to what can be quantified, understood, directly comprehended, and expressed by each individual in the organization. On the other hand, tacit knowledge, commonly referred to as know-how, is unique to each individual. It is composed, on the one hand, of their informal technical expertise and, on the other hand, of their personal beliefs and aspirations. According to the same previous reference, the KM process is articulated around five interacting facets centered on the notion of crucial knowledge: identification, preservation, valorization, updating and management. Each facet is the subject of sub-processes intended to provide a solution to all the problems concerned. In [5], the authors deal with the problem of identifying business processes sensitive to the level of the identification facet of crucial knowledge. Sensitive business processes require increased security management to reduce the risk of compromise and ensure business continuity. They propose a new methodology called SOPIM (Sensitive Organization's Process Identification Methodology) based on a multi-criteria decision-making approach to build a coherent family of criteria for evaluating and identifying these processes. The approach implemented mainly consists of two phases: (1) Construction of a model of decision-maker preferences and (2) Exploitation of the preference model (decision rules) to classify the "potentially sensitive organizational processes". Once sensitive organizational processes are identified, they will be subject to modeling, then execution and then exploration. In the context of this research article, we are interested in exploring processes.

With the advancement of information technology, a new generation of KM has emerged, known as IKM (Intelligent Knowledge Management) [6]. This new generation is characterized by the use of AI (Artificial Intelligence) tools that help to capitalize on the knowledge of experts. According to [7], AI plays a potential role in supporting KM activities. It allows to:

- Promote predictive analysis through self-learning analytical capabilities.
- Recognize previously unknown patterns.
- Sift through organizational data and discover relationships.
- Develop new declarative knowledge.
- Collect, classify, organize, store and retrieve explicit knowledge.
- Analyze and filter multiple content and communication channels.
- Facilitate the reuse of knowledge by teams and individuals.
- Connect people working on the same problems by promoting weak ties and "know who."
- Facilitate collaborative intelligence and shared organizational memory.
- Generate an overall perspective on knowledge sources and bottlenecks.
- Create more connected coordination systems between organizational silos.

- Improve the application of localized knowledge by searching and preparing knowledge sources.
- Promote fair access to knowledge without fear of social cost.
- Etc.

2.2 Business Process Management (BPM)

Regardless of its size, a business is built around processes that cover all its activities and services (e.g. Human Resources, Sales, Quality, Procurement, etc.). According to [8], Business Process Management (BPM) is a process-centered approach that effectively monitors the progression of activities within the organization with the goal of improving overall performance and, in turn, results. Utilizing a BPM tool provides real-time traceability of various exchanges that can occur between actors involved in a process and offers greater responsiveness through the generation of different indicators in the form of automatic alerts or notifications. This enables decision making to be facilitated, bottlenecks to be identified more quickly, deadlines to be met, and the cost of producing products and services to be controlled [9]. The BPM lifecycle consists of five phases: Design, modeling, execution, monitoring, and optimization. (See Fig. 1).

Fig. 1. Cycle de vie du BPM[1]

The first phase (Design) involves identifying existing processes, and then designing future processes. The second phase (Modeling) involves graphically representing the model as closely as possible to reality. The third phase (Execution) is the implementation phase of BPM. Business procedures are interpreted by an execution engine that implements and coordinates all types of interactions between users, system tasks, and computer resources. The fourth phase (Monitoring) encompasses the regulation of individual processes to easily consult information on their status and provide statistics on the performance of one or more processes. Finally, the fifth phase (Optimization) allows

[1] https://www.objetconnecte.com/bpm-presentation.

for the adjustment of a process in order to minimize costs and optimize efficiency. In the context of this research article, we are positioning ourselves at the level of the Monitoring phase.

The field of BPM has seen many technological advancements thanks to the integration of Artificial Intelligence. In this context, American consulting and research company Granter introduced the concept of iBPMS (Intelligent Business Process Management Suite) in 2012. iBPMS is considered the evolution of BPM, as it combines the use of predictive analysis, process intelligence, and emerging technologies with traditional BPM practices.

3 Predictive Monitoring of Business Processes

According to [10], predictive process monitoring aims to detect and anticipate potential issues in advance. It enables countermeasures to be applied to prevent problems from arising and aid in making appropriate decisions proactively. With real-time monitoring techniques, data is analyzed in real-time, which can assist in decision-making and optimizing processes that are currently in operation.

According to [11], predictive process monitoring is performed in real-time during the execution of instances. It should be noted that in the learning phase, the event log as well as additional information form the input data for the predictive monitoring method. These data go through an encoding step into feature vectors to be interpreted by the prediction algorithm. This algorithm generates a prediction model which is applied to instances currently in execution, in order to determine the predicted output value for that process instance. The majority of predictive monitoring techniques consist of an offline component (costly in computation) that handles the generation of the predictive model and an online component (fast) that handles predictions based on the generated model.

Event logs are the main source of input data for Process Mining techniques, particularly for predictive process monitoring. Each line in log files represents execution information for an activity, primarily consisting of the process instance ID and timestamp. According to references [12] and [13], the log may also include additional information such as the cost of the activity or the name of the resource executing it. The technical characteristics that describe event logs include the number of instances, number of activities, number of events, and number of process variants [14].

Predictive process monitoring with a temporal perspective is a research topic first addressed by [15]. The proposed prediction approach extracts a transition system from the event log with additional temporal information. This approach was a foundational reference for [12, 16–18]. According to [19], the transition system uses decision trees to predict the completion time and next activity of a process instance. In the two works of [16] and [17], the approach of [15] was extended by clustering traces from the event log based on contextual features.

According to [18], an approach is exploited by adding Naive Bayes and Support Vector Regressors models for annotating the transition system. Additional attributes have positively influenced the prediction quality. The disadvantage of these methods is that they assume that the event log used for learning contains all possible behaviors of the process. This assumption is generally not valid in reality.

The two approaches proposed by [20] and [19] are similar. They present a general framework for enriching the event log with derived information and discovering correlations between process characteristics using decision trees. The difference between the two approaches is that in the one proposed by [19], infrequent behaviors have been discarded. This latter approach is similar to the one proposed by [18], except that the process model is a reduced type of transition system. The problem posed by correlation is that numerical values need to be discretized, which reduces precision.

In the two works by [21] and [22], two probabilistic methods based on the Hidden Markov Model (HMM) are proposed. The probabilistic method allows predicting the probability of future activities, providing information on the advancement of the process. The approach proposed by [23] exploits a generic model based on decision trees which can provide decision criteria according to the actual goals of the process.

Other approaches from various fields have been proposed for deadline prediction. According to [24], process mining is based on queueing systems. This approach is based on building an annotated transition system and using non-linear regression algorithms for deadline prediction. According to [25], the proposed approach allows for predicting remaining time using expressive probabilistic models and only workflow information.

A decision tree-based predictive model is proposed by [26]. This model estimates the probability that an user-defined constraint is satisfied by the ongoing instances. The task is also encountered in the work of [27]. Traces are treated as complex symbolic sequences encoded by two methods: indexed encoding and HMM encoding. The same encoding is used by [28]. Clustering is used to divide the dataset, and a random forest-based predictor is trained for each cluster.

With reference [29], natural language processing (NLP) is combined with various classifiers to derive representative features of individual documents. Random forests were found to be the most effective predictors. Subsequently, approaches based on deep neural networks have emerged. Both the approach of [30] and [22] employ a recurrent neural network composed of two hidden RNN layers, utilizing basic LSTM cells to predict the next event. The approach of [22] also utilizes an LSTM-based neural network to predict the next activity and its execution time. Both of these approaches do not take into account additional attributes and are sensitive to hyperparameter selection and require a significant amount of training time. The approach of [10] utilizes artificial neural networks (ANNs) to predict whether a process instance exceeds the expected deadline. [31] presents a comparison between two machine learning models (Random Forest and SVM) and two deep learning models (LSTM and DNN). The results showed that the LSTM model is the best.

Based on our review of the literature, we did not find a standard scientific approach to follow for carrying out a Process Mining project. In our work, we follow the well-known CRISP-DM procedure in Data Science. Furthermore, most event logs used store business data. In this study, we deal with execution data from the workflow engine of a BPMS. We also notice that models are generally represented graphically in Petri net and less frequent transitions are eliminated. In our work, we add all paths followed by process instances to the database as additional attributes. Finally, we note that some approaches have tested more than one prediction model. In this work, we apply six models to choose the one with the best accuracy value.

4 BPETPM (Business Process Execution Time Prediction Method)

During its execution, a business process can go through several stages that represent all of its tasks (or activities). The sequence of these tasks constitutes a case (or instance). Therefore, all information related to an instance is saved by the BPMS in the form of an event log. For the same process, it is possible to have multiple instances with different paths. The execution time of a process depends on the sequence of its tasks and their execution times. However, if the execution time of a process or task exceeds a given threshold, it can be an indicator used to detect bottlenecks and help decision-makers to optimize and improve the performance of different business processes.

The CRISP-DM (CRoss Industry Standard Process for Data Mining) is a field-tested procedure that guides data exploration work. It represents a key to success in Data Science. According to [32], CRISP-DM is a popular process in practice and research. It is an organizational process model and is not limited to any technology. It includes descriptions of typical project phases and tasks included in each phase, and an explanation of the relationships between these tasks. It offers an overview of the data exploration lifecycle. According to [33], CRISP-DM breaks down the exploration process into six main steps: Process detection, data detection, data preparation, modeling, evaluation, and result released (See Fig. 2).

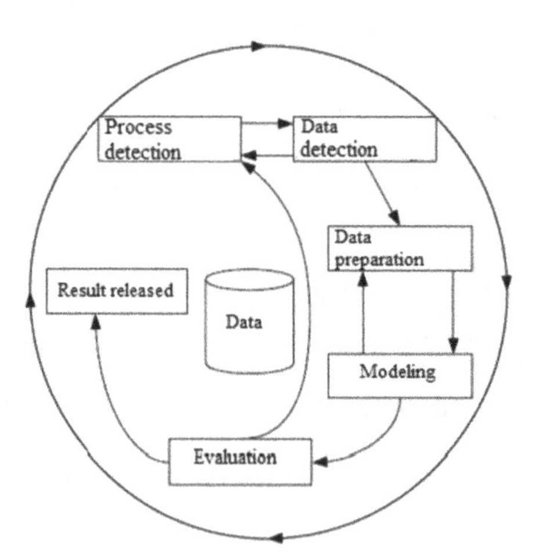

Fig. 2. CRISP-DM procedure [33] p. 216

In this context, and based on the CRISP-DM procedure, the BPETPM method (see Fig. 3) allows for the extraction of the different paths followed by process instances from the event log. The execution time of each path is then calculated. The goal of BPETPM is to predict the execution times of business processes. This prediction can be used to offer decision-makers, in the case of a semi-automatic process, or the system, in the case of an automatic process, the ability to choose the most optimal path.

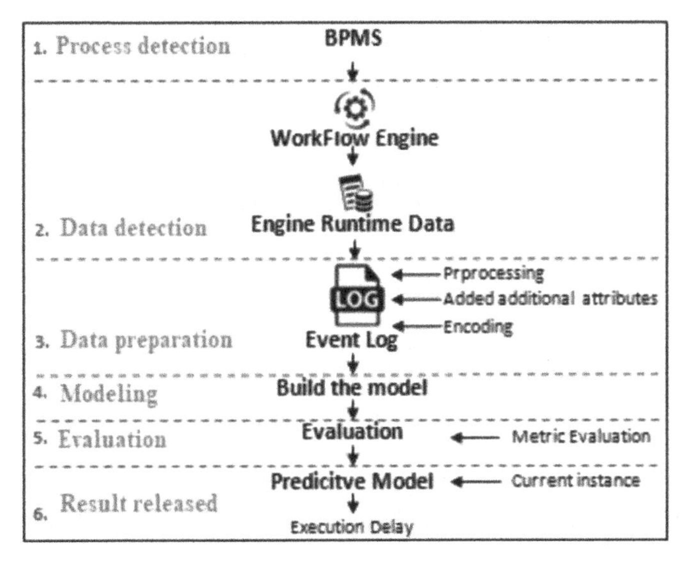

Fig. 3. BPETPM

4.1 Phase 1: Process Detection

Our method focuses on the business processes of the company and the flow of activities in order to improve overall performance and as a result. Each company is built around a set of processes that cover its activities. With technological advancements, the company strongly requires the use of intelligent software solutions to quickly respond to issues and improve competitiveness. Business Process Management Systems (BPMS) are a potential solution for managing and controlling these processes. As such, the implementation of workflow engines, which are a component of BPMS, allows for the separation of business data and process execution data.

4.2 Phase 2: Data Detcection

Data related to the execution of processes is saved by the workflow engine in event logs. The retrieval of process execution data from a BPMS workflow engine can be done in different ways: Either from a database integrated within the BPMS. Generally, the tables in this database are temporary, created in memory and lost upon server restart. Or through the configuration of REST APIs to connect to the integrated database and retrieve execution traces and save them to disk. Or through the configuration of a separate database (MySQL, Oracle, MongoDB, etc.) to the BPMS through connectors. In the context of our work, we are using two CSV-type event logs. The contents of these two files are extracted from a MySQL database configured to record execution data from a BPMS workflow engine.

4.3 Phase 3: Data Preparation

Having a log file containing the execution data of a workflow engine does not mean that the data is ready for modeling. However, this file may present redundancies, missing values, unnecessary values, categorical variables that require encoding, etc. Additionally, these data require enrichment by adding additional attributes. In the context of our BPETPM method, we propose the following approach for data preparation:

1. Remove empty columns.
2. Remove indexing columns because they are not useful for modeling.
3. Remove redundant columns.
4. Replace null values with appropriate values.
5. Sort the data based on the “date” column.
6. Determine, for each instance, the path followed by the task (human or automatic).
7. Add a column containing the path per instance.
8. Add a column containing the execution duration of each instance.
9. Calculate the interquartile range (IQR). The first quartile (Q1), the second quartile (Q2), and the third quartile (Q3) of the column containing the instance durations (see Fig. 4)
10. Add a column “description” containing the value “late” if the duration of the instance is greater than Q3, and the value “in time” otherwise.
11. Add all previously defined columns to a Data Frame.
12. Convert the Data Frame to a CSV file.
13. Verify if there are any missing values in the database.
14. Decompose the dataset into independent variables X, which encompasses: the process ID, the instance ID, the actor who started the instance, the actor who finished the instance, and the path followed by the instance, and a dependent variable y, which represents the “description” variable.
15. Verify the type of variables and encode categorical variables using LabelEncoder to transform them into numerical values. It should be noted that, based on our review of existing literature, the most commonly used encoding is OneHotEncoder. However, it may increase the size of the database and the learning time by adding as many columns as there are values per variable.
16. Divide the dataset (X; y) into a training set (X_train; y_train) and a test set (X_test; y_test).
17. Standardize the values of the X_train and X_test datasets.

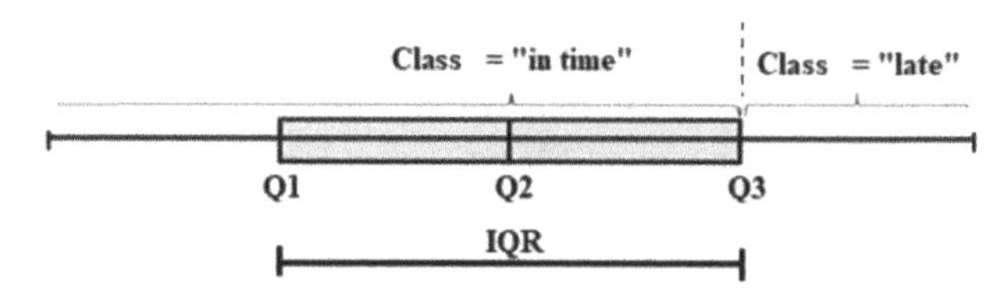

Fig. 4. Boxplot for class selection

4.4 Modeling

In terms of modeling, the first step is to determine the type of problem. In our context, we are trying to predict if a process instance is completed on time or late. Therefore, the prediction is a qualitative variable. As such, we can deduce that our problem is a supervised learning problem of the classification type. In order to choose the best supervised learning algorithm, we have chosen to compare the results of applying the following six algorithms: KNN (k-NearestNeighbors), DecisionTree, RandomForest, SVM (Kernel = linear), SVM (Kernel = RBF) and LogisticRegression.

4.5 Evaluation

In evaluating each model, we first predict the outcome y_pred for the independent variables in the test set X_test. Then, we compute the accuracy based on the confusion matrix (Table 1).

Table 1. Confusion matrix

		Predictive Class	
		Positive	Negative
Current Class	Positive	TP True Positive	FP False Positive
	Negative	FN False Negative	TN True Negative

Accuracy represents the rate of correct classification. It quantifies the extent of instances effectively classified compared to the aggregate instances.

$$\text{Accuracy} = \frac{TP + TN}{TP + TN + FP + FN} \quad (1)$$

4.6 Result Released

At the end of our work, we deploy the BPETPM method for the prediction of business process execution times to the end-users.

5 iBPMS4PET (Intelligent Business Process Management System for Prediction Execution Time)

In accordance with the steps of our BPETMP method, in this section we present an intelligent business process management system named iBPMS4PET (intelligent Business Process Management System For Prediction Execution Time) for predicting the execution times of a business process.

In this article, we have chosen the application framework of the intelligenceWay group (https://iway-tn.com/). We are focusing on the field of health insurance, specifically the management process of incoming mail. It is worth noting that the I-Way Group has a computer solution named I-Santé (https://i-sante.tn/). This solution provides an end-to-end management suitable for all health insurance professions (health fund, mutual insurance, and insurance company) based on a 100% digital platform and a highly secure health card. I-Santé meets the needs of insured persons, insurers, and health professionals through the use of advanced technologies such as the BPM Workflow Engine, GED, Rule Engine, etc. The business process being treated is the "Incoming mail" process (see Fig. 5) modeled on the Bonitasoft platform.

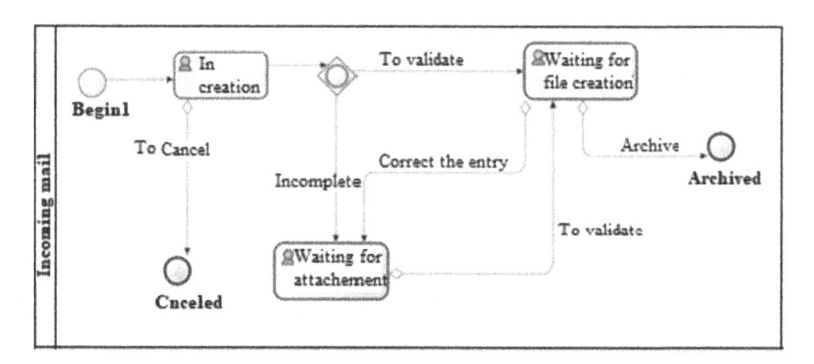

Fig. 5. Incoming mail process

In this work, we use two CSV type event logs extracted from an I-WAY configured MySQL database. They store the execution data of a BonitaSoft iBPMS workflow engine. The first log file "BN_PROC_INST.csv" saves the execution traces by process instance, and the second file "BN_ACT_INST.csv" saves the execution traces by task. Using the Python programming language, we process these two event logs following the steps previously described in the BPETPM method. In the data preparation step, we use the pandas, numpy, pylab, scipy-optimize, matplotlib and seaborn libraries. The result is a single file named "Result.csv" containing 4817 rows and 24 columns. The columns ready for modeling are: [PROCESS_UUID_, INST_UUID_, START_BY, END_BY, user_act0, user_act1, user_act2,..., useractnb, duration_INST, description] such as:

PROCESS_UUID_: represents the process identifier.

INST_UUID_: represents the instance identifier.

START_BY: represents the user who started the process instance.

END_BY: represents the user who completed the process instance.

user_act i: represents the performer (person or machine) of activity number i.

duration_INST: represents the execution time of the instance. Description: represents a description of the instance delay.

In terms of modeling, we start by importing the following classifiers: KNeighborsClassifier, DecisionTreeClassifier, RandomForestClassifier, SVM (kernel = linear), SVM (kernel = rbf), and LogisticRegression. For each model, training is done using standardized training variables. Then we test each model with the test variables and keep the prediction result in a "y_pred" variable. At the evaluation step, for each model, we import the "classification_report" module from the "sklearn.metrics" library and display the evaluation report. In terms of accuracy, the results are presented in Table 2.

Table 2. Evaluation

Classification Model	Accuracy
KNN	83%
RandomForest	81%
SVM (kernel = linear)	81%
SVM (kernel = rbf)	**84%**
LogesticRegression	80%
Decision Tree	78%

6 Conclusion and Perspectives

In the course of this research article, we have addressed the issue of predictive monitoring of business processes. This problem represents a current research topic in the field of BPM and plays an important role in process-oriented organizations. In this context, we have developed a method (BPETPM). The contributions made in this work can be summarized as follows:

- Following the CRISP-DM, a well-known Data Science procedure, to develop a method in Process Mining.
- Applying Process Mining techniques to event logs that record execution data from a BPMS workflow engine.
- Extracting all the paths followed by a business process from the event logs and using these paths as additional attributes with the data to be analyzed.
- Using the LabelEncoder method to encode categorical variables.
- Determining the prediction classes based on the third quartile of the durations of business process instances.
- Predicting the execution time of a business process based on the paths followed by its instances.

In the data preparation phase, we carried out cleaning of the data recorded in event logs. Then, we defined and added additional attributes, including the path relative to an instance, the execution time of an instance, and the execution time description of an instance. If an instance's execution time is less than or equal to the third quartile of the execution time column, it is considered to be executed on time (description = "on-time"), otherwise it is considered to be executed late (description = "late"). We encoded categorical variables using the LabelEncoder method. Then, we standardized all independent variables in order to reduce the gaps between their values.

In the modeling phase, we tested six models for predicting instance execution times. The goal of this prediction is to determine whether a new process instance will be completed on time or late. In terms of Accuracy, the evaluation showed that the SVM (kernel = rbf) model outperforms the Decision-Tree, Random-Forest, SVM (kernel = linear), KNN, and Logistic-Regression models.

At the implementation level, we applied the BPETPM method to create an intelligent process management system for predicting execution times called iBPMS4PET. This system was applied to an incoming mail management process in the healthcare mutual field.

In addition to the contributions made in this master's thesis, there are some points that merit further study. In the short term, it may be feasible to use other types of event logs and test other classification models based on neural networks. In the medium term, we plan to leverage business data contained in relational or NoSQL databases with workflow engine execution data to detect bottlenecks. Additionally, we plan to develop a monitoring system for visualizing performance indicators related to business processes.

References

1. Van der Aalst, W.M., Schonenberg, M.H., Song, M.: Time prediction based on process mining. Inf. Syst. **36**(2), 450–475 (2011)
2. van der Aalst, W. M. P.: Process mining: a 360 degree overview In: van der Aalst, W.M.P., Carmona, J. (eds.) Process Mining Handbook. LNBIP, vol. 448, pp. 3–34. Springer, Cham (2022). https://doi.org/10.1007/978-3-031-08848-3_1
3. van der Aalst, W. M. P.: Foundations of process discovery. In: van der Aalst, W.M.P., Carmona, J. (eds.) Process Mining Handbook. LNBIP, vol. 448, pp. 37–75. Springer, Cham (2022). https://doi.org/10.1007/978-3-031-08848-3_2
4. Morey, D., Maybury, M.T., Thuraisingham, B.M.: Knowledge Management: Classic and Contemporary Works. MIT Press, Cambridge (2002)
5. Turki, M., Saad, I., Gargouri, F., Kassel, G.: A business process evaluation methodology for knowledge management based on multicriteria decision-making approach. In: Information Systems for Knowledge Management, pp. 249–277 John Wiley & Sons, Ltd (2014)
6. Sanzogni, L., Guzman, G., Busch, P.: Artificial intelligence and knowledge management: questioning the tacit dimension. Prometheus **35**(1), 37–56 (2017)
7. Jarrahi, M.H., Askay, D., Eshraghi, A., Smith, P.: Artificial intelligence and knowledge management: a partnership between human and AI. Bus. Horiz. **66**(1), 87–99 (2023)
8. Dumas, M., La Rosa, M., Mendling, J., Reijers, H.A.: Process implementation with executable models. Presented at the (2018). https://doi.org/10.1007/978-3-662-56509-4_10
9. Ko, R.K., Lee, S.S., Wah Lee, E.: Business process management (BPM) standards: a survey. Bus. Process. Manag. J. **15**(5), 744–791 (2009)
10. Teinemaa, I., Dumas, M., Rosa, M.L., Maggi, F.M.: Outcome-oriented predictive process monitoring: review and benchmark. ACM Trans. Knowl. Discov. Data (TKDD) **13**(2), 1–57 (2019)
11. Márquez-Chamorro, A.E., Resinas, M., Ruiz-Cortés, A.: Predictive monitoring of business processes: a survey. IEEE Trans. Serv. Comput. **11**(6), 962–977 (2017)
12. van der Aalst, W.M.P.: data scientist: the engineer of the future. In: Mertins, K., Bénaben, F., Poler, R., Bourrières, JP. (eds.) Enterprise Interoperability VI Proceedings of the I-ESA Conferences, vol. 7, pp. 13–26. Springer, Cham (2014). https://doi.org/10.1007/978-3-319-04948-9_2

13. vom Brocke, J., Zelt, S., Schmiedel, T.: On the role of context in business process management. Int. J. Inf. Manag. **36**(3), 486–495 (2016)
14. Augusto, A., et al.: Automated discovery of process models from event logs: review and benchmark. IEEE Trans. Knowl. Data Eng. **31**(4), 686–705 (2019)
15. van der Aalst, W., et al.: Process mining manifesto. In: Daniel, F., Barkaoui, K., Dustdar, S. (eds.) BPM 2011. LNBIP, vol. 99, pp. 169–194. Springer, Heidelberg (2012). https://doi.org/10.1007/978-3-642-28108-2_19
16. Folino, F., Greco, G., Guzzo, A., Pontieri, L.: Mining usage scenarios in business processes: outlier-aware discovery and run-time prediction. Data Knowl. Eng. **70**(12), 1005–1029 (2011)
17. Folino, F., Guarascio, M., Pontieri, L.: Discovering context-aware models for predicting business process performances. In: Meersman, R., et al. (eds.) OTM 2012. LNCS, vol. 7565, pp. 287–304. Springer, Heidelberg (2012). https://doi.org/10.1007/978-3-642-33606-5_18
18. Polato, M., Sperduti, A., Burattin, A., de Leoni, M.: Data-aware remaining time prediction of business process instances. In: 2014 International Joint Conference on Neural Networks (IJCNN), Beijing, China, pp. 816–823 (2014)
19. Ceci, M., Lanotte, P.F., Fumarola, F., Cavallo, D.P., Malerba, D.: Completion time and next activity prediction of processes using sequential pattern mining. In: Džeroski, S., Panov, P., Kocev, D., Todorovski, L. (eds.) DS 2014. LNCS (LNAI), vol. 8777, pp. 49–61. Springer, Cham (2014). https://doi.org/10.1007/978-3-319-11812-3_5
20. de Leoni, M., van der Aalst, W.M.P., Dees, M.: A General framework for correlating business process characteristics. In: Sadiq, S., Soffer, P., Völzer, H. (eds.) BPM 2014. LNCS, vol. 8659, pp. 250–266. Springer, Cham (2014). https://doi.org/10.1007/978-3-319-10172-9_16
21. Lakshmanan, G.T., Shamsi, D., Doganata, Y.N., Unuvar, M., Khalaf, R.: A Markov prediction model for data-driven semi-structured business processes. Knowl. Inf. Syst. **42**(1), 97–126 (2015)
22. Tax, N., Verenich, I., La Rosa, M., Dumas, M.: Predictive Business process monitoring with LSTM neural networks. In: Dubois, E., Pohl, K. (eds.) CAiSE 2017. LNCS, vol. 10253, pp. 477–492. Springer, Cham (2017). https://doi.org/10.1007/978-3-319-59536-8_30
23. Ghattas, J., Soffer, P., Peleg, M.: Improving business process decision making based on past experience. Decis. Support. Syst. **59**, 93–107 (2014)
24. Senderovich, A., Weidlich, M., Gal, A., Mandelbaum, A.: Queue mining for delay prediction in multi-class service processes. Inf. Syst. **53**, 278–295 (2015)
25. Rogge-Solti, A., Weske, M.: Prediction of business process durations using non-Markovian stochastic Petri nets. Inf. Syst. **54**, 1–14 (2015)
26. Maggi, F.M., Di Francescomarino, C., Dumas, M., Ghidini, C.: Predictive monitoring of business processes. In: Jarke, M., Mylopoulos, J., Quix, C., Rolland, C., Manolopoulos, Y., Mouratidis, H., Horkoff, J. (eds.) CAiSE 2014. LNCS, vol. 8484, pp. 457–472. Springer, Cham (2014). https://doi.org/10.1007/978-3-319-07881-6_31
27. Leontjeva, A., Conforti, R., Di Francescomarino, C., Dumas, M., Maggi, F.M.: Complex symbolic sequence encodings for predictive monitoring of business processes. In: Motahari-Nezhad, H.R., Recker, J., Weidlich, M. (eds.) BPM 2015. LNCS, vol. 9253, pp. 297–313. Springer, Cham (2015). https://doi.org/10.1007/978-3-319-23063-4_21
28. Di Francescomarino, C., Ghidini, C., Maggi, F.M., Milani, F.: Predictive process monitoring methods: which one suits me best? In: Weske, M., Montali, M., Weber, I., vom Brocke, J. (eds.) BPM 2018. LNCS, vol. 11080, pp. 462–479. Springer, Cham (2018). https://doi.org/10.1007/978-3-319-98648-7_27
29. Teinemaa, I., Dumas, M., Maggi, F.M., Di Francescomarino, C.: Predictive business process monitoring with structured and unstructured data. In: La Rosa, M., Loos, P., Pastor, O. (eds.) BPM 2016. LNCS, vol. 9850, pp. 401–417. Springer, Cham (2016). https://doi.org/10.1007/978-3-319-45348-4_23

30. Evermann, J., Rehse, J.R., Fettke, P.: A deep learning approach for predicting process behaviour at runtime. In: Dumas, M., Fantinato, M. (eds.) BPM 2016. LNBIP, vol. 281, pp. 327–338. Springer, Cham (2017). https://doi.org/10.1007/978-3-319-58457-7_24
31. Kratsch, W., Manderscheid, J., Röglinger, M., Seyfried, J.: Machine learning in business process monitoring: a comparison of deep learning and classical approaches used for outcome prediction. Bus. Inf. Syst. Eng. **63**, 261–276 (2021)
32. Schröer, C., Kruse, F., Gómez, J.M.: A systematic literature review on applying CRISP-DM process model. Proc. Comput. Sci. **181**, 526–534 (2021)
33. Zhang, Y.: Sales forecasting of promotion activities based on the cross-industry standard process for data mining of e-commerce promotional information and support vector regression. J. Comput. **32**(1), 212–225 (2021)

An Accurate Random Forest-Based Action Recognition Technique Using only Velocity and Landmarks' Distances

Hajer Maaoui[1,2], Amani Elaoud[2(✉)], and Walid Barhoumi[2,3]

[1] Institut Supérieur des Sciences Appliquées et de Technologie de Mateur, Université de Carthage, 7030 Mateur, Tunisia

[2] Institut Supérieur d'Informatique, Research Team on Intelligent Systems in Imaging and Artificial Vision (SIIVA), LR16ES06 Laboratoire de recherche en Informatique, Modélisation et Traitement de l'Information et de la Connaissance (LIMTIC), Université de Tunis El Manar, 2 Rue Bayrouni, 2080 Ariana, Tunisia
amani.elaoud@fst.utm.tn

[3] Ecole Nationale d'Ingénieurs de Carthage (ENICarthage), Université de Carthage, 45 Rue des Entrepreneurs, 2035 Tunis-Carthage, Tunisia

Abstract. Human action recognition has gained significant attention recently as it can be adopted for various potential applications, notably for smart activity monitoring for surveillance and assisted living purposes. However, recognizing human activities is a challenging task because of the variety of human actions in daily life. In this work, we propose to extract skeletons from RGB images using deep learning architectures while adopting hand-crafted features for action recognition. The main idea is to extract few features from the skeleton data in order to ensure low feature dimensionality while leading to fast and accurate action recognition. In fact, our objective is to define hand-crafted features that are clearly explainable and related to the kinematics. Each joint is modeled by a single feature vector that encodes only the most essential kinematic information to characterize an action: the relative locations of the joints as well as the velocity of these joints. We have validated the proposed technique on the challenging UTD-MHAD multimodal action dataset and the preliminary obtained results are very encouraging.

Keywords: Human action recognition · Deep learning · Velocity · Skeleton · Landmarks

1 Introduction

The automatic recognition and interpretation of human behaviours from visual data have been an active research topic in the last years due to its potential application in a variety of domains [10]. In particular, a great interest has been granted to human action recognition, which has proved to be effective for understanding human actions for a number of real-world applications. Indeed, many

I. Saad et al. (Eds.): ICIKS 2023, LNBIP 486, pp. 129–144, 2024.
https://doi.org/10.1007/978-3-031-51664-1_9

research works have designed accurate human action recognition techniques that have been employed in many modern applications, such as video surveillance [28], environmental home monitoring [1], human-machine interaction [40], video storage and retrieval [36], and entertainment [29]. For instance, human action recognition is a fundamental task of intelligent video-surveillance systems that seek to understand what the subjects are doing in an input video [22]. The main objective is to automatically understand the human behaviours as well as the interactions with the environment. This could be very useful not only for various visual surveillance systems that look to identify dangerous human activities [11], but also for autonomous navigation systems in order to perceive human behaviors for safe operating [21]. In this context, human action recognition has recently become increasingly popular thanks to the availability of effective motion capture tools and accessible high quality visual data. Indeed, the ability to interpret such information has eventually bring computers closer to human capabilities. However, action recognition is still a major challenge in many real-world contexts because of the diversity of movements, the complexity of the captured motions, especially in crowded environments, the inter-subjects variability, and the difficulty of creating standard datasets that can be considered as realistic and relevant. For instance, the action of drinking can be done in several ways dependent on the individual actor, the context, the point of view, the environment, the style of movement, and many other parameters. Generally, each action can be performed in several ways, what complicates its modeling using characteristics describing this action; and how to capture and represent this action is another challenge. In fact, there are many challenges that can make it difficult to identify human actions in still images, comparatively to what can be performed using one or many videos. For instance, authors in [41] have proposed an effective action recognition from monocular RGB videos using a transformer-based framework in order to exploit temporal information. Furthermore, the authors in [33] have investigated RGB-based action recognition using multi-view videos. They have adopted a supervised contrastive learning framework in order to learn a feature embedding robust to changes in viewpoint. Overall, various action recognition methods have been proposed and many well-known action datasets are provided and some of them are publicly available. The two most common types of data that have been used for human action recognition are RGB videos and skeletons. Nevertheless, although that they cover full information of the human body, RGB videos data are computationally expensive and complex in practical usage. These data are also sensitive to illumination changes, camera calibration, lighting conditions, and background changing. However, the skeletons are not sensitive to illumination changes, lighting conditions and backgrounds. Indeed, skeleton-based techniques have proven their effectiveness to deal with human action recognition with less computational complexity while being less sensitive against environmental changes. The cost of processing skeleton information is, in fact, negligible, i.e. few milliseconds to process a whole activity [46]. But, most of the few publicly available skeletons' datasets have been designed under in-lab environment.

Thus, we propose in this work the recognition of human actions from RGB videos while extracting sequences of skeleton data using an accurate deep learning architecture. In fact, given an ordered sequence of human joint coordinates representing a person performing an action, the aim is to categorize this collection into one of the possible action categories. To deal with this issue, we have opted for few features (velocity and landmarks' distances) for indexing skeleton sequences that illustrate enough information including the whole body. Indeed, our main idea consists to explore few hand-crafted features, which are clearly explainable and related to the kinematics, in order to ensure low feature dimensionality while leading to fast and accurate action recognition. The proposed technique has been validated on the challenging UTD-MHAD multimodal action dataset and the preliminary results are very promising.

The rest of this paper is organised as follows. In the next section, we will briefly discuss the related work on human action recognition. In Sect. 3, we will present our contribution within the context of automated action recognition. Section 4 will show the obtained results by the proposed method. Finally, the concluding remarks and some suggestions for future works will be discussed in Sect. 5.

2 Related Work

Many research surveys have focused on human action recognition such as the works of [12] and [15], where comprehensive studies have been provided on traditional hand-crafted-based methods as well as deep learning-based ones for action recognition. In particular, for the hand-crafted approach, characteristics can be used to capture the joint trajectories in skeletons in order to construct features designed to distinguish human movement representations. In fact, hand-crafted features are extracted manually using designed algorithms, whereas the trainable features can be determined automatically from videos in the deep learning methods. Within the framework of hand-crafted methods, the work of [39] has investigated video action detection with relational dynamic-poselets. Domain and contextual knowledge has been widely exploited for human activity recognition in video streams [26]. Likewise, authors in [44] have used EigenJoints, as features that have been based on joint position differences. The method has captured many features from each frame, including posture, motion, and offset, and then Principal Component Analysis (PCA) and normalizing have been used to define the action description. Similarly, the joint points' trajectory has been employed in [13] while the histogram of oriented displacement (HOD) has been used to extract direction between two subsequent locations in the temporal dimension. The authors in [32] have proposed a spatio-temporal relationship matching method designed to measure structural similarity between feature sets extracted from two videos. Then, they created pairwise relationship between features and presented each video sequence using relationship histograms. Recently, authors in [9] have introduced weighted linear combination of distances within two manifolds for 3D human action recognition.

Within the framework of deep learning methods, the authors in [7] have presented a long-term recurrent convolutional network with a 2D CNN extracting frame-level from RGB features in order to generate the action label. Likewise, authors in [16] have used video frames as an input to a CNN architecture for video action identification. In the work of [20], a spatiotemporal LSTM network, which is able to effectively fuse the RGB features with the skeleton ones within the LSTM unit, has been suggested. Recently, the work of [8] has introduced PoseConv3D based on 3D-CNN-based architectures for skeleton-based action recognition, while investigating 3D heatmap volumes as input. Furthermore, a self-attention model for short-time pose-based classificcation has been adopted for human action recognition in [24]. It is worth mentioning that there are other methods using both hand-crafted and deep learning features such as the method of [14], where hand-crafted and deep learning based techniques have been combined for feature extraction in order to recognize a human action from a video containing complete action execution.

Furthermore, some works have been interested in analyzing the human body models. In fact, human body model-based action recognition methods made use of 2D or 3D information about human body parts, such as position and motion. Typically, the pose of the human body is recovered, and the action recognition is based on pose estimates, trajectories of body parts, joint locations, and/or landmarks. For instance, the authors in [45] have proposed a method to detect human actions in videos that were captured by uncalibrated moving cameras, based on the trajectories of human joint points. The work of [2] has also used common reference based on common reference trajectories. These trajectories were used as a representation of the nonlinear dynamical system that generated the action, and they were adopted thereafter in order to reconstruct the phase space of the appropriate dimension using a delayed embedding scheme. Similarly, the authors in [38] have investigated silhouettes in 3D by using the space-time occupancy patterns. For joint locations, the work of [37] has proposed an example of the Lie group representation of skeletons in order to facilitate the implementation of action-curved manifolds from the Lie group.

Overall, human landmarks have shown to be effective in 3D motion analysis, while human action dynamic representation use skeleton data to encode motion dynamics. This could contribute for ignoring human appearance information, illumination change, shadows and other factors. Several works have been proposed in the literature in order to deal with skeleton from Kinect and skeleton from RGB data. In fact, Kinect sensor has prompted intensive research efforts on 3D action recognition thanks to the extra dimension of depth. In fact, the information given by depth maps is insensitive to background clutter and includes rich 3D structural information of the scene. In particular, the depth information from Kinect camera can be effectively analyzed to better locate and extract the body joints, which form the human skeleton. For instance, authors in [18] have introduced learning skeleton information model for human action analysis using Kinect. Indeed, action recognition was performed using an input depth map data stream from a single Kinect sensor and skeleton-tracking algorithms

[27]. Likewise, authors in [17] have investigated a convolutional co-occurrence feature learning framework to use hierarchical methodology with different levels of contextual information. Inspired by the booming graph-based methods, the work of [42] has proposed a Spatial Temporal Graph Convolutional Networks (ST-GCN) for skeleton-based action recognition. Recently, authors in [43] have extracted features from depth and skeleton data using the spatial information of human motion with information entropy and the temporal information through stitching in order to distinguish different motions. Furthermore, given the lack of availability of 3D datasets, many recent studies in the computer vision community have tried to extract skeletons from RGB data and predict landmarks. For instance, the work of [23] has been based on predicting 3D joint locations for raw image pixels. Similarly, the model joint dependencies in the CNN via a max-margin formalism was investigated in [19]. Authors in [25] have presented also 3D pose estimation including human detection and 3D human root localization.

3 Proposed Method

In this work, we have investigated skeletons from RGB videos in order to recognize human action by focusing on 3D human joints. In the first step, we extract key-frames form the input 2D video, in order to reduce the problem complexity. Then, we extract the skeleton from each frame using a deep learning architecture. Finally, we use different features with a machine learning classifier in order to recognise the performed action within the input video (Fig. 1).

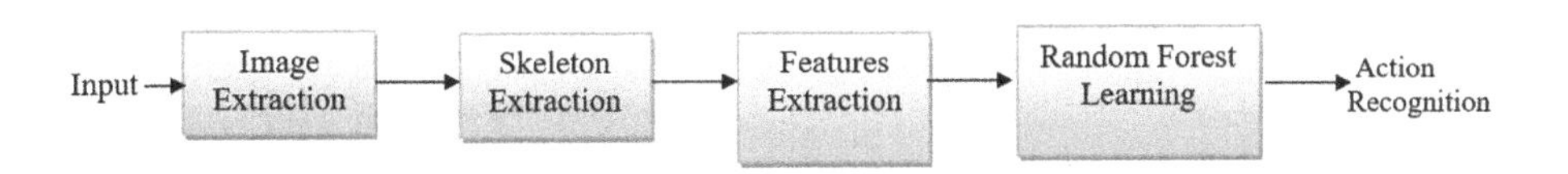

Fig. 1. Outline of the proposed action recognition technique.

3.1 Images Extraction

Compared to still images, video sequences afford more information about how objects and scenarios change over time. In fact, we start the proposed technique by extracting some relevant key-frames from the input video, in order to minimize the computational cost. The input of this step is the video sequence illustrating humans in actions. Then, we have applied a spatio-temporal sampling allowing to select some key frames without a considerable loss of information, since successive frame illustrate generally the same motion unit [3]. The output of this step is a set of RGB key-frames that will be used in the next step in order to extract the skeletons illustrating coarsely the human motion. The main idea is to recognize the input action in the video by using a frame-by-frame analysis based on kinematic patterns.

3.2 Skeleton Extraction

For many years, skeleton detection has remained a very difficult task in computer vision. The situation has changed after the introduction of affordable RGB-D sensors, such as Microsoft Kinect or Asus Xtion. Thus, skeleton detection became possible and many works proposed action recognition methods based on skeleton information, while achieving good results comparing to RGB-based methods. However, skeleton extraction algorithms based on RGB-D information fail when the filmed person is too far from the sensor and signal to noise ration drops down. Moreover, many skeleton detection algorithms have problems when persons are occluded for instance by furniture, which is often the case in daily-living surveillance scenarios. To deal with these issues, recent advancements in deep-learning show promising results of skeleton detection on RGB videos.

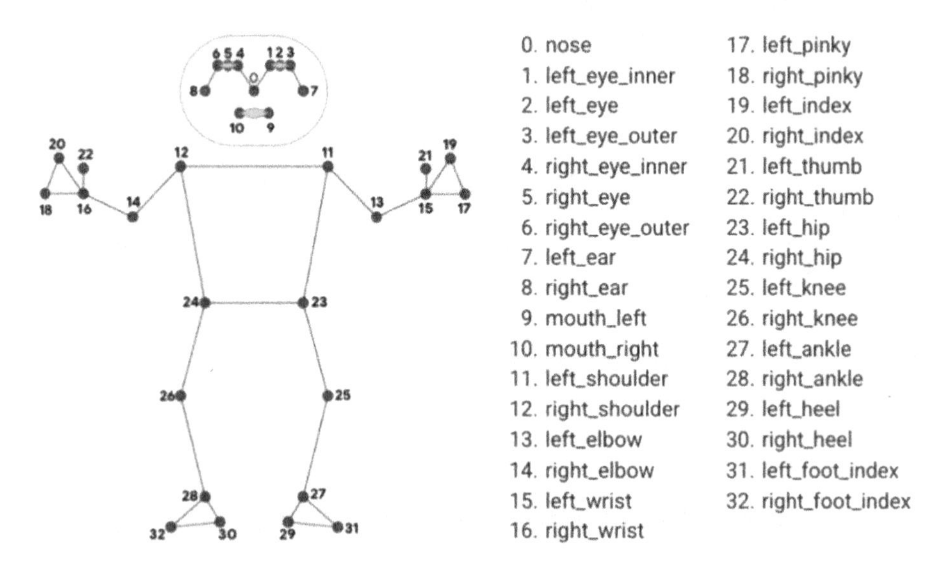

Fig. 2. The keypoint topology used [4].

In this work, we use the Blazepose model [4] for detecting human pose and extracting key points. The model consists of a lightweight body pose detector followed by a pose tracker network and can be easily implemented through a very helpful library such as 'mediapipe'. The current standard for human body pose is the COCO topology, which consists of 17 landmarks across the torso, arms, legs, and face. Nevertheless, the COCO keypoints only localize to the ankle and wrist points, missing scale and orientation data for hands and feet, which is vital for practical applications like fitness and dance. Thus, the inclusion of more keypoints seems essential for various application of domain-specific pose estimation models, like those for hands, face, or feet. Indeed, the model can estimate the skeleton in cases where either the whole person is visible, or where hips and shoulders keypoints can be confidently annotated while supporting heavy occlusions. The main idea of the adopted pose estimation algorithm [4] is to predict keypoint coordinates for the person on the current frame using combined heatmap and offset. It worth noting that the use of the heatmap and offset loss is only in the training phase and the corresponding output layers are removed from the model before running the inference. In fact, the network architecture combined heatmap, offset, and regression approach. Thus, the heatmap has been effectively used to supervise the lightweight embedding, which was utilized by the regression encoder network. In this work, we use the key-frames extracted from the studied video as input, while the joints body are extracted from each image using deep learning. The pose estimation algorithm is applied on the set of RGB frames extracted from videos to acquire the skeleton data. Using BlazePose, we extract 33 human body keypoints, which is a superset of COCO, BlazeFace and BlazePalm topologies (Fig. 2). We obtain 33 keypoints or joints with coordinate (x, y, z) (Fig. 3). With BlazePose, a topology of 33 human body keypoints is presented, which is a superset of COCO, BlazeFace and BlazePalm topologies in order to determine body semantics from pose prediction alone that is consistent with face and hand models. In fact, skeleton sequences can provide a very informative encoding of the human joints' trajectories for human motion through spatial and temporal dependencies. These results will be used to calculate the Euclidean distance between the adjacent joints as well as the velocity as shown in the next step.

3.3 Features Extraction

The skeleton sequences represent a sparse representation of human action videos. It has been proved previously that this sparse representation still representative of the human action [6]. In order to deal with undesirable variability of skeleton sequences, we propose to extract Kinematic features from each skeleton. In fact, for each skeleton, we investigate the distances between the adjacent joints (Fig. 4) as well as the velocity between each couple of successive skeletons. Thus, we compute lengths for skeleton segments with different actions obtained from different subjects (1).

$$d_i = \sqrt{(x_1 - x_2)^2 + (y_1 - y_2)^2 + (z_1 - z_2)^2} \tag{1}$$

The main idea of this work is to use rich structured data like 3D pose which not require sophisticated feature modeling or learning. Moreover, we reduce pose data over time to velocity informations. Some existing studies for skeleton-based action recognition used velocities to model actions [34]. Instead, the velocity at an instant t of each joint j, is considered as a very discriminative kinematic feature. Given the locations of a joint j at two successive frames F_i and $F_{(i+1)}$, we define the velocity vector (Fig. 5). The input of motion velocities is started by the second frame for each skeleton to obtain a feature vector at each time. Thus, we compute velocity between landmarks of successive frames in order to obtain a global vector of velocity (2). We have $(n - 1)$ values of velocities and

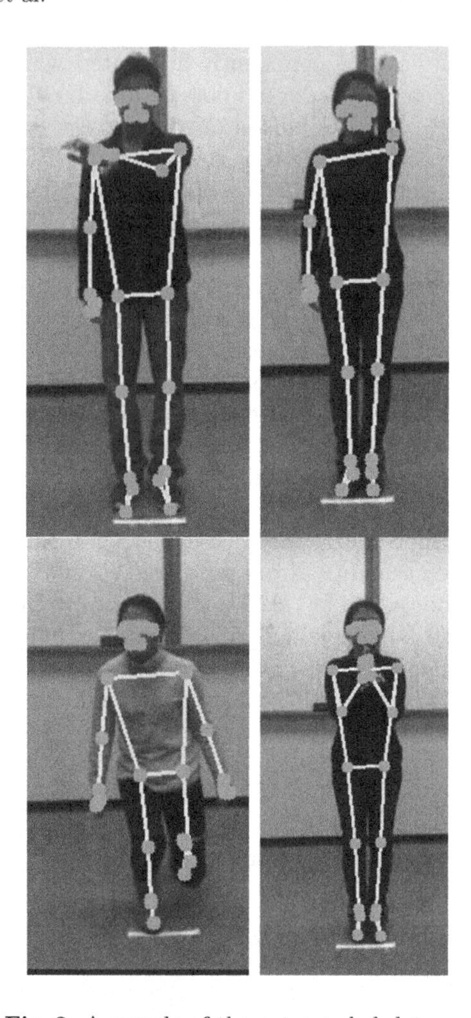

Fig. 3. A sample of the extracted skeletons.

$(n \times k)$ joint distances (where n denotes the number of frames within the input video sequence and k is the number of computed distances per frame). In fact, we have as an input for the training set of actions a vector that contains (n-1) values of velocities with $(n \times k)$ joint distances.

$$V = F_{(i+1)} - F_i \tag{2}$$

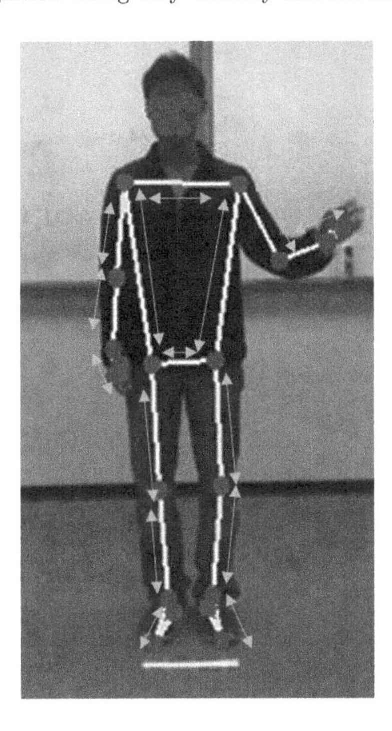

Fig. 4. Local features represented by the lengths of different parts of the skeleton.

3.4 Random Forest Classifier

In the final step, we use the random forest as a machine learning classifiers in order to predict the label of the action in the input video [31]. Random forest is a statistical and machine learning algorithm for prediction and classification. It illustrates an ensemble of weak learners which are decision trees [30]. Each tree contains split nodes and leaf nodes. A binary classification based on the value of a particular feature is performed by each split node. The action is reserved in the left partition or the right partition depend on value of a threshold. If the classes are linearly separable, action class will get a leaf node. Each tree independently predicts its label and a majority voting scheme is used in order to predict the final label of the feature vector. In fact, random decision forests are fast and effective multi-class classifiers. The main contribution of the proposed method resides in adopting machine learning for the distances of adjacent joints and velocity according to each action. The objective of this work is to use smaller feature dimensionality while training a random decision forest in order to lead to fast and accurate feature extraction and action recognition. In fact, the input for the machine learning process consists only of vectors contains landmarks' distances

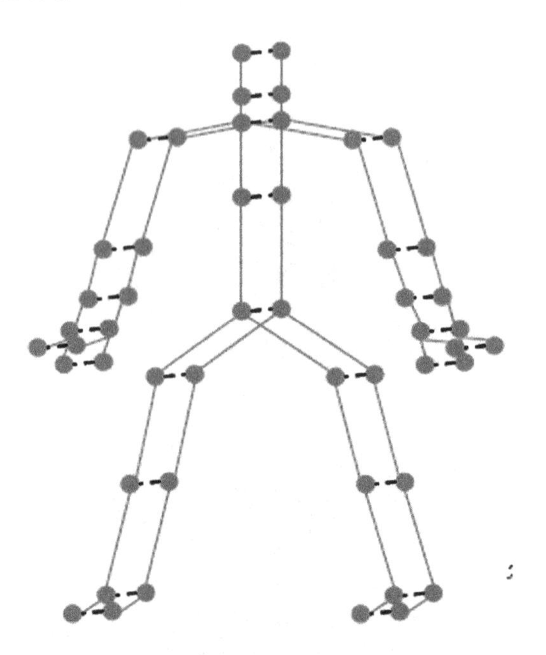

Fig. 5. The motion velocity representation.

and velocities. To this end, a random forest model is trained in order to learn each feature according to his action, while selecting empirically the number of decision trees in the forest as well as the number of features considered by each tree when splitting a node has been set to. This study has employed an empirical approach for threshold determination.

4 Results

In order to demonstrate the effectiveness of the proposed action recognition technique, we have used the challenging UTD-MHAD multimodal action dataset [5]. This dataset is composed of 27 actions performed by 8 subjects (4 females and 4 males) such that each subject performs each action 4 times (Fig. 6). The used dataset contains 861 data sequences collected using Kinect V1 camera and a wearable inertial sensor. The actions are: "right arm swipe to the left", "right arm swipe to the right" .... It covers sport actions (e.g. "bowling", "tennis serve" and "baseball swing"), hand gestures (e.g. "draw X", "draw triangle", and "draw circle"), daily activities (e.g. "knock on door", "sit to stand" and "stand to sit") and training exercises (e.g. "arm curl", "lung" and "squat"). The cross-subjects protocol is used as in [5], namely, the data from the subjects number 1, 3, 5, 7 are used for training while the subjects 2, 4, 6, 8 are used for testing. More precisely, we have employed in this work the training data and the testing data using landmark's distances and velocities in order to identify the studied action. In fact, we have applied the random forest in order to predict the action, according to few kinematic features characterizing temporally as well as spatially the motion performed during the input RGB video.

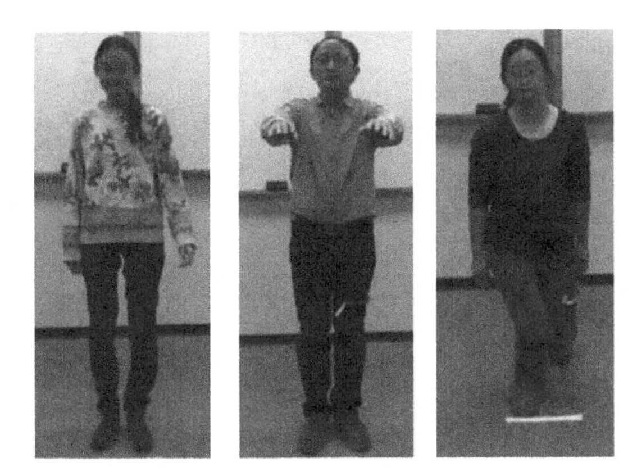

Fig. 6. A sample of the dataset used.

We have evaluated the proposed technique on the standard UTD dataset using the accuracy of the system, which is the most used evaluation metric in classification problems, and the F1-score which is a combination of precision and recall. Table 1 presents the effect of several training parameters on machine learning random forest with accuracy. The first one is the number of trees resulting in a forest. The second one is the depth which is the maximum depth of trees used (the number of levels in the decision tree). We try different number of trees and depths to find the best accuracy. In fact, increasing the number of trees increases the training and execution time without significant accuracy benefits (Fig. 7).

Figure 8 shows F1-score using different values of number of trees and samples split which represents the minimum number of samples required to split an internal node. This can vary between considering at least one sample at each node to considering all of the samples at each node. This figure shows different cases, such as case 1 where the number of trees is equal to 100 and the split is equal to 10 (the red curve illustrates the chosen hyper-parameters). Thus, we have used only 100 trees with depth equal to 10 in order to obtain flexible cost-efficient method with very high accuracy.

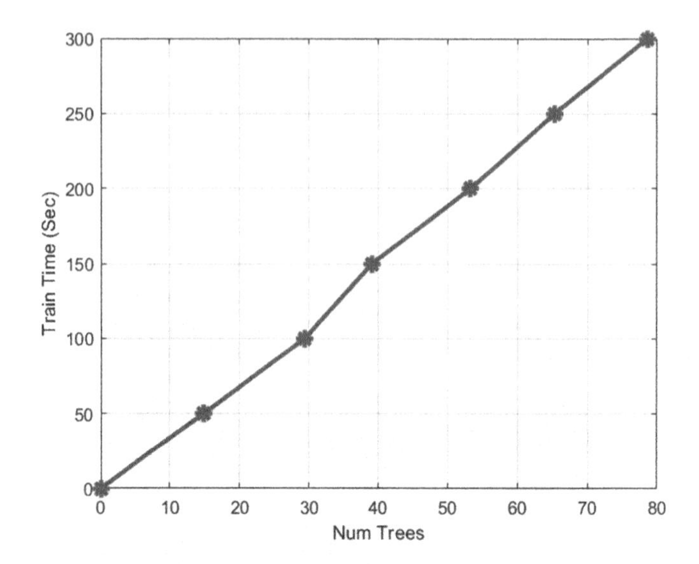

Fig. 7. Evolution of the training CPU time according to the number of trees.

Table 1. Comparison of all models

Accuracy (%)	Number of trees	Depth
97.45	100	10
95.3	100	20
90.46	400	10
62	50	10

We show also the confusion matrices in Fig. 9, while the rows of the confusion matrix correspond to the true class and the columns correspond to the predicted class. It is worth noting that each action has ben performed four times by each of the four subjects. The accuracy of our technique, using a random forest classifier trained with the best configuration, is equal to 97.45%. We present the optimal selection of the parameters on machine learning random forest. The proposed method outperforms the state-of-the-art methods with a value of accuracy of 93.45% against 95.37% for the hand-crafted method of [9] and 91% for the deep learning method introduced in [35]. The results of the case studies present time efficiency both at training and test time and develop an accurate activity recognition model without requiring large datasets, and especially while avoiding the overfitting risk of deep learning-based techniques while remaining explainable thanks to the adopted kinematic features.

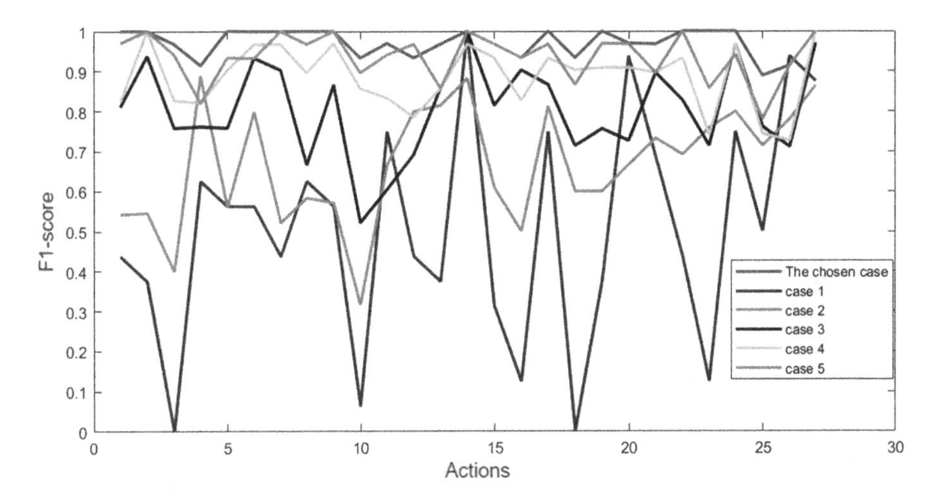

Fig. 8. F1-score by action for different combinations of many random forest parameters.

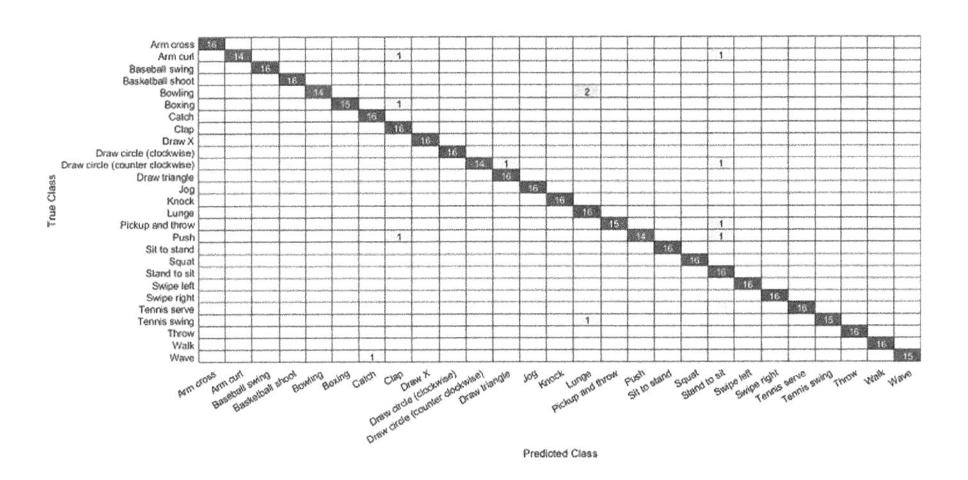

Fig. 9. The confusion matrix for the UTD dataset.

5 Conclusion

In this paper, we have presented a fast and accurate human action recognition based on skeleton information provided by RGB data. In fact, RGB cameras can provide accurate joint kinematic estimation and the skeleton information can be independent of clothes, pose and illumination changes. We have thus presented a framework for human action recognition using segments connecting adjacent anatomical landmarks and motion velocities. Then, we have adapted the random forest classifier in order to predict the action label depending on the nature of the studied action. We use as input for the machine learning classifier a vector illustrating the human representation through the human key joints' locations and velocities. The proposed technique proved to be accurate and achieving a fast testing speed, but it is largely dependent on the accuracy of the skeleton estimation. This dependency could be minimized by considering color and/or texture features around the human main landmarks. In future work, it might be interesting to consider distances per region (legs, arms, head, ...), even if it increases the number of attributes.

References

1. Aggarwal, J.K., Ryoo, M.S.: Human activity analysis: a review. ACM Comput. Surv. (CSUR) **43**(3), 1–43 (2011)
2. Ali, S., Basharat, A., Shah, M.: Chaotic invariants for human action recognition. In: 2007 IEEE 11th International Conference on Computer Vision, pp. 1–8. IEEE (2007)
3. Barhoumi, W., Zagrouba, E.: On-the-fly extraction of key frames for efficient video summarization. AASRI Procedia **4**, 78–84 (2013)
4. Bazarevsky, V., Grishchenko, I., Raveendran, K., Zhu, T., Zhang, F., Grundmann, M.: BlazePose: on-device real-time body pose tracking. arXiv preprint arXiv:2006.10204 (2020)
5. Chen, C., Jafari, R., Kehtarnavaz, N.: UTD-MHAD: a multimodal dataset for human action recognition utilizing a depth camera and a wearable inertial sensor. In: 2015 IEEE International Conference on Image Processing (ICIP), pp. 168–172. IEEE (2015)
6. Chen, Y.T., Fang, W.H., Dai, S.T., Lu, C.C.: Skeleton moving pose-based human fall detection with sparse coding and temporal pyramid pooling. In: 2021 7th International Conference on Applied System Innovation (ICASI), pp. 91–96. IEEE (2021)
7. Donahue, J., et al.: Long-term recurrent convolutional networks for visual recognition and description. In: Proceedings of the IEEE Conference on Computer Vision and Pattern Recognition, pp. 2625–2634 (2015)
8. Duan, H., Zhao, Y., Chen, K., Lin, D., Dai, B.: Revisiting skeleton-based action recognition. In: Proceedings of the IEEE/CVF Conference on Computer Vision and Pattern Recognition, pp. 2969–2978 (2022)
9. Elaoud, A., Barhoumi, W., Drira, H., Zagrouba, E.: Weighted linear combination of distances within two manifolds for 3D human action recognition. In: VISIGRAPP (5: VISAPP) (2019)
10. Elaoud, A., Barhoumi, W., Drira, H., Zagrouba, E.: Person re-identification from different views based on dynamic linear combination of distances. Multimed. Tools Appl. **80**, 17685–17704 (2021)
11. Fisher, R.: BEHAVE: computer-assisted prescreening of video streams for unusual activities. The EPSRC project GR S 98146 (2007)
12. Girdhar, P., et al.: Vision based human activity recognition: a comprehensive review of methods & techniques. Turk. J. Comput. Math. Educ. (TURCOMAT) **12**(10), 7383–7394 (2021)
13. Gowayyed, M.A., Torki, M., Hussein, M.E., El-Saban, M.: Histogram of oriented displacements (HOD): describing trajectories of human joints for action recognition. In: Twenty-third international joint conference on artificial intelligence (2013)
14. Herath, S., Harandi, M., Porikli, F.: Going deeper into action recognition: a survey. Image Vis. Comput. **60**, 4–21 (2017)

15. Jegham, I., Khalifa, A.B., Alouani, I., Mahjoub, M.A.: Vision-based human action recognition: an overview and real world challenges. Forensic Sci. Int.: Digit. Invest. **32**, 200901 (2020)
16. Karpathy, A., Toderici, G., Shetty, S., Leung, T., Sukthankar, R., Fei-Fei, L.: Large-scale video classification with convolutional neural networks. In: Proceedings of the IEEE Conference on Computer Vision and Pattern Recognition, pp. 1725–1732 (2014)
17. Li, C., Zhong, Q., Xie, D., Pu, S.: Co-occurrence feature learning from skeleton data for action recognition and detection with hierarchical aggregation. arXiv preprint arXiv:1804.06055 (2018)
18. Li, G., Li, C.: Learning skeleton information for human action analysis using Kinect. Signal Process.: Image Commun. **84**, 115814 (2020)
19. Li, S., Zhang, W., Chan, A.B.: Maximum-margin structured learning with deep networks for 3D human pose estimation. In: Proceedings of the IEEE International Conference on Computer Vision, pp. 2848–2856 (2015)
20. Liu, J., Shahroudy, A., Xu, D., Kot, A.C., Wang, G.: Skeleton-based action recognition using spatio-temporal LSTM network with trust gates. IEEE Trans. Pattern Anal. Mach. Intell. **40**(12), 3007–3021 (2017)
21. Lu, M., Hu, Y., Lu, X.: Driver action recognition using deformable and dilated faster R-CNN with optimized region proposals. Appl. Intell. **50**, 1100–1111 (2020)
22. Mabrouk, A.B., Zagrouba, E.: Abnormal behavior recognition for intelligent video surveillance systems: A review. Expert Syst. Appl. **91**, 480–491 (2018)
23. Martinez, J., Hossain, R., Romero, J., Little, J.J.: A simple yet effective baseline for 3D human pose estimation. In: Proceedings of the IEEE International Conference on Computer Vision, pp. 2640–2649 (2017)
24. Mazzia, V., Angarano, S., Salvetti, F., Angelini, F., Chiaberge, M.: Action transformer: a self-attention model for short-time pose-based human action recognition. Pattern Recogn. **124**, 108487 (2022)
25. Moon, G., Chang, J.Y., Lee, K.M.: Camera distance-aware top-down approach for 3D multi-person pose estimation from a single RGB image. In: Proceedings of the IEEE/CVF International Conference on Computer Vision, pp. 10133–10142 (2019)
26. Onofri, L., Soda, P., Pechenizkiy, M., Iannello, G.: A survey on using domain and contextual knowledge for human activity recognition in video streams. Expert Syst. Appl. **63**, 97–111 (2016)
27. Papadopoulos, G.T., Axenopoulos, A., Daras, P.: Real-time skeleton-tracking-based human action recognition using Kinect data. In: Gurrin, C., Hopfgartner, F., Hurst, W., Johansen, H., Lee, H., O'Connor, N. (eds.) MMM 2014. LNCS, vol. 8325, pp. 473–483. Springer, Cham (2014). https://doi.org/10.1007/978-3-319-04114-8_40
28. Patrikar, D.R., Parate, M.R.: Anomaly detection using edge computing in video surveillance system. Int. J. Multimed. Inf. Retrieval **11**(2), 85–110 (2022)
29. Poquet, O., Lim, L., Mirriahi, N., Dawson, S.: Video and learning: a systematic review (2007–2017). In: Proceedings of the 8th International Conference on Learning Analytics and Knowledge, pp. 151–160 (2018)
30. Quinlan, J.R.: Induction of decision trees. Mach. Learn. **1**, 81–106 (1986)
31. Rahmani, H., Mahmood, A., Huynh, D.Q., Mian, A.: Real time action recognition using histograms of depth gradients and random decision forests. In: IEEE Winter Conference on Applications of Computer Vision, pp. 626–633. IEEE (2014)
32. Ryoo, M.S., Aggarwal, J.K.: Spatio-temporal relationship match: video structure comparison for recognition of complex human activities. In: 2009 IEEE 12th International Conference on Computer Vision, pp. 1593–1600. IEEE (2009)

33. Shah, K., Shah, A., Lau, C.P., de Melo, C.M., Chellappa, R.: Multi-view action recognition using contrastive learning. In: Proceedings of the IEEE/CVF Winter Conference on Applications of Computer Vision, pp. 3381–3391 (2023)
34. Song, Y.F., Zhang, Z., Shan, C., Wang, L.: Constructing stronger and faster baselines for skeleton-based action recognition. IEEE Trans. Pattern Anal. Mach. Intell. **45**(2), 1474–1488 (2022)
35. Usmani, A., Siddiqui, N., Islam, S.: Skeleton joint trajectories based human activity recognition using deep RNN. Multimed. Tools Appl. **82**, 46845–46869 (2023)
36. Van Gemert, J.C., Jain, M., Gati, E., Snoek, C.G., et al.: APT: action localization proposals from dense trajectories. In: BMVC, vol. 2, p. 4 (2015)
37. Vemulapalli, R., Arrate, F., Chellappa, R.: Human action recognition by representing 3D skeletons as points in a lie group. In: Proceedings of the IEEE Conference on Computer Vision and Pattern Recognition, pp. 588–595 (2014)
38. Vieira, A.W., Nascimento, E.R., Oliveira, G.L., Liu, Z., Campos, M.F.M.: STOP: space-time occupancy patterns for 3D action recognition from depth map sequences. In: Alvarez, L., Mejail, M., Gomez, L., Jacobo, J. (eds.) CIARP 2012. LNCS, vol. 7441, pp. 252–259. Springer, Heidelberg (2012). https://doi.org/10.1007/978-3-642-33275-3_31
39. Wang, L., Qiao, Yu., Tang, X.: Video action detection with relational dynamic-poselets. In: Fleet, D., Pajdla, T., Schiele, B., Tuytelaars, T. (eds.) ECCV 2014, Part V. LNCS, vol. 8693, pp. 565–580. Springer, Cham (2014). https://doi.org/10.1007/978-3-319-10602-1_37
40. Wen, W., Imamizu, H.: The sense of agency in perception, behaviour and human-machine interactions. Nat. Rev. Psychol. **1**(4), 211–222 (2022)
41. Wen, Y., Pan, H., Yang, L., Pan, J., Komura, T., Wang, W.: Hierarchical temporal transformer for 3D hand pose estimation and action recognition from egocentric RGB videos. arXiv preprint arXiv:2209.09484 (2022)
42. Yan, S., Xiong, Y., Lin, D.: Spatial temporal graph convolutional networks for skeleton-based action recognition. In: Thirty-second AAAI Conference on Artificial Intelligence (2018)
43. Yang, T., Hou, Z., Liang, J., Gu, Y., Chao, X.: Depth sequential information entropy maps and multi-label subspace learning for human action recognition. IEEE Access **8**, 135118–135130 (2020)
44. Yang, X., Tian, Y.L.: Eigenjoints-based action recognition using naive-bayes-nearest-neighbor. In: 2012 IEEE Computer Society Conference on Computer Vision and Pattern Recognition Workshops, pp. 14–19. IEEE (2012)
45. Yilmaz, A., Shah, M.: Recognizing human actions in videos acquired by uncalibrated moving cameras. In: Tenth IEEE International Conference on Computer Vision (ICCV 2005), vol. 1, pp. 150–157. IEEE (2005)
46. Yue, R., Tian, Z., Du, S.: Action recognition based on RGB and skeleton data sets: a survey. Neurocomputing **512**, 287–306 (2022)

Efficient Topic Detection Using an Adaptive Neural Network Architecture

Meriem Manai[1(✉)] and Sadok Ben Yahia[2]

[1] LIPAH-LR11ES14, Faculty of Sciences of Tunis, University of Tunis El Manar, 2092 Tunis, Tunisia
meriem.manai@fst.utm.tn

[2] Department of Software Science, Tallinn University of Technology, Akadeemia tee 15a, Tallinn, Estonia

Abstract. Topic detection is the process of identifying the underlying themes or topics present in a set of documents. It has become more critical due to the increase of information electronically available and the necessity to process and filter it. In this respect, we introduce a new approach to detecting topics called ClusART. Thus, we created a three-phase approach: a first phase during which lexical preprocessing was conducted. The second phase pays heed to the construction and the generation of vectors representing the documents carried out. In the topic detection phase, the FuzzyART algorithm is used for the training phase, and a classifier based on ParagraphVector is used for the test phase. The comparative study of our approach on the 20 Newsgroups dataset showed that our method could detect almost relevant topics.

Keywords: Topic detection · Text clustering · FuzzyART · PragraphVector classifier

1 Introduction

In the last fifteen years, the Internet has exploded, and many tools have sprung up to help people share information and get their opinions [20]. With the proliferation of data in various fields, users have the opportunity to access a multitude of data points. The latter can be stored in several forms. In our case, we deal with semi-structured textual data.

These data contain much helpful information, but humans can only extract some relevant information because of their limited ability to assimilate it. The recovery of information from this data and its processing are challenging when it is semi-structured. Thus, many techniques have been presented to process this type of data. However, the actual challenge is managing this latter according to its content, particularly its thematic content. For this reason, the detection and classification of topics evoke much interest in research dealing with different documents. Topic detection, also known as "topic analysis," "topic modeling," or "topic extraction," is the task of detecting previously unknown topics in the

I. Saad et al. (Eds.): ICIKS 2023, LNBIP 486, pp. 145–157, 2024.
https://doi.org/10.1007/978-3-031-51664-1_10

system [21]. In the field of natural language processing, there are various tasks with wide-ranging applications. These tasks include text classification, information retrieval, and text summarization. For example, topics can be used to provide brief summaries of documents that revolve around similar news or subjects. Topic detection methods must be capable of extracting useful information from this data to avoid losing relevant information when representing the topic of a text document. Different methods have been presented to solve this task in the past years.

One of the methods for performing topic detection is using adaptive neural networks, specifically the Fuzzy Adaptive Resonance Theory (ART) algorithm. ART is an unsupervised learning algorithm that automatically identifies and categorizes patterns in a data set. In topic detection, Fuzzy ART can be used to determine the latent topics present in a collection of documents. The algorithm creates a prototype for each topic and updates these prototypes as new documents are processed. The prototypes can be thought of as centroids or representations of each topic, and the documents are then assigned to the closest prototype based on their similarity [18].

Overall, using ART for topic detection is a flexible and efficient way to find the hidden structure in a text corpus, and it works well for analyzing large amounts of text data. This paper introduces ClusART, a new approach for topic detection using adaptive neural networks. We use the Fuzzy Adaptive Resonance Theory algorithm as a topic detection algorithm. The proposed method relies on the frequency of words in documents to generate vectors, which are our input for topic detection. We proceed to the training phase during this latter phase, using the Fuzzy Adaptive Resonance Theory algorithm (Fuzzy ART). We then obtain a set of documents classified by topic. After that, we use a classifier based on paragraph vectors for the test phase. At the end of this process, we get a set of documents classified according to their similarity to the topics.

The remainder of this paper is organized as follows. In Sect. 2, we review the related work and pinpoint the takeaways that will back the introduction of a new approach up. Section 3 thoroughly describes the new topic detection approach called ClusART. Section 4 extensively reports on the obtained results from the carried out experiments on a dataset composed of 20 Newsgroups. ClusART, which we introduce, has been shown to outperform its pioneering competitors, namely KNN and LDA. Issues of future work are given in Sect. 5.

2 Related Work

Topic detection is a scorching field of interest, as would witness the number of wealthy approaches that have been developed recently [7,22–24]. The remainder of this section thoroughly reviews the pioneering methods that paid heed to this topic.

Bigi et al. [3], five statistical methods for topic identification (including unigrams, TF-IDF, cache modeling, topic perplexity, and a weighted model) were tested on both newspaper and email corpora. The study found that the performance of these techniques was greatly influenced by the specific characteristics of the data collection.

Schonhofen [17] introduced a method for topic identification in Wikipedia articles using titles and categories. The approach measured relatedness between titles, categories, and a given document. Categories with the highest weights were selected as the document's topics. This method offers a concise and effective way to identify article topics.

Denecke et al. [6] utilized Latent Dirichlet Allocation (LDA) for sentence-level topic detection in the presence of noisy data. Their study involved an experimental evaluation on three real-world datasets, showcasing the practicality and effectiveness of their proposed solution. The results were promising, demonstrating the successful application of LDA in determining topics, even with short and noisy sentences.

Sayyadi et al. [16] proposed a topic detection analytical approach based on graphs. They convert text data into a term graph based on co-occurrence relations between terms using a KeyGraph algorithm. Thereafter, they partition the graph using a community detection approach. They performed an extensive evaluation to show KeyGraph's superiority in terms of accuracy and runtime performance.

Abainia et al. [2] compared the Support Vector Machine (SVM) classifier with the K-nearest neighbors (KNN) classifier for Arabic text classification. Their findings demonstrated the superior performance of SVM compared to KNN in classifying Arabic text.

Zhang et al. [25] presented a hybrid approach for topic detection, integrating semantic and co-occurrence relations. They combined various relations into a unified term graph using the idea discovery theory in topic modeling. Co-occurrence relationships were extracted using the IdeaGraph algorithm, while semantic relationships were obtained using an LDA-based approach. The enriched graph was then subjected to graph analytical methods for topic identification. Their study showcased improved topic detection quality and the ability to identify significant topics through this integrated approach.

Patel et al. [13] proposed an approach for topic detection and tracking in news articles by combining multiple learning techniques. Their method involved using agglomerative clustering with average linkage for topic detection, cosine similarity for measuring topic similarity, and the KNN classifier for topic tracking.

Jamil et al. [10] proposed a rule generation algorithm for topic identification called TopId, that filters important terms from the pre-processed text and applies a term weighting system to solve the problem of synonyms. They conducted a comparative study with a Rough Set and three experts in the Quran and Hadith research fields. The proposed topic identification method consistently identified topics similar to Rough Set and the experts have given. The result of the comparison proved that the proposed method was able to be used to capture topics for textual documents.

Liu et al. [11] proposed a topic detection and tracking method based on event ontology on news texts; this provides a hierarchical structure of event classes following the common sense of the domain.

Dieng et al. [8] proposed an embedded topic model (ETM), an LDA variant that includes the principle of word embedding on newsgroups. Each word is modeled with a categorical distribution with the inner product between the word's embedding and an assigned theme embedding as a natural parameter.

Proto et al. [15] proposed a method called ToPIC for clustering textual data collections into correlated document groups through LDA after choosing the proper configurations. The method has been validated on real collections of textual documents.

Recently, Mamo et al. [12] proposed a new approach that combines document pivot and feature pivot techniques with online clustering. They validated their proposed method on football match tweets. The authors in [9] investigated how Twitter content related to COVID-19 in both English and Portuguese reflects the topics and sentiments expressed by users. They utilized LDA (Latent Dirichlet Allocation) independently for each language and month. Subsequently, the authors manually labeled the topics by examining the most frequently occurring words and tweets within each topic.

Takeaways from the Scrutiny of the Related Work: In the context of Topic Detection and Tracking (TDT), there is a need for methods that can adapt quickly to changes in an adaptive environment. However, most of the cited works lack this adaptability. Interestingly enough, we can pinpoint that a new approach is a compelling challenge. The latter is expected to allow swift and stable learning, which is required to achieve a better result in detecting subjects. To alleviate this shortage, we introduce, in this paper, a new approach leveraging an artificial neural network to solve the plasticity-stability conundrum. The proposed approach aims to respond to environmental changes while maintaining previously acquired knowledge proactively. In neural network-based approaches, the document to be classified is presented as input, and the output layer represents all possible classes. The activated output layer values indicate the potential categories of the document.

3 The Proposed Topic Detection Method

In this section, we thoroughly describe our proposed topic detection method. The global architecture of our approach is shown in Fig. 1. The workflow starts with preprocessing our data. The latter is an important step as it enables us to eliminate irrelevant and redundant information from our data, which does not significantly contribute to the underlying topics or themes present in the collection of documents. Our second step is document representation, allowing us to represent the document for our next step conveniently. Finally, we move to the topic detection step.

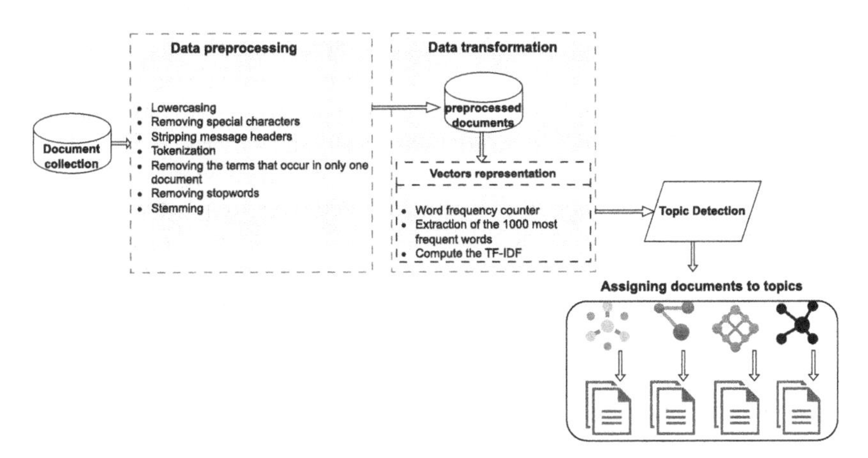

Fig. 1. The framework at a glance of the proposed approach

All these steps are detailed in the following.

3.1 Preprocessing

Before we apply our algorithm, the dataset's text documents must be preprocessed. The data are cleaned up during this step, so they no longer have to deal with irrelevant and redundant information or noisy and unreliable data. We achieve minimal preprocessing, which includes the following tasks:

- lowercasing: It involves converting all the text in a document or dataset to lowercase, ensuring consistent treatment of words irrespective of their original case.
- removing special characters such as hashtags and punctuation
- stripping message headers such as emails, IDs, and signatures to exclude newsgroups information
- tokenization: Is an essential NLP process that divides the text into smaller units called tokens. Its purpose is to break down text into meaningful units for further analysis and processing.
- removing the terms that occur in only one document
- removing stopwords: It includes removing articles (e.g., "a," "an," "the"), prepositions (e.g., "in," "on," "at"), and other frequently used words that do not contribute much to the overall understanding or analysis of the text.
- applying the Porter stemmer: is a widely used algorithm in NLP for reducing words to their base or root form, known as the stem. [14]

3.2 Document Representation

After preprocessing, we must represent our documents suitably to reduce their complexity and make them easier to work with. We transformed our documents from the full-text version to a document vector describing the documents' content. We then proceeded as follows: First of all, we applied the word "frequency counter." Terms with a high frequency are considered strong or confident words. We analyzed the distribution of word frequencies and determined the threshold where the frequency distribution no longer provided substantial meaningful information. Our findings showed that the top 1,000 most frequent words encompassed a significant portion of the overall vocabulary, whereas the less frequent words had a lesser impact on the overall topics. Consequently, we extracted and saved those 1,000 most frequent words. And at last, we computed the TF-IDF for each document in our dataset.

3.3 Topic Detection

Once representing documents as vectors is carried out, we move forward to topic detection. First, during the training phase, we used the FuzzyART algorithm.

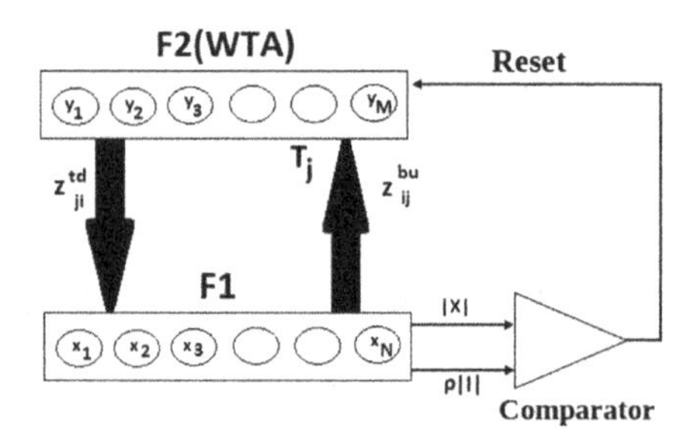

Fig. 2. Topological structure of Fuzzy ART's architecture [18]

FuzzyART [5] is an adaptive neural network that incorporates fuzzy logic to improve generalizability in recognizing ART forms. Complement coding is an optional and useful feature of FuzzyART that handles the absence of characteristics in model classifications, preventing unnecessary category proliferation. While FuzzyART is sensitive to noise, it consists of two layers: F1 (input layer) and F2 (category layer), as shown in Fig. 2. The adjustable vigilance parameter $\rho \in [0, 1]$ controls the vigilance subsystem. The input layer maps input vectors with close proximity based on a specific similarity measure to the same cluster, creating new clusters for unmatched patterns.

During the training phase, we initiate by configuring the network parameters. Next, we feed the document vectors into the network. The network is trained using the Fuzzy ART algorithm as part of the third step. Upon completion of the training phase, we save the network status and the topic structure for future reference.

In the test phase, we proceeded by loading the most recent topic structure and a collection of test documents. Utilizing an algorithm rooted in paragraph vectors, we assigned the documents to their respective topics, allowing for the possibility of multiple topic assignments for a single document. Consequently, we obtained classified documents based on their membership in the pre-established clusters.

The testing phase involves several steps. Initially, we configure the network parameters. In the second step, we load the previous cluster structure. Moving on to the third step, we present the document to be tested to the classifier algorithm based on Paragraph Vectors. Lastly, in the fourth step, we assign a similarity measure of the document to the pre-identified topics.

4 Results and Discussion

For our experiments, we utilized the 20 Newsgroups dataset, a widely recognized collection of text documents frequently employed in natural language processing (NLP) research and classification tasks, as both our training and testing dataset [1]. The common information retrieval metrics then assess the performance, i.e., precision, recall, and F1-measure (c.f., Eqs. (1), (2), and (3)).

$$\text{precision} = \frac{\text{relevant documents} \cap \text{retrieved documents}}{\text{retrieved documents}} \tag{1}$$

$$\text{recall} = \frac{\text{relevant documents} \cap \text{retrieved documents}}{\text{relevant documents}} \tag{2}$$

$$\text{F}_{\text{measure}} = 2 * \frac{\text{precision} * \text{recall}}{\text{precision} + \text{recall}} \tag{3}$$

Some qualitative results are discussed in the remainder:

4.1 Study of the Impact of the Vigilance Parameter ρ

The selection parameter, training parameter, and vigilance parameter are essential in controlling the functioning of a fuzzyART network. These parameters influence different aspects of the network's behavior.

The selection parameter α determines the selection functions in a network, with higher values leading to more classes and reduced impact of vigilance. It is generally advantageous to keep α small, and in this study, it is set to 0.2. The training parameter β influences the speed of weight updates, with higher values resulting in faster changes. In this study, β is assigned a value of 0.4. The specific values of α and β were determined through empirical observations, experimentation, and a thorough analysis of the dataset, evaluating performance by systematically varying the parameters through iterations.

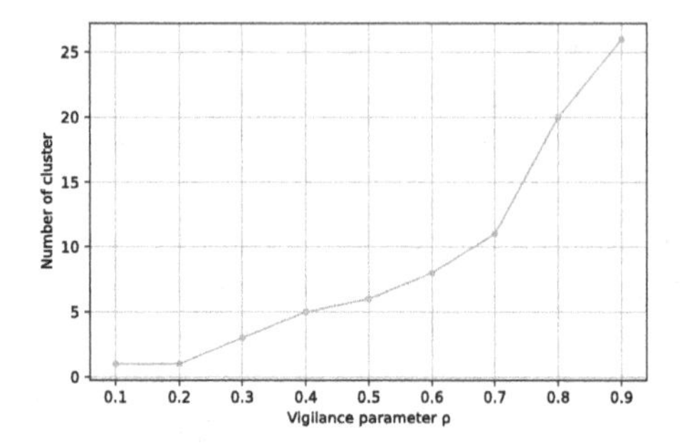

Fig. 3. Number of classes detected according to the values of the vigilance parameter

The vigilance parameter determines whether the class (the winning neuron) chosen by the network is accepted. The smaller α is, the coarser the classes created. Whereas the larger α, the more accurate the classes created are.

Therefore, we tried to study this vigilance parameter's impact on the number of classes created. Figure 3 shows the variation in the number of classes created according to the values assigned to the vigilance parameter. These tests led us to set the value of this parameter to 0.8. To assess the performance of the proposed approach to document topic detection, we conducted a comparative study with some pioneering approaches.

1. KNN: The k-Nearest Neighbor clustering algorithm is a popular clustering solution for TDT tasks (topic detection and tracking) [19].
2. LDA: It is a three-level hierarchical Bayesian model in which each collection element is represented as a finite mixture on an underlying set of topics [4].

4.2 Measurements for 5000 Documents

The recall, precision, and F-measure values for 5000 documents are reported in Fig. 4 and Table 1. It is worth noting that, according to the histograms sketched, we can highlight that ClusART outperforms both KNN and LDA. Their recall values are much lower than the ones obtained by our approach. However, the precision of KNN and LDA slightly exceeds our approach. The performance of our approach is confirmed in terms of the F-measure since it also outperforms both KNN and LDA.

Table 1. Values of recall, precision, and F-measure obtained for 5000 documents

	Precision	Recall	F-measure
KNN	0,59	0,49	0,53
LDA	**0,69**	0,62	0,65
ClusART	0,59	**0,76**	**0,66**

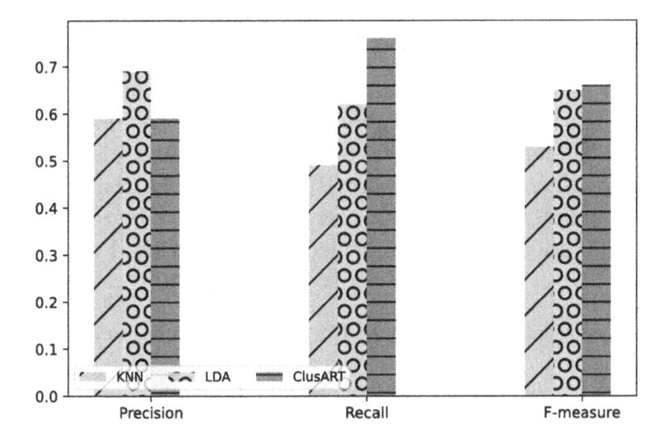

Fig. 4. Performance measurement for 5000 documents

4.3 Measurements for 10000 Documents

In the following, we increased the number of documents to 10,000 documents. Figure 5 and Table 2 show us the recall, precision, and F-measure values for our 10, 000 documents. The histograms indicate that, with the increased number of documents, ClusART always outperforms KNN and LDA. Their recall values are much lower than the ones obtained by our approach. The performance of our approach is always confirmed in terms of F-measure; it also exceeds KNN and LDA.

Table 2. Values of recall, precision, and F-measure obtained for 10, 000 documents

	Precision	Recall	F-measure
KNN	0,51	0,42	0,46
LDA	**0,53**	0,45	0,48
ClusART	0,39	**0,72**	**0,50**

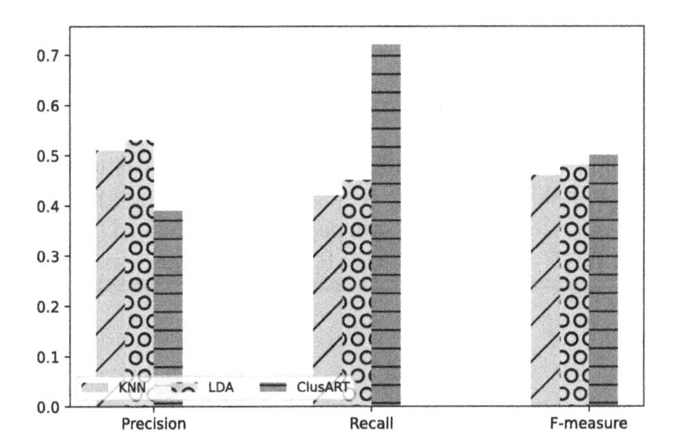

Fig. 5. Performance measurement for 10000 documents

4.4 Measurements for 15000 Documents

In the remainder of our tests, we further increased the number of documents to 15000. According to Fig. 6 and Table 3, we can see that despite the decrease in precision, recall, and F-measure values, ClusART exceeds KNN and LDA. Even if KNN and LDA have higher precision values than the ones obtained by ClusART, the latter surpasses them in terms of recall. The performance of our approach for 15000 documents is confirmed in terms of the F-measure metric.

Table 3. Values of recall, precision, and F-measure obtained for 15000 documents

	Precision	Recall	F-measure
KNN	0,48	0,37	0,41
LDA	**0,50**	0,43	0,46
ClusART	0,37	**0,69**	**0,48**

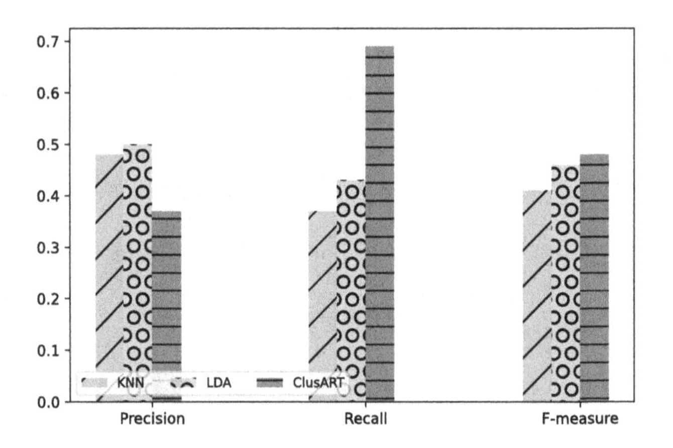

Fig. 6. Performance measurement for 15000 documents

4.5 Measurements for 20000 Documents

Figure 7 and Table 4 depict the recall, precision, and F-measure recorded values when considering 20000 documents. According to the histograms sketched, we can again point out that ClusART surpasses KNN and LDA. With a higher recall value than KNN and LDA. ClusART's performance is always confirmed in terms of F-measure.

Table 4. Values of recall, precision, and F-measure obtained for 20, 000 documents

	Precision	Recall	F-measure
KNN	0,46	0,34	0,39
LDA	**0,47**	0,41	0,43
ClusART	0,36	**0,62**	**0,45**

To sum up, for all the tests performed, we notice that the more we increase the number of documents, the more the three precision measurements (recall and F-measure) decrease for KNN, LDA, and ClusART. For precision measurements, KNN and LDA outperform ClusART. Regarding recall, the values obtained by ClusART are higher than those obtained by KNN and LDA. We can conclude that ClusART is more efficient than KNN and LDA, and this is confirmed by the F-measure values obtained where ClusART exceeds KNN and LDA. In this case, our approach can detect almost all relevant topics.

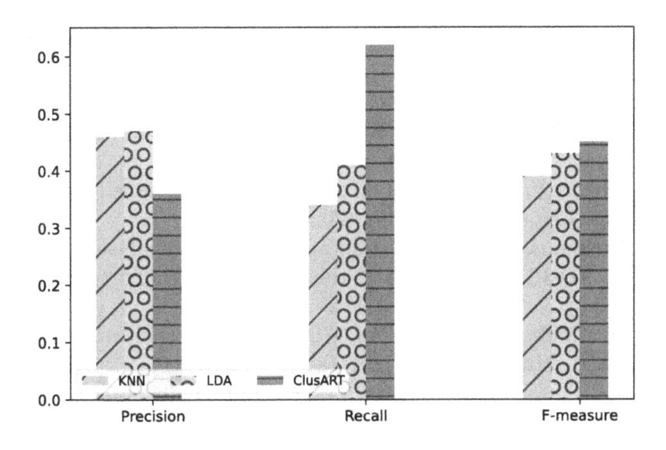

Fig. 7. Performance measurement for 20000 documents

5 Conclusion

This paper introduces a new approach for topic detection based on adaptive neural networks. The experimental evaluation showed that the presented approach outperformed the KNN and the LDA. Avenues for future work include the consideration of synonyms. It would be interesting to represent words in a document that are synonyms by a single word to avoid the vocabulary mismatch problem. The selection of nouns: it would be interesting to use an algorithm to filter out unnecessary text characteristics and ensure that only the name is selected. Noun-based text clustering can be efficient since the nouns may represent particular incidents and general events in texts and produce more good topics than verbs. And finally, it would be interesting to implement our approach in a multilingual environment.

References

1. 20 Newsgroups. http://qwone.com/~jason/20Newsgroups/
2. Abainia, K., Ouamour, S., Sayoud, H.: Topic identification of Arabic noisy texts based on KNN. In: 2015 International Conference on Information and Communication Technology Research, ICTRC 2015, pp. 92–95. IEEE (2015)
3. Bigi, B., Brun, A., Haton, J.P., Smaili, K., Zitouni, I.: A comparative study of topic identification on newspaper and e-mail. In: Proceedings - Eighth International Symposium on String Processing and Information Retrieval, SPIRE 2001, pp. 238–241. IEEE (2001)
4. Blei, D.M., Ng, A.Y., Jordan, M.I.: Latent Dirichlet allocation. J. Mach. Learn. Res. **3**, 993–1022 (2003)
5. Carpenter, G.A., Grossberg, S., Rosen, D.B.: Fuzzy art: fast stable learning and categorization of analog patterns by an adaptive resonance system. Neural Netw. **4**(6), 759–771 (1991)
6. Denecke, K., Brosowski, M.: Topic detection in noisy data sources. In: 2010 Fifth International Conference on Digital Information Management (ICDIM), pp. 50–55. IEEE (2010)
7. Di Corso, E., Proto, S., Vacchetti, B., Bethaz, P., Cerquitelli, T.: Simplifying text mining activities: scalable and self-tuning methodology for topic detection and characterization. Appl. Sci. **12**(10), 5125 (2022)
8. Dieng, A.B., Ruiz, F.J., Blei, D.M.: Topic modeling in embedding spaces. Trans. Assoc. Comput. Linguist. **8**, 439–453 (2020)
9. Garcia, K., Berton, L.: Topic detection and sentiment analysis in twitter content related to COVID-19 from brazil and the USA. Appl. Soft Comput. **101**, 107057 (2021)
10. Jamil, N.S., Ku-Mahamud, K.R., Din, A.M.: Topic identification method for textual document. J. Multidisc. Eng. Sci. Technol. (JMEST) **4**(2), 6643–6647 (2017)
11. Liu, W., Jiang, L., Wu, Y., Tang, T., Li, W.: Topic detection and tracking based on event ontology. IEEE Access **8**, 98044–98056 (2020)
12. Mamo, N., Azzopardi, J., Layfield, C.: Fine-grained topic detection and tracking on twitter. In: KDIR, pp. 79–86 (2021)
13. Patel, S., Suthar, S., Patel, S., Patel, N., Patel, A.: Topic detection and tracking in news articles. In: Satapathy, S.C., Joshi, A. (eds.) ICTIS 2017. SIST, vol. 84, pp. 420–426. Springer, Cham (2018). https://doi.org/10.1007/978-3-319-63645-0_48
14. Porter, M.F.: An algorithm for suffix stripping. Program **14**(3), 130–137 (1980)
15. Proto, S., Di Corso, E., Ventura, F., Cerquitelli, T.: Useful topic: self-tuning strategies to enhance latent dirichlet allocation. In: 2018 IEEE International Congress on Big Data (BigData Congress), pp. 33–40. IEEE (2018)
16. Sayyadi, H., Raschid, L.: A graph analytical approach for topic detection. ACM Trans. Internet Technol. (TOIT) **13**(2), 4 (2013)
17. Schonhofen, P.: Identifying document topics using the Wikipedia category network. Web Intell. Agent Syst. Int. J. **7**(2), 195–207 (2009)
18. Serrano-Gotarredona, T., Linares-Barranco, B., Andreou, A.G.: Adaptive Resonance Theory Microchips: Circuit Design Techniques, vol. 456. Springer, New York (2012). https://doi.org/10.1007/978-1-4419-8710-5

19. Soucy, P., Mineau, G.W.: A simple KNN algorithm for text categorization. In: Proceedings 2001 IEEE International Conference on Data Mining. ICDMm, vol. 63, pp. 647–648. IEEE (2001)
20. Wankhade, M., Annavarapu, C.S.R., Verma, M.K.: CBVoSD: context-based vectors over sentiment domain ensemble model for review classification. J. Supercomput. **78**(5), 6411–6447 (2022)
21. Wayne Charles, L.: Topic detection & tracking (TDT)-overview & perspective. In: Workshop held at the University of Maryland on, vol. 27, p. 28. Citeseer (1997)
22. Xiao, M., et al.: Hierarchical interdisciplinary topic detection model for research proposal classification. IEEE Trans. Knowl. Data Eng. (2023)
23. Yang, J., Lu, W., Hu, J., Huang, S.: A novel emerging topic detection method: a knowledge ecology perspective. Inf. Process. Manag. **59**(2), 102843 (2022)
24. Yang, S., Tang, Y.: News topic detection based on capsule semantic graph. Big Data Mining Analytics **5**(2), 98–109 (2022)
25. Zhang, C., Wang, H., Cao, L., Wang, W., Xu, F.: A hybrid term-term relations analysis approach for topic detection. Knowl.-Based Syst. **93**, 109–120 (2016)

Exploiting Machine Learning Technique for Attack Detection in Intrusion Detection System (IDS) Based on Protocol

Olomi Isaiah Aladesote[1](✉), Johnson Tunde Fakoya[1], and Olutola Agbelusi[2]

[1] Department of Computer Science, Federal Polytechnic, Ile Oluji, Ondo State, Nigeria
{isaaladesote,johfakoya}@fedpolel.edu.ng

[2] Department of Software Engineering, Federal University of Technology, Akure, Ondo State, Nigeria
aolutola@futa.edu.ng

Abstract. An intrusion detection system (IDS) can be either software or hardware that computerizes the process of keeping track of and evaluating network or computer system activity for indications of security issues. IDS is a crucial component of the security infrastructure of many organizations due to an increase in the frequency and intensity of attackers over the past decades. The study proposes machine learning techniques for the classification and detection of normal and attack traffics using protocol types records of the NSL-KDD dataset. Three sets of datasets were extracted from NSL-KDD datasets based on ICMP, UDP, and TCP. The experiment was conducted on WEKA 3.8.5 using KNN, KStar, LWL, BayesNet, Naïve Bayes, and PART algorithms. The results indicated that the PART algorithm has the highest performance rating while NaiveBayes has the lowest performance rating utilizing the Correlation-based feature selection (CFS) using the Ranking Filter approach. It is concluded that the PART algorithm performs well across the dataset while NaiveBayes does not perform well across the dataset.

Keywords: Correlation-based feature selection · Intrusion Detection System · NSL-KDD dataset

1 Introduction

The exponential increase in the number of internet users has resulted in an increase in security attacks over the internet, which are dangerous to the smooth operation of the internet and thereby prevent the 3 triads (Confidentiality, Integrity, and Availability) of information security from reality. Furthermore, the advancement of information technology (IT) has resulted in network security challenges [1]. As a result of this development, attackers are looking for ways to get access to the network through the back door. Hence, a system or approach to network security is required to keep attackers at bay [2]. Intrusion detection systems are a type of security technology that protects whole networks from unauthorized access. In this context, intrusion can be defined as any act that violates information security policies or the CIA trinity in a network [3].

I. Saad et al. (Eds.): ICIKS 2023, LNBIP 486, pp. 158–167, 2024.
https://doi.org/10.1007/978-3-031-51664-1_11

Intrusion detection is an important aspect of network defense since it alerts network administrators and personnel to hostile behavior. Low false-alarm rates and high intrusion detection rates would be the results of an effective Network Intrusion Detection System (NIDS) [4]. When regular traffic is mistakenly classified as hostile, a false alarm arises. A reliable and standard benchmark evaluation dataset is essential to develop an effective Intrusion Detection System (IDS). An IDS is introduced to guarantee the network security from various forms of threats and attacks [5]. Also, the performance of any IDS is totally dependent on the efficiency and accuracy of the dataset, hence selecting an appropriate dataset leads to an efficient and effective IDS [6].

Machine learning is a branch of computer science that arose from artificial intelligence's pattern recognition and computational learning theory. It investigates the design and development of algorithms that can learn from and predict data. When training data is labeled, machine learning is defined as supervised learning. Machine learning is unsupervised learning when training data is unlabelled. Semi-supervised learning is when training data is a mixture of labeled and unlabelled data. In recent times, many machine learning algorithms have been invented and refined to handle network attacks problem. Hence, we employed the following algorithms: KNN, KStar, LWL, BayesNet, Naïve Bayes and PART in detecting all forms of attacks and normal traffic.

We summarized the contributions of this paper as follows:

- Extract protocol-based records from NSL-KDD dataset
- Classify the data using machine learning techniques
- carry out a performance evaluation

2 Related Works

Numerous studies that were conducted over the past ten years concentrated on network intruder detection issues. While some of them employed conventional machine learning models, the other lately embraced deep learning models. A lot of research has been done on the methods and difficulties of employing conventional machine learning algorithms in intruder detection systems. [7] used the Weka to assess and contrast the performance of six supervised machine learning models with the records of NSL-KDD with emphasis along various dimensions, feature selection, hyper-parameter tuning sensitivity, and class mismatch issues. They measured the efficacy of the evaluated classifiers using seven performance metrics. The results showed that J48 and IBK perform better with the complete datasets and when carried out feature selection methods, respectively.

To discover anomalies in the KDD CUP 99 and NSL-KDD datasets, Meena et al. also presented a review article on different classification methods in WEKA. The study used two machine learning approaches (J48 and Naïve Bayes) and the metrics used are inaccuracy, time to build the models, sensitivity. The results revealed that J48 performs better than Naïve Bayes, though it required more time to build the model [8].

Different machine-learning methodologies were used in the research to evaluate the performance of the KDD '99 and NSL-KDD datasets. According to the findings, ANN outperforms other algorithms across all datasets, and classifiers trained on the KDD '99 dataset have greater accuracy than those trained on the NSL KDD dataset [9]. The authors studied the performance of KDD '99 and NSL-KDD datasets using five machine learning approaches (RFC, Naïve bayes, J48, Bayes Net and SVM). The following metrics (accuracy, recall, F1-score, and precision) were used to evaluate their performance. The results showed that RFC outperforms other techniques and also, R2L and U2L are not showing very encouraging outcomes because of its few numbers for the training set [10].

The preprocessing stage used the equal breadth binning method, and the sequential floating forward selection (SFFS) feature selection method produced 26 features. The study used a multi-agent, 2-layer categorization method on the NSL-KDD dataset. Tested and contrasted among the various classifications were: NB and DT, specifically NBTree, BFTree, J.48, and RF Tree. Compared to RF and J.48, NBTree and BFTree produced superior outcomes. When it came to categorizing common and DoS attacks against the recognizing R2L and U2R attacks, MLP performed well, this makes it the most suitable and efficient classifier for boosting recognition efficiency [11]. Table 1 presents the summary of the related works.

Table 1. Summary of the related works.

Author	Research Gap /Limitation	Performance Metrics	Dataset/ Features	Classification Type	Classifiers Evaluated
[7]	The study does not consider a case of class imbalance	Accuracy, TPR, FPR, Precision, Recall, F-measure, ROC area	NSL-KDD/ 14	Normal, DoS, Probe, R2L, U2R	Logistic, NB, MLP, SMO, IBK, J48
[8]	There is no discussion on the basis for feature extraction, training, and testing. Moreso, the study could have been conducted with many classification algorithms	Time to build model, Accuracy, Inaccuracy, F1 Score, and Precision	KDD '99 & NSL-KDD / NA	Normal, DoS, Probe, R2L, U2R	J48, Naïve Bayes
[9]	There is a need for classification algorithms to expand the categorization	Accuracy, Precision, Recall, F1-Score,	KDD '99 & NSL-KDD	Normal, DoS, Probe, R2L, U2R	Naïve Bayes, Support Vector Machines, Random Forests, and Artificial Neural Networks

(continued)

Table 1. *(continued)*

Author	Research Gap /Limitation	Performance Metrics	Dataset/ Features	Classification Type	Classifiers Evaluated
[10]	The novel approach/ technique is proposed to address the need for advancement in the intrusion detection system	Accuracy, F1-Score, Precision, Recall	KDD '99 & NSL-KDD	Normal, DoS, Probe, R2L, U2R	Random Forest, J48, Support Vector Machine, BayesNet, Naïve Bayes
[11]	The classifier appears excellent but may not be consistent in this circumstance because no performance measurements were utilized to evaluate the classifiers in this class imbalance condition	Accuracy, FP Rate, TP Rate, FN Rate, Precision, Recall, F-Score	NSL-KDD	Normal, DoS, Probe, R2L, U2R	Naïve Bayes, BF Tree, J48 Multilayer Perceptron, NB Tree RFT

3 Methodology

3.1 Dataset

The dataset used for the research is NSL-KDD, which is an improved version of the KDD '99 dataset. We extracted the dataset based on the protocol types (TCP, UDP, and ICMP). The number of TCP records is 102,689; the number of UDP records is 14,993, while the number of ICMP records is 8291. Table 1 depicts the attack types and Normal traffic based on protocol. WEKA 3.8.5 was used to conduct the research's experimental analysis (Waikato Environment for Knowledge Analysis). WEKA is a Java-based open-source machine learning scripting software created by Waikato University in New Zealand [12].

Table 2 revealed that all four attack types (DoS, R2L, U2R & Probe) are perpetrated via TCP; three out of the four attack types also carry out attacks via UDP protocol, and only two attack types (DoS & Probe) can take place via ICMP. Normal traffic occurs via all the protocols.

Table 2. Attack and Normal Traffics Based on Protocol.

	TCP	UDP	ICMP
DOS	42188	892	2847
R2L	993	-	-
U2R	51	3	-
PROBE	5857	1664	4135
NORMAL	53600	12434	1309
	102,689	**14,993**	**8291**

3.2 Preprocessing Stage

In this stage, the researchers considered any feature with at least 95% similar values redundant and consequently removed the feature. Out of the 44 attributes of the NSL-KDD dataset, 17 attributes, 27 attributes, and 29 attributes were eliminated from the data based on TCP, UDP, and ICMP, respectively.

3.3 Feature Selection Phase

Correlation-based feature selection (CFS) using Ranking Filter approach to eliminate irrelevant or non-significant attributes. Seven significant features were selected from ICMP based records, 8 records from UDP based records while 11 records were selected from TCP based records (Fig. 1).

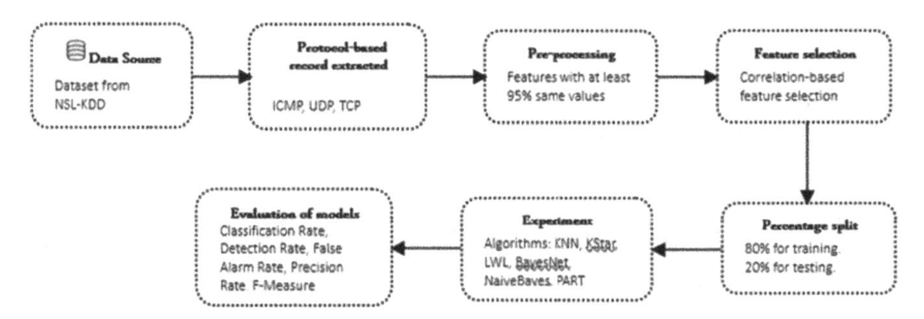

Fig. 1. The System Model.

3.4 Metrics

Equations 1 to 5 present the metrics used in the study, where TP is the True Positive, TN is the True Negative, FP is the False Positive, and FP is the False Negative.

$$\text{Classification Rate} = \frac{TP + TN}{TP + TN + FP + FN} \quad (1)$$

$$\text{Recall} = \frac{TP}{TP + FN} \tag{2}$$

$$\text{False Positive Rate} = \frac{FP}{FP + TN} \tag{3}$$

$$\text{Precision Rate (PR)} = \frac{TP}{TP + FP} \tag{4}$$

$$\text{F - measure (FM)} = \frac{2}{\frac{1}{PR} + \frac{1}{DR}} \tag{5}$$

4 Result and Discussion

4.1 ICMP

This section presents the results of ICMP-based dataset. In Table 3 and Fig. 2, PART exhibits outstanding performance in classification and detection rates. PART and BayesNet demonstrate high efficiency in accurately classifying normal traffic. Additionally, PART outperforms other algorithms in terms of precision rate and F-measure.

Table 3. Performance Metrics of ICMP-based Dataset

Algorithm	Classification Rate	Detection Rate	False Alarm Rate	Precision Rate	F-Measure
KNN	99.46	99.64	1.57	99.71	99.68
KStar	98.55	99.07	4.30	99.21	99.14
LWL	95.78	95.58	3.14	99.41	97.46
BayesNet	99.28	99.36	1.18	99.79	99.57
Naïve Bayes	92.05	98.65	44.31	92.46	95.45
PART	99.60	99.93	1.18	99.79	99.86

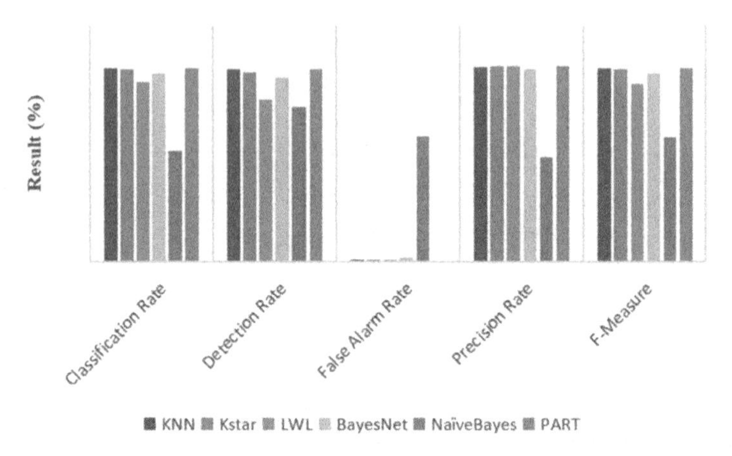

Fig. 2. Graphical Representation of Performance Metrics of ICMP-Based Dataset.

4.2 UDP

This section presents the results of UDP-based dataset. In Table 4 and Fig. 3, PART exhibits an excellent classification rate performance, while BayesNet surpasses other algorithms in detection rates. KNN, KStar, and PART.

Table 4. Performance Metrics of UDP-based Dataset

Algorithm	Classification Rate	Detection Rate	False Alarm Rate	Precision Rate	F-Measure
KNN	99.00	96.31	0.44	97.83	97.06
KStar	98.73	94.76	0.44	97.80	96.25
LWL	95.00	87.38	3.42	84.11	85.71
BayesNet	96.77	97.86	3.46	85.42	91.22
Naïve Bayes	91.47	86.46	7.48	70.62	77.74
PART	99.07	96.70	0.44	97.84	97.27

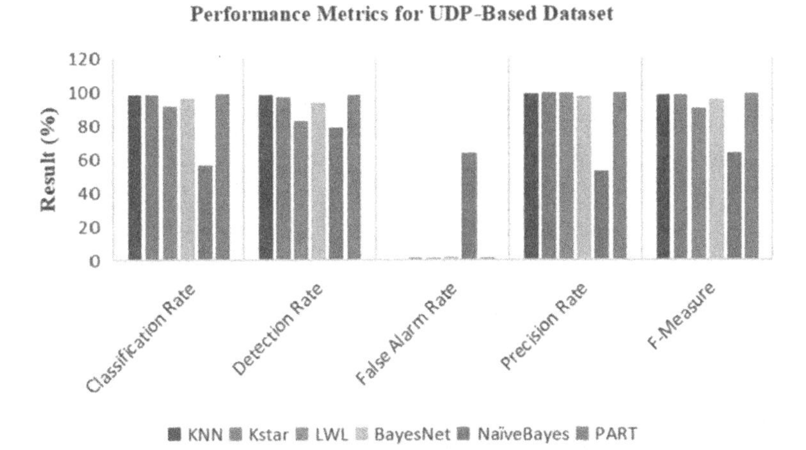

Fig. 3. Graphical Representation of Performance Metrics of UDP - Based Dataset.

4.3 TCP

This section presents the results of TCP-based dataset. Table 5 and Fig. 4 show that PART performs exceptionally well in classification rate. In detection rates, PART and KNN outperform the others. KStar and LWL work exceptionally well at accurately classifying regular traffic. KStar performs better than other algorithms in precision rate, while PART performs better in F-measure.

Table 5. Performance Metrics of TCP-based Dataset

Algorithm	Classification Rate	Detection Rate	False Alarm Rate	Precision Rate	F-Measure
KNN	98.50	97.98	1.02	98.87	98.42
KStar	98.11	96.59	0.49	99.44	97.99
LWL	91.50	82.76	0.49	99.36	90.31
BayesNet	95.89	93.58	2.00	97.72	95.61
Naïve Bayes	56.55	78.94	63.96	53.06	63.46
PART	98.70	97.98	0.63	99.30	98.64

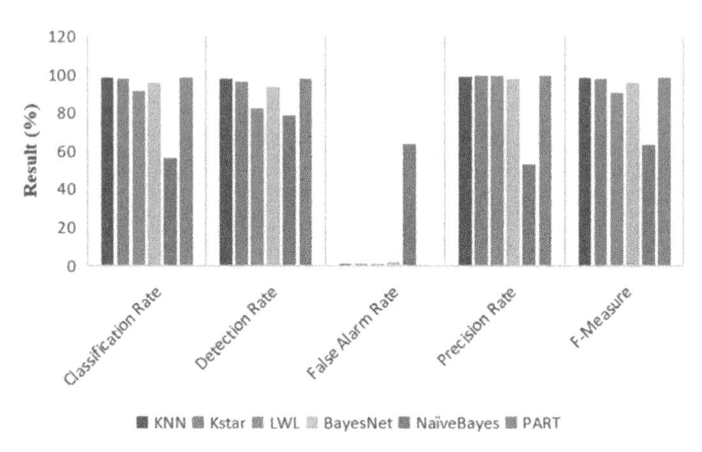

Fig. 4. Performance Metrics for TCP-Based Dataset

5 Conclusion

Table 6 is an extraction from Tables 3, 4 and 5, such that an algorithm with the best performance in each metrics is given a point and where more than an algorithm has the same best performance, a point is allocated to each of the algorithms. PART algorithm has the highest performance rating while NaiveBayes has the lowest performance rating. We therefore deduce that PART algorithm performs well across the dataset while NaiveBayes does not perform well across the dataset. We therefore recommend the use of other machine learning algorithms and ensemble models for further study.

Table 6. Performance Rating of the Algorithm

Algorithm	Classification Rate	Detection Rate	False Alarm Rate	Precision Rate	F-Measure	Performance Rating
KNN		1	1			2
KStar			2	1		3
LWL			1			1
BayesNet		1	1	1		3
Naïve Bayes						0
PART		1	1			2

References

1. Amudha, P., Karthik, S., Sivakumari, S.: A hybrid swarm intelligence algorithm for intrusion detection using significant features. Sci. World J. **2015** (2015). https://doi.org/10.1155/2015/574589
2. Jaiganesh, V., Sumathi, P., Mangayarkarasi, S.: An analysis of intrusion detection system using back propagation neural network. In: 2013 International Conference on Information Communication Embedded System ICICES, pp. 2013. 232–236 (2013). https://doi.org/10.1109/ICICES.2013.6508202
3. Aladesote, O.I., Alese, B.K., Dahunsi, F.: Intrusion detection technique using hypothesis testing. Lect. Notes Eng. Comput. Sci. **1**, 393–397 (2014)
4. Zuech, R., Khoshgoftaar, T.M., Seliya, N., Najafabadi, M.M., Kemp, C.: A new intrusion detection benchmarking system. In: Proceedings of 28th International Florida Artif. Intell. Res. Soc. Conf. FLAIRS 2015, pp. 252–255 (2015)
5. Agarwal, A., Sharma, P., Alshehri, M., Mohamed, A.A., Alfarraj, O.: Classification model for accuracy and intrusion detection using machine learning approach, pp. 1–22 (2021). https://doi.org/10.7717/peerj-cs.437
6. Imran, H.M., Abdullah, A.B., Palaniappan, S.: Towards the low false alarms and high detection rate in intrusions detection system. Int. J. Mach. Learn. Comput. **3**, 332–336 (2013). https://doi.org/10.7763/ijmlc.2013.v3.332
7. Mahfouz, A.M., Venugopal, D., Shiva, S.G.: Comparative analysis of ML classifiers for network intrusion detection. In: Yang, X.-S., Sherratt, S., Dey, N., Joshi, A. (eds.) Fourth International Congress on Information and Communication Technology. AISC, vol. 1027, pp. 193–207. Springer, Singapore (2020). https://doi.org/10.1007/978-981-32-9343-4_16
8. Meena, G., Choudhary, R.R.: A review paper on IDS classification using KDD 99 and NSL KDD dataset in WEKA. In: 2017 International Conference on Computer, Communications and Electronics (COMPTELIX), pp. 553–558 (2017).https://doi.org/10.1109/COMPTELIX.2017.8004032
9. Sapre, S., Ahmadi, P., Islam, K.: A Robust Comparison of the KDDCup99 and NSL- KDD IoT Network Intrusion Detection Datasets Through Various Machine Learning Algorithms (2019)
10. Nehra, D., Kumar, K., Mangat, V.: Pragmatic analysis of machine learning techniques in network based IDS. In: Luhach, A.K., Jat, D.S., Hawari, K.B.G., Gao, X.-Z., Lingras, P. (eds.) ICAICR 2019. CCIS, vol. 1075, pp. 422–430. Springer, Singapore (2019). https://doi.org/10.1007/978-981-15-0108-1_39
11. Amira, A.S., Hanafi, S.E.O., Hassanien, A.E.: Comparison of classification techniques applied for network intrusion detection and classification. J. Appl. Log. **24**, 109–118 (2017). https://doi.org/10.1016/j.jal.2016.11.018
12. Hall, M., Frank, E., Holmes, G., Pfahringer, B., Reutemann, P., Witten, I.H.: The WEKA data mining software. ACM SIGKDD Explor. Newsl. **11**, 10–18 (2009). https://doi.org/10.1145/1656274.1656278

Robust Aggregation Function in Federated Learning

Rahim Taheri[1(✉)], Farzad Arabikhan[1], Alexander Gegov[1], and Negar Akbari[2]

[1] University of Portsmouth, School of Computing, Portsmouth, UK
{rahim.taheri,farzad.arabikhan,alexander.gegov}@port.ac.uk
[2] University of Portsmouth, Portsmouth Business School, Portsmouth, UK
negar.akbari@port.ac.uk

Abstract. Maintaining user data privacy is a crucial challenge for machine learning techniques. Federated learning is a solution that enables machine learning models to be trained using data residing on different devices without centralizing the data. This training method offers several advantages: Firstly, federated learning helps preserve user privacy by storing data on separate devices rather than transferring it to a central location for training. Secondly, training machine learning models on a diverse set of devices through federated learning improves their robustness, ensuring optimal performance in a wide range of real-world scenarios. Finally, federated learning can promote scalability by enabling simultaneous training on a vast number of devices. So, it can increase the scale of training and enable the development of more sophisticated models. The aggregation function is one of the main steps in federated learning and is used on the server side to aggregate local models sent from the client side. The most widely used aggregation function is Krum, which, despite the research done to improve its robustness, is still vulnerable to adversarial samples. In this paper, a method is proposed to improve the robustness of the Krum aggregation function. The results confirm that the proposed method is more robust against adversarial samples than the original version of the Krum aggregation function.

Keywords: Federated Learning · Robustness · Krum Aggregation

1 Introduction

Today, various machine learning (ML)-based models trained on data are widely used. However, there are many concerns from users about maintaining their privacy when sharing data to train these models. Federated learning (FL) is an increasingly popular ML approach and data analysis due to its ability to facilitate collaboration among multiple devices with confidential data without violating privacy. FL is a technique that enables multiple clients to collaborate on training a neural network while keeping their data privacy. By keeping data decentralized, FL allows devices to work together to enhance the accuracy of a ML model while simultaneously safeguarding the privacy and security of the data. This framework has been proposed for various fields, including healthcare, finance, sensor networks, and mobile applications. For example, FL can be utilized to develop ML models with medical data gathered from multiple medical

I. Saad et al. (Eds.): ICIKS 2023, LNBIP 486, pp. 168–175, 2024.
https://doi.org/10.1007/978-3-031-51664-1_12

institutions while ensuring the privacy of patients. In mobile applications, FL can be used to enhance the user experience by tailoring application functionalities based on individual user data.

1.1 Federated Learning

The usual approach for FL algorithms involves a synchronized communication process between the server and clients, where each client performs some local computation based on its own data. The resulting updates are then aggregated and used to update the server model. The most commonly used training algorithm is FedAvg [1], which operates as follows.

- The server initiates a global model and distributes it to m clients. The global model is denoted as $w^{(t)}$, which represents the parameters of this model.
- Upon receiving $w^{(t)}$, each client i treats it as a local model, denoted as $w_{i,0}^{(t)}$, and proceeds to update it using local data and the gradient descent algorithm.

$$w_{i,j+1}^{(t)} = w_{i,j}^{(t)} - \gamma \nabla F_i \left(w_{i,j}^{(t)} \right) \tag{1}$$

  In Eq. 1, $j = 0, \ldots, n-1$ where n is number of gradient descent steps and γ is learning rate.
- Every client i sends a vector, $w_i^{(t+1)}$, representing the updated local model to the server. Assuming that the number of clients in each round t is S_t, the global model is obtained using the FedAvg method. This method employs the Eq. 2, which computes the weighted average of the local models [2].

$$w^{(t+1)} = \frac{\sum_{i \in S_t} \alpha_i w_i^{(t+1)}}{\sum_{i \in S_t} \alpha_i} \tag{2}$$

The weight of client i is represented by α_i in this equation. Recently, there has been growing interest in exploring alternative aggregation algorithms, beyond FedAvg, to improve robustness.

1.2 Motivation

Despite its potential in real-world applications, FL is susceptible to degraded model performance resulting from corrupted updates. Such updates can be caused by malicious individuals inserting inaccurate data into the system or by hardware malfunctions in budget-friendly devices. These corruptions can result in unreliable and imprecise ML models, creating uncertainty in the model's output and diminishing trust in its reliability. FL is susceptible to several types of attacks, including data and model poisoning. FL is vulnerable to attacks where a fraction of clients can be controlled by the attackers, who manipulate the local data and model updates to harm the performance of the global model or introduce a backdoor. One example of this is the backdoor attack, where an attacker can label images with a 'trash' label for certain brands of automobiles, contaminating the global model. Thus, it is crucial to protect against such attacks to ensure that FL training is robust. Some researchers have suggested the need for a robust aggregation function that can adapt to these attacks and ensure robust FL.

The defence against adversarial attacks in FL is complicated by the fact that the server can only rely on model updates submitted by clients, while attackers have greater design flexibility. Furthermore, the presence of diverse data distribution in FL adds complexity to the situation. As a result, the primary focus of defence is on developing a robust aggregation technique for model updates. In this technique, the model update is treated as a matrix, with the goal of filtering out modified update vectors from attackers and retaining only the legitimate update vectors submitted by benign clients.

To address this challenge, researchers have introduced the Krum aggregation approach in FL. This technique is designed to counteract the negative effects of flawed or malicious updates during the aggregation process. Krum aggregation utilizes a technique that identifies a subset of updates that are less susceptible to corruption. This process involves selecting the update that is closest to the center of the other updates while excluding those that are farthest from the center. By employing this method, Krum aggregation diminishes the influence of malicious or defective updates while still incorporating the input of valid updates. The Krum technique has been implemented in a variety of contexts to aggregate local models in FL.

1.3 Contribution

The innovations presented in this paper can be summarized as follows:

- Introducing a robust aggregation method for FL.
- The proposed method modifies the Krum algorithm, which has been widely used and studied in previous research.
- The proposed method has a relatively low computational complexity.
- The performance of the proposed method is compared with that of the Krum algorithm.

1.4 Paper Organization

This paper is structured by these sections. The Sect. 2 containing a thorough review of the literature. Section 3 outlines the proposed method for aggregation function in federated learning. Section 4 reviews the simulation environment and its results, while Sect. 5 presents the final conclusion.

2 Related Works

Various aggregation functions have been proposed in the literature to protect FL against adversarial attacks. One of the first functions used for aggregation was FedAvg, which has been investigated in many research studies for its robustness against malicious data and adversarial attacks. Most of the proposed methods to improve robustness of FedAvg, consider the distribution of local models in their computations [3,4].

While classical median-based robust aggregations, such as coordinate-wise and geometric have been proposed by Yin et al [5], their performance is impacted by the heterogeneous distribution of data in FL. The authors of the study introduced a median aggregation function for FL, where the server sorts all client-provided parameters and chooses the median as the aggregate parameter. In cases where clients have partitions of varying sizes, the efficiency of aggregation rules may be compromised, as these rules assume that all clients have equal amounts of data during training. Additionally, the computational complexity of calculating the median increases with the number of clients and model parameters.

The multiKrum [6] is another robust aggregation method that involves selecting one or more local models that are highly similar to each other. The Krum function assigns a score to each client based on the sum of Euclidean distances between that client and the other clients. multiKrum selects and aggregates the local models with the smallest sum of distances, and then applies a standard FedAvg only to the selected models. Although this approach restricts the global server to only collecting similar models, it may impact the training process's performance by removing some models. Additionally, the computational complexity of multiKrum can increase with the number of clients and model parameters.

In [7], a robust aggregation mechanism designed to rank the gradient estimators of clients is proposed. The approach only considers the gradients of the top clients with the highest ratings for updating the model at each iteration. However, the number of selected clients, must be predetermined.

In [8], the proposed approach aims to detect and exclude malicious client updates at each iteration, utilizing data from all trustworthy clients to compute the model update. This eliminates the need for a predefined parameter as number of clients and enables us to leverage information from all clients, particularly in situations where there are minimal or no malicious clients. The authors introduced an adaptive federated aggregation (AFA) algorithm that uses a *Hidden Markov Model* to detect and block malicious clients from sending bad model updates. Preventing model destruction reduces the computational and communication costs associated with retraining and redistributing the model. This method calculates the similarity between the client model updates and the aggregated model parameters to select a cluster of clients with similar scores to the global model. This method assumes that benign clients make up more than half of the total number of clients, so their updates are retained while discarding the rest. The algorithm includes customizable hyper-parameters that allow for accepting or rejecting local updates based on the heterogeneity and scenario of the dataset. In each iteration, the model parameters are updated using only the updates from the selected benign clients, which are clustered based on their similarity to the global model parameters.

3 Proposed Method

Devising a robust aggregation approach is crucial because attackers may continually develop new methods to overcome current defenses. Therefore, it is essential to have an aggregation approach that can easily adapt to defend against such challenging attacks. In this section, we will propose a framework for this type of aggregation approach. We propose a method for detecting attacks during the aggregation of update vectors in FL. Our goal is to develop a model based on data that can detect corrupt update vectors, specifically by identifying their vulnerable areas susceptible to various types of attacks. FL is a decentralized ML approach, in which a global model is trained on a server with the participation of n clients. The global model is denoted as W_g, and for each client i, their local model is represented as $W_1^{(i)}$. First, the local models are initialized with the global model. Then, each client i trains their local model with local data and sends the update vector to the server (Eq. 3).

$$\boldsymbol{u}_i = W_1^{(i)} - W\mathrm{g} \tag{3}$$

Upon receiving the update vectors, the server combines them using an aggregation function to obtain a combined vector,$g\left(\{\boldsymbol{u}_i\}\right)$. This aggregated vector is then utilized to update the global model in round i (Eq. 4). After updating the global model, it is returned to the clients for use in the next round of training.

$$W_{\mathrm{g}} = W_{\mathrm{g}} + g\left(\{\boldsymbol{u}_i\}\right) \tag{4}$$

A robust aggregation is essential to prevent adversarial attacks during FL training. Such attacks can target either local data or update vectors directly. Nevertheless, since the server lacks access to the clients' local training data or the number of locally trained samples, it can only utilize the update vectors for aggregation. In this context, the robustness of the aggregation function becomes crucial.

Assuming that $\mathbb{B}$ represents a collection of valid update vectors from benign clients with $|\mathbb{B}|$ value and $\mathbb{D}$ represents a collection of malicious update vectors from attackers with $|\mathbb{D}|$ value, defining the robust mean, $\mathbb{M}_{\mathrm{R}}$, as the average of the legitimate update vectors results in Eq. 5:

$$\mathbb{M}_{\mathrm{R}} = \sum_{i=1}^{n} \frac{\mathbb{F}(\text{Number of legitimate } u_i \in \mathbb{B})}{|\mathbb{B}|} \boldsymbol{u}_i, \tag{5}$$

In Eq. 5, $\mathbb{F}$ is a function that represents the number of benign update vectors in the dataset $\mathbb{B}$. The objective of a robust aggregation approach $g(\cdot)$ is to find the robust mean $\mathbb{M}_{\mathrm{R}}$ by minimizing the Eq. 6:

$$\underset{g}{arg\,min} \|g\left(\{\boldsymbol{u}_i\}\right) - \boldsymbol{M}_{\mathrm{R}}\| \tag{6}$$

As proposed in [9], a probability-based method can be used to estimate the $\mathbb{F}$ function in Eq. 5. Therefore, we use $p\left(\boldsymbol{u}_i \in \mathbb{B} \mid q_r\right)$ as an estimation. This estimate represents the probability that u_i is a valid update vector for a benign client, while q_r represents a robust estimate. We define q as an estimate for the robust krum $\mathbb{M}_{\mathrm{R}}$.

$$q_0 = \mathrm{Krum}\left(\{\boldsymbol{u}_i\}\right), \tag{7}$$

$$q_{r+1} = \sum_{i=1}^{n} \frac{p\left(\boldsymbol{u}_i \in \mathbb{B} \mid q_r\right)}{|\mathbb{B}|} \boldsymbol{u}_i, \tag{8}$$

In Eq. 7, Krum is an aggregation function that is used in FL. Krum operates by having individual clients utilize their respective local data to calculate their model updates and transmit them to the central server. The server selects a group of clients whose model updates closely resemble the global model, and averages the updates from these chosen clients to derive an updated global model. This updated global model is then distributed to all clients.

Algorithm 1. Robust Krum

Input: u_i,R,k,ε
Output: q,w_i

```
1:  q_0 ← Krum({u_i}, k)
2:  for r in R do
3:      for i in n do
4:          s_i = |u_i| / |Σ_{j=1}^{n} u_j ∈ B|
5:          if (s_i > ε) then
6:              w_i ← 1
7:          else
8:              w_i ← 0
9:          end if
10:     end for
11:     q_{r+1} = Σ_{i=1}^{n} w_i u_i
12:     q.append(q_{r+1})
13: end for
14: return q
```

In [9], the authors used an attention function to estimate the probability. In our work, we opted to utilize the ratio of the size of the update vector i to the total update vectors in order to obtain the value of w for this purpose.

$$s_i = \frac{|u_i|}{|\sum_{j=1}^{n} \boldsymbol{u}_j \in \mathbb{B}|} \tag{9}$$

Finally, we use a threshold limit to determine the selected w_i. We use this method by checking whether the value of s_i exceeds the epsilon threshold limit; if so,w_i is as 1 effective; otherwise, it is deemed as 0.

The proposed approach is presented in Algorithm 1.

3.1 System Setting

The paper's experiments were conducted using the Cifar10 dataset, with 10 clients and either 5, 10, or 15 rounds running in the FL algorithm. The proposed aggregation algorithm exhibited strong accuracy in two cases, with and without attack, as evidenced in Tables 1, 2, and 3. The experiments employed the label-flipping attack from [10] as an adversarial attack. The results demonstrate that accuracy increases as the number of rounds increases, with the potential for accuracy to exceed 90% with around 100 rounds. Additionally, a simple neural network with two hidden layers was used as the client model in all implementations.

4 Results

In this section, we compare the accuracy of the proposed method with different settings. To achieve this, we have compared the accuracy of four different implementation scenarios.

- **KA**: FL with Krum aggregation
- **RKA**: FL with Robust Krum aggregation
- **KA-WA**: FL with Krum aggregation (With Attack)
- **RKA-WA**: FL with Robust Krum aggregation (With Attack)

The results of implementing these methods are reported in Tables 1, 2, and 3 for 5, 10, and 15 rounds in FL.

Table 1. Accuracy of FL by different aggregation function after 5 rounds-Cifar10 dataset

Algorithm	Accuracy
KA	41.86
RKA	42.11
KA-WA	44.23
RKA-WA	43.52

Table 1 presents the results of running the FL algorithm on the Cifar10 dataset for 5 rounds. As shown in the table, the accuracy of the model in the Robust Krum algorithm is comparable to that of the Krum algorithm in both attack and non-attack scenarios. Likewise, Table 2 compares the results of FL execution with 10-rounds. It is apparent that the accuracy has improved compared to the 5-round execution, and the accuracy of the proposed method remains comparable to that of Krum's method.

Table 2. Accuracy of FL by different aggregation function after 10 rounds-Cifar10 dataset

Algorithm	Accuracy
KA	45.73
RKA	43.96
KA-WA	44.76
RKA-WA	42.89

Table 3, like the previous tables, shows the results of running the proposed method and compares it to Krum's aggregation method for 15 rounds. The improved accuracy observed with this method highlights the success of the proposed approach.

Table 3. Accuracy of FL by different aggregation function after 15 rounds-Cifar10 dataset

Algorithm	Accuracy
KA	64.83
RKA	63.19
KA-WA	49.23
RKA-WA	51.71

5 Conclusions

This study introduces a novel aggregation function for robust federated learning that utilizes a neural network as client models. Our aggregation approach is the first of its kind to employ an attack-adaptive Krum-based strategy, which learns to defend against diverse attacks in a data-driven manner. The experimental results demonstrate that our proposed method is mostly superior to Krum's method, delivering comparable outcomes in many cases.

References

1. McMahan, B., Moore, E., Ramage, D., Hampson, S., y Arcas, B A.: Communication-efficient learning of deep networks from decentralized data. In: Artificial Intelligence and Statistics, pp. 1273–1282, PMLR (2017)
2. Pillutla, K., Kakade, S.M., Harchaoui, Z.: Robust aggregation for federated learning. IEEE Trans. Signal Process. **70**, 1142–1154 (2022)
3. Liu, Z., Guo, J., Yang, W., Fan, J., Lam, K.Y., Zhao, J.: Privacy-preserving aggregation in federated learning: a survey. IEEE Trans. Big Data (2022)
4. Taheri, R., Shojafar, M., Alazab, M., Tafazolli, R.: FED-IIoT: a robust federated malware detection architecture in industrial IoT. IEEE Trans. Ind. Inform. **17**(12), 8442–8452 (2020)
5. Yin, D., Chen, Y., Kannan, R., Bartlett, P.: Byzantine-robust distributed learning: towards optimal statistical rates. In International Conference on Machine Learning, pp. 5650-5659. PMLR (2018)
6. Blanchard, P., El Mhamdi, E.M., Guerraoui, R., Stainer, J.: Machine learning with adversaries: Byzantine tolerant gradient descent. In: Advances in Neural Information Processing Systems, vol. 30 (2017)
7. Xie, C., Koyejo, S., Gupta, I.: Zeno: distributed stochastic gradient descent with suspicion-based fault-tolerance. In International Conference on Machine Learning, pp. 6893-6901. PMLR (2019)
8. Mu noz-Gonz alez, L., Co, K.T., Lupu, E.C.: Byzantine-robust federated machine learning through adaptive model averaging, arXiv preprint arXiv:1909.05125 (2019)
9. Wan, C.P., Chen, Q.: Robust federated learning with attack-adaptive aggregation, arXiv preprint arXiv:2102.05257 (2021)
10. Jebreel, N.M., Domingo-Ferrer, J., Sanchez, D., Blanco-Justicia, A.: Defending against the label-flipping attack in federated learning, arXiv preprint arXiv:2207.01982 (2022)

Knowledge Management, Context and Ontology

Should I Share or Should I Go?
A Study of Tacit Knowledge Sharing Behaviors in Extended Enterprises

Léontine Lando and Pierre-Emmanuel Arduin(✉)

Université Paris-Dauphine, PSL, CNRS DRM, Paris, France
pierre-emmanuel.arduin@dauphine.psl.eu

Abstract. This paper proposes to study the management of tacit knowledge within extended enterprises and particularly multinational technology corporations, *i.e.*, companies working on digital software and Internet-related services with a presence in more than one state. Indeed, due to the particular nature of tacit knowledge, human behaviors deployed to share it or on the contrary to retain it are of particular interest in that kind of organizations with not only an international cultural context, but also a highly digital environment.

We argue that tacit knowledge management within organizations can be improved if we better understand practices, habits, and behaviors when arises to employees the question of "Should I Share or Should I Go?".

In this paper we present a study involving a sample of 88 participants who expressed their tacit knowledge behaviors as employees working in a multinational technology corporation. We identified trends on the importance of tacit knowledge as well as on the role of management and technology in sharing behaviors. Particularly, we observed that the presence of established procedures does not imply positive tacit knowledge sharing behaviors. We finally conclude by sketching future research directions on the observation of this phenomenon.

Keywords: Tacit knowledge · Knowledge sharing · Extended Enterprises · Multinational organizations · Technological tools

1 Introduction

Where competition increases within companies but also between companies, tacit knowledge is a real way to stand out and gain in competitiveness [28]. Tacit knowledge, by nature, is a resource difficult to replicate and that may be the reason why individuals and organizations create added value [18]. However, if this resource is difficult to replicate, how can it be identified and shared among members of an organization fostering its own growth?

In the case of companies with an international presence, the problem is particular because the distance means that there are fewer opportunities for direct observation, direct group learning, and direct face-to-face. Nonaka and Takeuchi [18] argued that companies create knowledge through exchange and combination,

I. Saad et al. (Eds.): ICIKS 2023, LNBIP 486, pp. 179–196, 2024.
https://doi.org/10.1007/978-3-031-51664-1_13

but tacit knowledge is individual, making its formalization and communication difficult [23]. Close interactions among members are then necessary and this is where technology comes in, but how and how efficiently?

If tacit knowledge refers to know-how in practices, methods, and the understanding of certain situations, the challenge for a company is to ensure its sharing so that it can benefit to all the subsidiaries. The benefits can be economic (*e.g.*, efficiency gain in daily actions), strategic (*e.g.*, more understanding of international issues), or even human (*e.g.*, better management and work organization) [17]. Moreover, as tacit knowledge is directly linked to the interpretation and experience of each individual, the challenge is to know how to share this knowledge in order to retain it within the company.

In the second section of this paper, background theory and assumptions are presented: First, a literature review on tacit knowledge and its management; Second, a focus on extended enterprises and tacit knowledge management within them; Third, a discussion on the impact of language and technology in international tacit knowledge management; Fourth, a review of practices and indicators of tacit knowledge management within organizations. Then in the third section of this paper the field study is exposed: First, a description of the respondents and methodology; Second, an analysis of hypotheses and collected data; Third, a discussion of the results. Finally the fourth section presents the limits of this work as well as the conclusions and perspectives.

The aim of this paper is to improve the understanding of tacit knowledge sharing practices within organizations and particularly within extended enterprises, where individuals have different cultures, languages, and interpretative frameworks. Such a study allows to highlight what arises to employees when they have to share their tacit knowledge and are asking themselves: "Should I Share or Should I Go?".

2 Background Theory and Assumptions

2.1 Tacit Knowledge and Its Management

According to Tsuchiya [26], knowledge is created during two cognitive processes. The first, *sense-giving*, transforms data into information: "When datum is sense-given through interpretative framework, it becomes information" [26, p. 88]. Individuals therefore transform the data they receive into information as soon as they interpret it in order to communicate it in their own way. The second, *sense-reading*, transforms information into knowledge: "When information is sense-read through interpretative framework, it becomes knowledge" [26, p. 88]. In other words, information becomes knowledge when it is interpreted and internalized by the individual who is receiving it.

These interpretations are made a priori at the individual level, according to the mental models specific to each individual [3]. These mental models are the source of the interpretations made by individuals facing the reality which is external to them, and where they perceive data and information. This then explains why from the same inputs, *i.e.*, data and information received, as well as within a common context, different individuals may create different tacit knowledge. This phenomenon is called *meaning variance* [2]. This concept emphasizes the fact that the meaning read in the same stimulus may vary from one individual to another. Consequently, it may seem complicated to ensure that knowledge is transmitted and acquired in the same way by different individuals within organizations. This is particularly the case when we refer to tacit knowledge. The personal quality of tacit knowledge makes its formalization and communication difficult according to Polanyi [23], interactions should then be close among employees seeking to share tacit knowledge within their organization.

According to Nonaka and Takeuchi [18], knowledge is converted from made-explicit to tacit and vice-versa through a four-stages spiral:

- **Socialization**: from an individual's tacit knowledge to another individual's tacit knowledge;
- **Externalization**: from an individual's tacit knowledge to made-explicit knowledge;
- **Combination**: from existing made-explicit knowledge combined to create new made-explicit knowledge;
- **Internalization**: from made-explicit knowledge interpreted and appropriated to an individual's tacit knowledge.

In this sense, the management of tacit knowledge is not completely opposed to the one of made-explicit knowledge. Tacit knowledge management does not only concern socialization, but also the externalization of knowledge, as well as the combination and the internalization of this knowledge.

Tacit knowledge is mainly carried by individuals and is not transferable through a medium. However, within an organization, tacit knowledge is the basis of organizational knowledge [26, p. 93]: "Knowledge of individuals becomes organizational knowledge only when it is regarded to be legitimate.". This legitimization relies on initiatives coming from the organization which must establish processes, events or other means so that members of the organization exchange (as discussed in Sect. 2.4). Indeed, organizations create knowledge through exchange and combination [18].

Interestingly, without sufficient trading volume, tacit knowledge cannot be sufficiently shared to build significant organizational knowledge. The organization is then positioned as if it was in charge of establishing the conditions to create the organizational knowledge. The organization has thus the additional role of fostering a "common interpretative framework" to all its employees, in order to go beyond mental models and reduce the effects of meaning variance [2]. Individuals exchanging information must be able to understand, interpret and assimilate information in the most similar way possible. Such an idea is highlighted in the concept of *interpretative frameworks commensurability* [3,10,26].

Grundstein [10] talks about codes that are assimilated through a common academic education: people who have obtained the same diploma will have shared common codes. A connection can here be done with the definition of secondary socialization given by Berger and Luckmann [6] in the field of sociology: "Secondary socialization is the internalization of institutional or institution-based 'sub-worlds'." [6, p. 158]. Individuals who have acquired the codes of the same sub-worlds have more chance to integrate knowledge in the same way.

Thus, the organization should promote the commensurability of interpretative frameworks: create, share, and disseminate to members the codes, practices and values relating to the sub-world that is the organization. One of the roles of the organization would therefore be to foster shared or team mental models, encouraging them to become organizational knowledge [5]. Mental models having then both an individual and collective dimension, we consider that tacit knowledge management within organizations in general and within extended enterprises in particular is then not only an individual but also a collective mater.

2.2 Tacit Knowledge Management in Extended Enterprises

Through the various subsidiaries they have in multiple countries around the world, extended enterprises can take different strategic orientations, which will influence how they are managed [24]. Table 1 summarizes the four different internationalization strategies [8].

Table 1. Extended enterprises internationalization strategies (adapted from [8,24])

Adaptation		Integration: Low	Integration: Strong
	Strong	**Multinational Strategy** Structure with geographical divisions	**Transnational Strategy** Worldwide Matrix, geographical and product divisions
	Low	**International Strategy** International and export divisions	**Global Strategy** Worldwide functional structure

Global and transnational strategies, once in place, make it possible to standardize practices, management types, routines, and other resources or practices in all the organizational entities [24]. These strategies also make it possible to have a quite integrated culture within all the organizational entities, *i.e.*, beyond their respective national cultures. Global and transnational organizations have employees sharing common values. This can potentially be the sign of a high commensurability between employees' interpretative frameworks [3], leading to better tacit knowledge sharing between employees. Nevertheless, due to their high degree of centralization, the creation of new tacit knowledge can be mitigated because of too similar points of view [1].

International and multinational strategies may imply more difficulties to install common interpretative frameworks between the employees, because their cultures are strategically more oriented towards their respective local spaces. There are less shared values, even if it does not preclude information sharing on the local activity that can be used to create knowledge adapted to the contexts of other countries, since interpretation is done at the individual level and is influenced by experience. The more the points of view are complementary, the more employees can co-create tacit knowledge [1].

Indeed, individuals can build mental models similar when they are in a configuration favoring exchanges in order to perceive and interpret the information in the same way [15]. The organization's environment must allow its members to share their experiences and learnings so that others can observe and practice in turn [5]. This is all the more true as tacit knowledge cannot be materialized and strongly relies on socialization. One of the ways to create these patterns of common interpretations is to foster an environment leading to the creation and preservation of common and shared mental models. Authors such as Orasanu and Salas [19] began quite early to study the extent to which individuals share and construct similar mental models when interacting together in a team setting. Others such as Langan-Fox *et al.* [15] go deeper when they state that "in order to work together successfully, teams must perceive, encode, store, and retrieve information in similar ways. Thus the quality of a team's output will depend not only on the information available to individual team members, but also on the shared or team mental model." [15, p. 243].

Particularly, by insisting on the importance of tacit knowledge, the extended enterprise as introduced by Grundstein *et al.* [12] is presented as increasingly developing its activities in a planetary space with three dimensions (Fig. 1):

1. **A global space:** this space covers the set of the enterprise that are the geographic places of localization.
2. **A local space:** this space corresponds to the subset of the enterprise located in a given geographic zone.
3. **An influence space:** this space is the field of interactions of the enterprise with the others organizations.

This extended enterprise is placed within an unforeseeable environment that leads towards uncertainty and doubt. Two kinds of digital information networks overlap: (1) a formal information network between the internal or external entities in which circulate data and codified knowledge, and (2) an informal information network between members, nomadic or sedentary employees, in which information exchanges and tacit knowledge sharing take place. This network is implemented by using digital communication technologies in the planetary space, where the language also impacts such an international tacit knowledge management.

2.3 The Impact of Language and Technology in International Tacit Knowledge Management

In extended enterprises, as presented in the previous section, employees have not only different national cultures, but also languages that might be different. All do not have the same level in a common working language such as English and

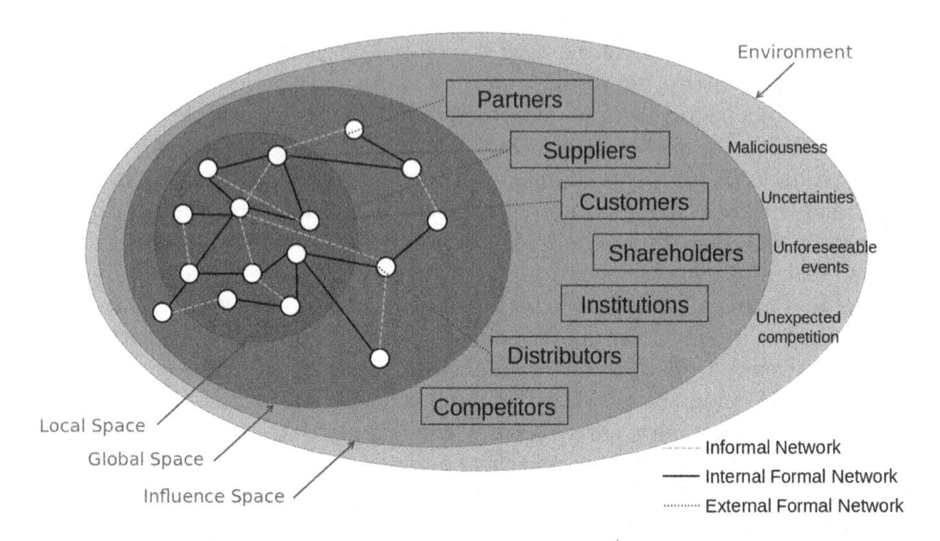

Fig. 1. The formal and informal networks within the extended enterprise (adapted from [12])

this can have an impact on the meaning variance [2]. If tacit knowledge is shared mainly through direct exchanges and observation, how do multinational teams manage to collaborate, socialize, and actually share tacit knowledge?

Authors such as Tenzer *et al.* [25] worked on the influence of language in communication within extended enterprises and made several findings. Indeed, these authors observed that an insufficient level in the working language is a real obstacle to communication because the information communicated no longer depends primarily on what individuals really want to say, but more on what they are able to say, even if they are extremely competent in their field. This leads individuals of minority nationality with less abilities of the working language in the extended enterprise to be often less integrated. Such an obstacle to their socialization within a group mitigates their tacit knowledge sharing. Particularly, even if the words and vocabulary used are understood, the meaning given to these words might not be the same [25].

The use of technological tools as a means of extending and maintaining a network helps to manage tacit knowledge, especially in extended enterprises where teams are dispersed [9]. Face-to-face communication is for example no longer considered as the principal way of sharing tacit knowledge, particularly when experts are not always geographically co-located [20]. Additionally, such a use of technological tools will depend on the internationalization strategy adopted by the extended enterprise (see Table 1 in Sect. 2.2). It is mainly in organizations with a transnational and multinational strategy that experts are precisely distributed in certain places [8,24]. For example, in transnational organizations, geographical areas are assimilated to product or business skill centers.

Even if language is a basis for international tacit knowledge management, several technological solutions position themselves as facilitators of exchanges within extended enterprises. Some authors have began in the early 2010s to study the various facilitators and inhibitors of knowledge management [22]. From their point of view technology essentially plays a secondary rather than a primary role in knowledge management and even more in the case of tacit knowledge. It must therefore be stressed that technology plays a role of facilitator, but that the main lever is social-cultural and human. Moreover, if we consider that tacit knowledge sharing also includes the passage from made-explicit to tacit knowledge (internalization), and from tacit to made-explicit knowledge (externalization) [18], then the technology also makes it possible to indirectly share tacit knowledge. Indeed, if made-explicit knowledge is disseminated, interpreted and integrated by individuals, then these individuals will have shared tacit knowledge. The same goes for externalization: it is necessary to master tacit knowledge in order to externalize it through technology. Then technology, such as Intranets for example, also makes it possible to indirectly contribute to the creation, capture and sharing of tacit knowledge, in addition to made-explicit knowledge [22]. When employees for example benefit from even made-explicit knowledge that allows them to better understand information, their interpretative frameworks might evolve, leading them to become able to interpret new tacit knowledge [3].

Actually some practices and indicators used within organizations can be quite instructive to design a study of tacit knowledge sharing behaviors within extended enterprises.

2.4 Practices and Indicators of Tacit Knowledge Management

Since the early 2000s, several authors considered practical solutions to handle the question of tacit knowledge management within organizations [28]. We observed that these practical solutions can be categorized into three plans: (1) organizational, (2) managerial, and (3) technological.

Organizational Plan

1. **Establishing communities of practice** [14]: A community of practice consists of a group of people who share good practices on one or more subjects of interest. Due to the reciprocity and motivation of their members, the communities of practice allow a socialization framework leading its members to interpret the information in the same way, *i.e.,* mitigating meaning variance [2].
2. **Reorganize work time and space** [4]: Time slots devoted to the creation and sharing of knowledge on employee planning make it possible to formalize the importance of a knowledge, particularly tacit knowledge, sharing approach. These time slots can be an opportunity to share feedback with others, so that employees share the same level of information as it is the case in After Action Reviews (AAR) [16].

Managerial Plan

1. **Appoint a knowledge manager** [4]: This person is responsible for the implementation and proper execution of initiatives dedicated to knowledge sharing and creation, including tacit knowledge. S/he must then improve the procedures and possibly measure them through frameworks allowing to analyze the level of tacit knowledge consideration for example [3].
2. **Reward employees significantly contributing to knowledge sharing** [14]: These rewards help to motivate employees and can be of different kinds. They are most often informal (congratulations), but can also be formalized (recognition of skills), or even honorary (symbolic prize). A common practice is to reward the most involved employees by offering them an access to chosen trainings opening specific career plans, which can produce a virtuous circle for the enterprise. Rewards may also be pecuniary, but do not perpetuate the motivation.

Technological Plan

1. **Use online collaboration tools** [22]: As digital technology is a facilitator that complements human relations, it makes it possible to replicate what is happening at the local level in a digital, *i.e.,* more global, space. Experts with rare skills or local knowledge can benefit therefore be contacted throughout the organization [1].
2. **Videoconferences** [20]: This kind of software is now well established in our daily lives and allows to achieve more general public such as colleagues in the global rather than in the local space (see Fig. 1 in Sect. 2.2). In addition, some software allows to do instant translations, overcoming then certain language barriers.
3. **Record lectures or podcasts** [20]: This allows access to the information created and shared during the videoconferences above mentioned, in a delayed way. This is particularly useful for extended enterprises with a multinational strategy (see Table 1 in Sect. 2.2), where employees might have planning problems with time differences, for example.
4. **Create blogs, wikis and Frequently Asked Questions (F.A.Q.)** [20]: On this type of media, the information is in fact tacit knowledge that has been made-explicit. After having lived and understood certain experiences (tacit knowledge), employees share them. In the case of wikis, the contribution of different individuals creates knowledge, both made-explicit (externalization and combination in the sense of [18]) and tacit (socialization and internalization in the sense of [18]).

Some authors studied for example the impact of tacit knowledge on innovations and observed that knowledge tacitness improves the usefulness of interactions for developing innovations [21]. The question is then to find indicators of a sharing level and an adoption level of these approaches, some might be [4,14,27]:

- The size of the knowledge management teams;
- The budget allocated to knowledge management, not only to finance training, but also financial rewards and employees in charge of managing knowledge management initiatives;
- Lists and demographics of employees connected to some events;
- Lists of employees contributing to blogs or wikis and of those who have accessed them;
- Employees' feedback on the existing initiatives;
- The number of communities of practice.

It might seem hard to know whether the adoption of these approaches is a fact, an illusion, or a nightmare in extended enterprises. To see if and how they really apply in the daily lives of employees within extended enterprises, the work presented in this paper rely on a field study.

3 Field Study

The objective of this analysis is to study tacit knowledge sharing behaviors in the international context of extended enterprises. The impact of language and technology in international tacit knowledge management is regarded through the lens of the practices and indicators of tacit knowledge management. Particularly, this analysis studies the implemented approaches and their impact on employees' willingness to share their tacit knowledge.

3.1 Description of the Respondents and Methodology

A questionnaire has been disseminated on several networking platforms, social media, and professional social networks. The questionnaire was initially intended for all working people, but it has been redirected to an audience that had already worked in extended enterprises, named "international companies" in the questionnaire. The other responses received have therefore been excluded in order to carry out the analyzes that will follow. Of the 121 responses received, 88 came from employees who had already worked in extended enterprises.

Table 2 shows that the data collected comes from relatively young and qualified people (average age 25.32 years, 56 female, 32 male, 3 PhD, 20 MSc, 54 BSc, 5 without degree, 6 no answer). This explains why the majority of respondents did not have a significant work experience (average of 2 years in the company).

The expected responses to the main survey questions are ratings on a Likert-type scale at 5 levels, in order to measure the opinions, feelings and expectations of the respondents. The answers evolve between 1 (strongly disagree) and 5 (strongly agree). This choice was made in order to avoid biases that could have influenced responses and the numbers here are qualitative and not quantitative: they only represent ideas and judgments.

Table 2. Distribution of respondents

	Age	Seniority (years)	
		in the company	on the post
Average	25.32	2	1.59
Median	23	1	1
Third quartile	25	2	2
Maximum	58	14	9
Minimum	20	0.5	0.5

3.2 Hypotheses and Data Analysis

In order to see if the theories of the studied literature are verified in the considered sample of respondents, three hypotheses have been designed: first, on the similitudes between made-explicit and tacit knowledge management within organizations (H1); second, on the impact of language on information assimilation (H2); and third, on the relationship between rewards and initiative in tacit knowledge management (H3). Each of these hypotheses relies on the literature presented in Sect. 2 and on survey questions (ranging from $Q1$ to $Q15$).

Hypothesis 1 (H1). *Organizations that manage made-explicit knowledge (*$Q8$*) also take into account tacit knowledge (*$Q1, Q5$*).*

→ Such an hypothesis allows to see if there is a correlation between the made-explicit and tacit knowledge management initiatives.

Hypothesis 2 (H2). *There is a relationship between the language level and the ease of assimilating correctly information in this language (*$Q15$*).*

→ Such an hypothesis allows to understand if there is a positive relationship between the language level and the ease to assimilate information in such a language.

Hypothesis 3 (H3). *There is a relationship between rewards (*$Q10$*) and employees' initiative to go further in tacit knowledge management (*$Q4$*).*

→ Such an hypothesis allows to understand if the most motivated employees are motivated by rewards or not.

Questions 1 to 4 concern the experience of the respondents in the organization regarding to the international aspect of their businesses. The following questions are about their experiences in the subsidiary in which they work, *i.e.*, at a rather local level of the extended enterprise. As the respondents are French, it is important for non-French speaking readers to be made aware that "know-how" (*savoir-faire*) is specifically identified with tacit-knowledge in French.

Looking at the maximal values (answer "strongly agree", value = 5), it is interesting to notice that the extended enterprises considered in the sample seem to

Table 3. Most frequent answers (Mode) per question. (1 = Strongly disagree, ..., 5 = Strongly agree)

Question	Mode
Q1: My company implements means to promote the sharing of knowledge and know-how between employees of different countries.	5;3
Q2: I take myself the initiative to discuss with my colleagues abroad to better do or understand my missions, as well as the overall activity of my company.	4
Q3: I can easily apply the working methods of my colleagues abroad to my position, in my local office.	4
Q4: I would like to participate in a general mentoring program at all the subsidiaries of my company.	4
Q5: My company implements means to promote the sharing of knowledge and know-how with the people in my local team.	2;5
Q6: I prefer to chat and observe others in person to learn rather than relying solely on formalized procedures.	5
Q7: I spontaneously share my know-how with my colleagues.	4
Q8: The methods and procedures related to my position are clearly written and formalized.	3
Q9: With digital tools (videoconferencing and collaborative tools), I find it more complicated to exchange and acquire knowledge from other colleagues.	3
Q10: My company encourages the transmission of know-how through rewards or another recognition system.	1
Q11: I prefer to learn in a group and an ad hoc basis rather than in a 1-to-1 relation with follow-up (mentoring or coaching type)	2
Q12: I prefer to learn through a medium (written or recorded) rather than only by observation and discussion.	2
Q13: I usually ask for advice to my colleagues or hierarchical superiors.	4
Q14: I often use another working language than French.	1
Q15: I assimilate information as well in this language than in French.	4

consider and implement tacit knowledge management initiatives: for questions 1 ("*My company implements means to promote the sharing of knowledge and know-how between employees of different countries*") and 5 ("*My company implements means to promote the sharing of knowledge and know-how with the people in my local team*"), the respondents seem to strongly agree. Initiatives are implemented both locally ($Q5$) and globally ($Q1$).

On the contrary, the minimal values (answer "strongly disagree", value = 1) have been recorded for question 10 ("*My company encourages the transmission of know-how through rewards or another recognition system*") and 14 ("*I often use another working language than French*"). Comparing these results with the previous ($Q1$ and $Q5$), as well as with the literature, there seems to be a paradox here: rewards should be a way to encourage knowledge management initiative when they are actually implemented. However, a weak mode for the question concerning rewards ($Q10$) highlights that they are not always integrated. This could imbalance the knowledge management initiatives, making them not sustainable for the organizations, as raised in some of the literature [7,13].

An analysis of the results presented in Table 3 and of detailed modes per question leads to notice that the extended enterprises in which the respondents work seem to have implemented means to promote exchanges between employees from the same ($Q5$) or different ($Q1$) countries. It is indeed the role of the organization to promote a framework fostering tacit knowledge sharing. On the other hand, it seems that there is no reward system in the implemented initiatives (Q10). Knowledge sharing is not promoted or not promoted enough to reward the employees of these extended enterprises.

3.3 Discussion of the Results

Each hypothesis will now be tested using graphical analyzes comparing the number of responses values for each question of the survey.

H_1**:** Organizations that manage made-explicit knowledge (by formalizing the procedures for example, $Q8$) also take into account tacit knowledge (by implementing means to promote knowledge sharing for example, $Q1$ and $Q5$).

The number of respondents stating that their organization have poorly formalized procedures evolve in a positive way with the number of respondents whose organizations did not implement tacit knowledge management (range from 1 to 4 in Fig. 2). On the other hand, where organizations seem to formalize quite well, the approaches to foster tacit knowledge management appear to be less present and even fall (range 4 and 5 in Fig. 2).
→ **H1 is rejected**: The relation between made-explicit and tacit knowledge management implementation does not seem to be always true.

H_2**:** There is a relationship between the language level and the ease of assimilating correctly information in this language ($Q15$).

The two factors: language level (Fig. 3a) and ease of assimilation (Fig. 3b) evolve in a similar way, the two graphs having globally the same form. Actually, the language level seems to affect the way in which individuals capture and assimilate information, and therefore potentially the tacit knowledge they create through their interpretative frameworks (see Sect. 2.1).
→ **H2 is validated**: Language level seems to impact the ease of assimilation and consequently the tacit knowledge creation.

H_3**:** There is a relationship between rewards ($Q10$) and the initiative taken by employees to go further in tacit knowledge management ($Q4$).

Here, the relationship between these two questions seems clearly reversed on the extremes (ranges 1–2 and 4–5 in Fig. 4) and confused when the respondents are more undecided (central responses with a value of 3 in Fig. 4).
→ **H3 is rejected**: Rewards do not seem to impact employees initiative in tacit knowledge management.

Table 4 presents the data if the extreme values are aggregated: answers 1 and 2, *i.e.*, "strongly disagree" and "disagree", as well as 4 and 5, *i.e.*, "strongly agree" and "agree". Such a table allows to observe the positioning of the sample in percentage for each question. The detailed frequencies of answers in collected data are presented in Appendix.

Table 4. Repartition of responses' frequency per question when negative (strongly disagree and disagree) as well as positive (strongly agree and agree) values are aggregated (see also Appendix)

Question	Strongly disagree or disagree	Undecided	Strongly agree or agree
Q1	32%	24%	44%
Q2	50%	16%	33%
Q3	41%	25%	34%
Q4	15%	23%	63%
Q5	32%	22%	47%
Q6	6%	16%	78%
Q7	9%	19%	72%
Q8	41%	33%	26%
Q9	35%	28%	36%
Q10	56%	22%	23%
Q11	40%	26%	34%
Q12	56%	31%	14%
Q13	11%	18%	70%
Q14	42%	17%	41%
Q15	27%	25%	48%

Quite a half of the employees in the sample do not take the initiative to discuss with their colleagues abroad ($Q2 : 50\%$) and their way of working is not the same ($Q3 : 41\%$). These two answers are perhaps complementary: respondents could not take initiatives to exchange because their working methods are not the same abroad. Paradoxically, the literature stated that diversity enables knowledge creation [1], a point that seems then to have been neglected by the organizations in this sample of respondents.

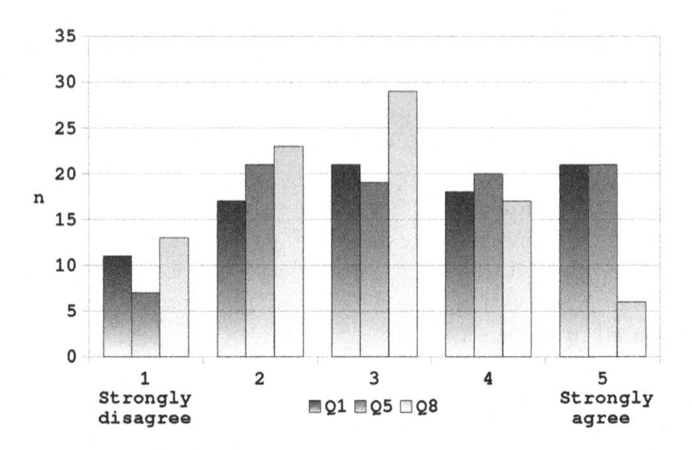

Fig. 2. Response values to questions 8 (made-explicit knowledge is managed), 1 and 5 (tacit knowledge is managed)

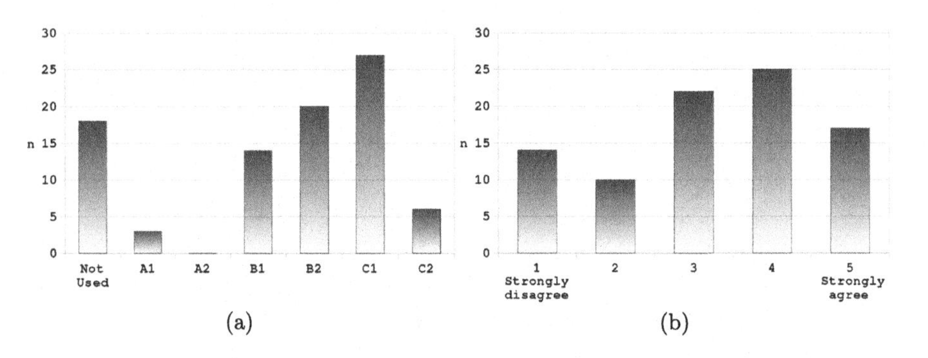

Fig. 3. Number of respondents in each language level (a) and to the question "I assimilate information as well in this language than in French" (b)

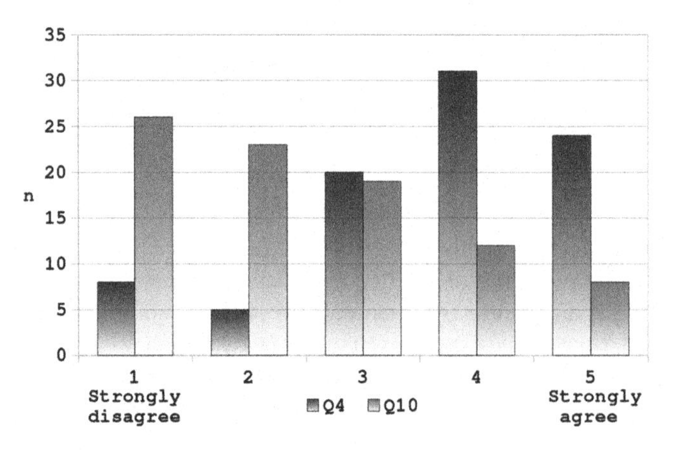

Fig. 4. Response values to questions 4 (employees' initiative in tacit knowledge management) and 10 (impact of rewards)

Moreover, one should notice that the preferred method for learning is a rather social way ($Q4, Q6, Q7, Q13 \mapsto \oplus$, whereas $Q11, Q12 \mapsto \ominus$). It is as if there were a discrepancy between possibilities, desires, and reality in terms of tacit knowledge management within extended enterprises.

4 Limits, Conclusions and Perspectives

Through a graphical analysis of the data obtained, some of the hypotheses seem to be verified within the framework of the considered sample of respondents, although some relationships are not marked enough and would need further analyses. The study presented in this paper raises a point on tacit knowledge management, which seems not considered as much as made-explicit knowledge management.

Indeed, it is complicated to quickly and easily identify tacit knowledge, being given that it is hold by individuals. Evidence of this knowledge can be identified through observation, but also the valuation of the impacts it might have on the activity. Through promotion, most often, visibility is for example given to best practices and experiences that can enrich the team as a whole. Tacit knowledge is held not only by individuals, but also collectively through a shared interpretation of events [21].

A limit of this study is that the sample is not diverse enough to identify whether certain parameters directly related to individuals (age, gender, professional experience, etc.) could have an impact on some tacit knowledge management practices. Future research will study whether voluntarily taking part in knowledge management initiatives is linked to demographics, to experience, or even to the type of job occupied by the respondent.

Moreover, there is no real comparison between what is implemented within an organization, its internationalization strategy, and the actual use made by employees. Future research could study these points to identify why certain tacit knowledge management procedures and tools are not working as expected. The question would then be to analyze personal characteristics by crossing them with organizational data. Also, the notion of measure and measurement in the questionnaire should be more addressed in future research, except for the question of reward which is indirectly linked to measures.

The interest of knowledge management being generally the adoption and adequacy of practices within organizations and relevant changes to go towards a knowledge-based management [11], future research could study such relevancy, adequacy and adoption of tacit knowledge management approaches within organizations.

Appendix

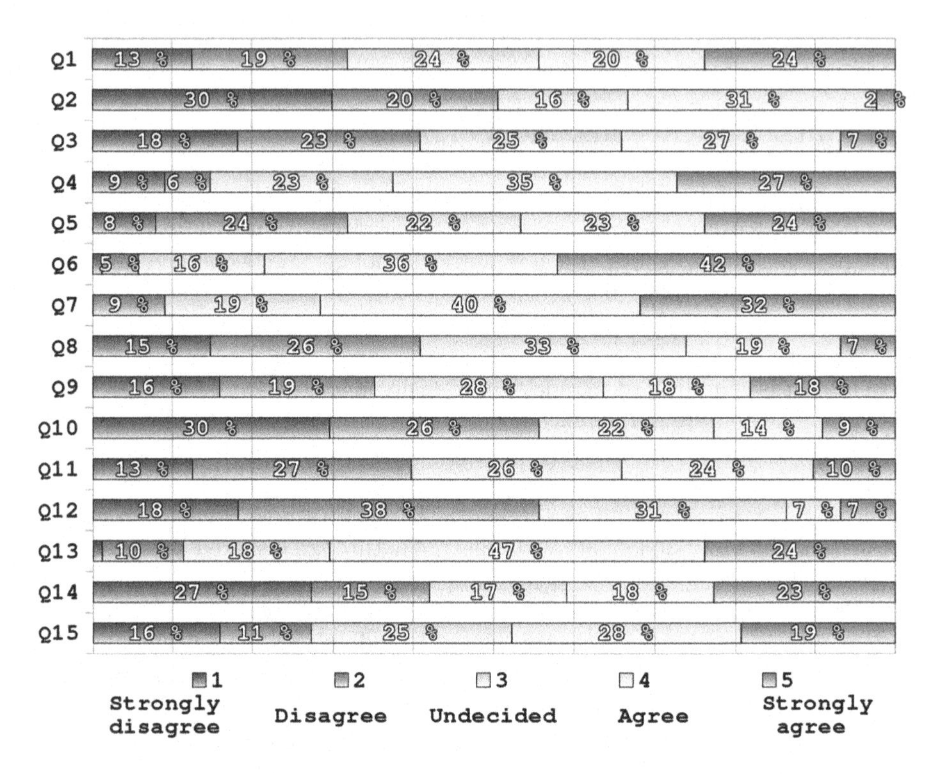

Fig. 5. Repartition of response frequencies per question

References

1. Adenfelt, M., Lagerström, K.: Enabling knowledge creation and sharing in transnational projects. Int. J. Project Manage. **24**(3), 191–198 (2006)
2. Arduin, P.-E.: On the use of cognitive maps to identify meaning variance. In: Zaraté, P., Kersten, G.E., Hernández, J.E. (eds.) GDN 2014. LNBIP, vol. 180, pp. 73–80. Springer, Cham (2014). https://doi.org/10.1007/978-3-319-07179-4_8
3. Arduin, P.E., Rosenthal-Sabroux, C., Grundstein, M.: Considering tacit knowledge when bridging knowledge management and information systems for collaborative decision-making. Inf. Syst. Knowl. Manag. 131–158 (2014)
4. Arfi, W.B.: Partage des connaissances: articulation entre management de l'innovation et management des connaissances: cas des plateformes d'innovation d'un groupe leader du secteur agroalimentaire en Tunisie. Ph.D. thesis, Université de Grenoble (2014)
5. Autissier, D., Vandangeon-Derumez, I., Vas, A.: Conduite du changement: concepts clés. Dunod (2010)

6. Berger, P., Luckmann, T.: The Social Construction of Reality: A Treatise in the Sociology of Knowledge. Open Road Media, New York (2011)
7. Bock, G.W., Kim, Y.G.: Breaking the myths of rewards: an exploratory study of attitudes about knowledge sharing. Inf. Resour. Manag. J. **15**(2), 14–21 (2002)
8. Donaldson, L., Joffe, G.: Fit-the key to organizational design. J. Organ. Des. **3**(3), 38–45 (2014)
9. Fang, Y., Neufeld, D., Zhang, X.: Knowledge coordination via digital artefacts in highly dispersed teams. Inf. Syst. J. **32**(3), 520–543 (2022)
10. Grundstein, M.: Three postulates that change knowledge management paradigm. In: New Research in Knowledge Management, Models and Methods, pp. 1–26 (2012)
11. Grundstein, M.: Knowledge-based management: the creative power of tacit knowledge in the "age of new normal". In: Recent Advances in Knowledge Management. IntechOpen (2022)
12. Grundstein, M., Arduin, P.E., Rosenthal-Sabroux, C.: From information system to information and knowledge system. In: Proceedings of the XI Conference of the Italian Chapter of the Association for Information Systems (itAIS), Genoa, Italy, 21–22 November (2014)
13. Ho, L.A., Kuo, T.H.: How system quality and incentive affect knowledge sharing. Ind. Manag. Data Syst. **113**(7), 1048–1063 (2013)
14. Janus, S.S.: Becoming a Knowledge-Sharing Organization: A Handbook for Scaling Up Solutions Through Knowledge Capturing and Sharing. World Bank Publications, Washington, DC (2016)
15. Langan-Fox, J., Code, S., Langfield-Smith, K.: Team mental models: techniques, methods, and analytic approaches. Hum. Factors **42**(2), 242–271 (2000)
16. Massingham, P.: An evaluation of knowledge management tools: Part 2-managing knowledge flows and enablers. J. Knowl. Manag. **18**(6), 1101–1126 (2014)
17. Ndlela, L.T., Du Toit, A.: Establishing a knowledge management programme for competitive advantage in an enterprise. Int. J. Inf. Manage. **21**(2), 151–165 (2001)
18. Nonaka, I., Takeuchi, H.: The Knowledge-Creating Company: How Japanese Companies Create the Dynamics of Innovation. Oxford University Press, Oxford (1995)
19. Orasanu, J., Salas, E.: Team decision making in complex environments. In: Klein, G., Orasanu, J., Calderwood, R., Zsambok, C. (eds.) Decision Making in Action: Mental Models and Methods, pp. 327–345. Ablex, Norwood (1993)
20. Panahi, S., Watson, J., Partridge, H.: Towards tacit knowledge sharing over social web tools. J. Knowl. Manag. **17**(3), 379–397 (2013)
21. Pérez-Luño, A., Alegre, J., Valle-Cabrera, R.: The role of tacit knowledge in connecting knowledge exchange and combination with innovation. Technol. Anal. Strateg. Manag. **31**(2), 186–198 (2019)
22. Pinho, I., Rego, A., e Cunha, M.P.: Improving knowledge management processes: a hybrid positive approach. J. Knowl. Manag. **16**(2), 215–242 (2012)
23. Polanyi, M.: The Tacit Dimension. Anchor Day, New York (1966)
24. Porter, M.E.: Competition in Global Industries. Harvard Business Press, Brighton (1986)
25. Tenzer, H., Pudelko, M., Zellmer-Bruhn, M.: The impact of language barriers on knowledge processing in multinational teams. J. World Bus. **56**(2), 101184 (2021)

26. Tsuchiya, S.: Improving knowledge creation ability through organizational learning. In: ISMICK 1993 Proceedings, International Symposium on the Management of Industrial and Corporate Knowledge, pp. 87–95 (1993)
27. Van Buren, M.E.: A yardstick for knowledge management. Train. Dev. **53**(5), 71–78 (1999)
28. Von Krogh, G., Ichijo, K., Nonaka, I., et al.: Enabling Knowledge Creation: How to Unlock the Mystery of Tacit Knowledge and Release the Power of Innovation. Oxford University Press on Demand (2000)

Designing a User Contextual Profile Ontology: A Focus on the Vehicle Sales Domain

Ngoc Luyen Le[1,2(✉)], Marie-Hélène Abel[1], and Philippe Gouspillou[2]

[1] Université de technologie de Compiègne, CNRS, Heudiasyc (Heuristics and Diagnosis of Complex Systems), CS 60319, 60203 Compiègne Cedex, France
ngoc-luyen.le@hds.utc.fr

[2] Vivocaz, 8 B Rue de la Gare, 02200 Mercin-et-Vaux, France

Abstract. In the digital age, it is crucial to understand and tailor experiences for users interacting with systems and applications. This requires the creation of user contextual profiles that combine user profiles with contextual information. However, there is a lack of research on the integration of contextual information with different user profiles. This study aims to address this gap by designing a user contextual profile ontology that considers both user profiles and contextual information on each profile. Specifically, we present a design and development of the user contextual profile ontology with a focus on the vehicle sales domain. Our designed ontology serves as a structural foundation for standardizing the representation of user profiles and contextual information, enhancing the system's ability to capture user preferences and contextual information of the user accurately. Moreover, we illustrate a case study using the User Contextual Profile Ontology in generating personalized recommendations for vehicle sales domain.

Keywords: Ontology · User Modeling · Knowledge base · Contextual Profile

1 Introduction

Understanding and customizing experiences for users interacting with various systems and applications is essential in today's digital landscape. To accomplish this, individual user profiles are integrated with their corresponding contextual information, resulting in the creation of a user contextual profile [14]. A user profile is a compilation of personal information and preferences about a particular user, including demographic data, interests, past behavior, and other attributes that define the user's identity within a system or application [16]. For example, the user Henri wants to buy a vehicle for professional use and another for his family with two children. Therefore, the user Henri can have two distinct profiles, each catering to different needs and preferences. While contextual information for each profile refers to the situational factors surrounding a user's interaction

I. Saad et al. (Eds.): ICIKS 2023, LNBIP 486, pp. 197–212, 2024.
https://doi.org/10.1007/978-3-031-51664-1_14

with a system or application [30]. This may include location, time, device being used, and even the user's current activity or emotional state. In the example, each of the two profiles of the user Henri has unique contextual information related to Henri's preferences, needs, and interactions within the application. Consequently, a user contextual profile combines the user profile with the contextual information specific to that profile, creating a more comprehensive understanding of the user's needs and preferences in various situations. For instance, when searching for a professional vehicle, the user contextual profile may prioritize factors such as fuel efficiency, compact size, and reliability, while for a family vehicle, it may focus on safety features, ample seating, and cargo space.

In order to effectively model user contextual profiles, utilizing ontology-based approaches can be advantageous. Ontologies are formal, structured representations of knowledge within a specific domain that enable the organization and sharing of data in a machine-readable format. Once the ontology is designed and implemented, it can be integrated into applications and systems, allowing them to reason over the user contextual profiles and infer new knowledge based on the available data [12,29,32]. This can improve the accuracy and relevance of recommendations, as well as facilitate more efficient data integration and interoperability among different applications and services.

A considerable number of studies in the field of user modeling have primarily focused on either user profiles or user contexts [25,28]. The aspects of contextual information related to each profile of the user have not been thoroughly examined. This underexplored area offers an opportunity for further research, focusing on the integration of contextual information with different profiles of the user to develop a deeper understanding of users' needs and preferences in various situations. Consequently, our work aims to address this gap by designing a User Contextual Profile Ontology (UCPO) with a focus on the vehicle sales domain. This ontology considers both user profiles and contextual information on each profile, utilizing a standard ontology development methodology. Furthermore, we provide an illustration of the use of UCPO through a case study on generating personalized recommendations in the vehicle sales domain.

The remainder of this article is organized as follows: Sect. 2 introduces works from the literature on which our approach is based. Section 3 presents our main contributions, outlining the methodology for designing a user contextual profile ontology with a focus on the vehicle sales domain. In Sect. 4, we provide a case study on personalized recommendations based on our developed ontology. Finally, we conclude and discuss the perspectives.

2 Related Work

In this section, we investigate the modeling and integration of user profiles and contextual information for specific profiles, as well as examine various approaches to ontology design. The related works' analysis will inform and support the development of a user contextual profile ontology.

2.1 Ontological User Modeling

Ontologies have proven to be highly effective in modeling user profiles and contexts. They provide a thorough depiction of a particular domain of interest, simplifying browsing and query refinement. Specifically, ontologies have been observed to outperform other methods in user modeling when compared to other methods utilized [26]. Ontological user modeling refers to the use of ontologies for representing and managing user profiles, preferences, and contextual information in a structured and machine-readable format. User-related information such as profiles, preferences, and context are represented using concepts, relationships, and attributes within a formal, hierarchical structure. This structured representation enables more effective processing, reasoning, and inferencing by computer systems.

Ontological user modeling has previously been investigated across numerous research areas, demonstrating its adaptability and potential to enhance user experiences in diverse domains such as e-learning [6], e-commerce [4], recommender systems [8–11], personalized web services [24], and context-aware applications [22], among others [5,13,17]. Specifically, Golemati et al. [16] designed an ontology that incorporates concepts and properties essential for modeling user profiles. Their model predominantly focuses on static user characteristics and provides a foundation for the development of a more general, comprehensive, and extensible user model. Adewoyin et al. [1] proposed a versatile user modeling architecture for smart environments. They represented the characteristics of users using the behavior concept, which encompasses all relevant aspects that can aid in effective behavioral modeling and monitoring. Sutterer et al. [28] put forth a user profile ontology designed specifically to depict situation-dependent sub-profiles. This ontology can be utilized by context-aware adaptive service platforms for mobile communication and information services, facilitating automatic and situation-dependent personalization of such services. Skillen et al. [25] developed a User Profile Ontology for personalizing context-aware applications in mobile environments. Their focus lies on user behavior and characterizing the needs of users for context-aware applications. Contrasting with the work in [16], their ontological user modeling approach emphasizes dynamic components for application usage.

The works mentioned highlight numerous advantages of ontological user modeling, which is capable of capturing both static and dynamic user characteristics related to permanent, temporary information, and user evolution. Moreover, user contexts play a crucial role in user modeling, contributing significantly to the creation of personalized and adaptive systems. Most of these studies integrate user contexts and user profiles. Nonetheless, the contextual information can be dependent on the profile and its primary objectives. Therefore, it is necessary to distinguish contextual information for the user from that for the profile clearly. The development of a user contextual profile ontology is essential for clearly organizing and distinguishing the contextual information related to users and their respective profiles. In the following section, we will examine ontology development methodologies to achieve this objective.

2.2 Ontology Development Methodology

Ontology development methodologies provide a structured and systematic approach to the creation, maintenance, and evolution of ontologies. These methodologies are specifically designed to ensure the quality, coherence, and practicality of the resulting ontology, while also promoting its consistency and usability. By employing ontology development methodologies, researchers and practitioners can effectively enhance the overall performance and applicability of the ontologies they create, making them more reliable and valuable for their intended use cases.

Various ontology development methodologies have been proposed in the literature, each featuring its own unique set of guidelines and principles. Some widely adopted methodologies include: Methontology [3], Ontology Development 101 (OD101) [19], NeON Methodology [27], Unified Process for Ontology Building (UPON) [2], Simplified Agile Methodology for Ontology Development (SAMOD) [20], Modular Ontology Modeling (MoMo) [23], and others. Firstly, Methontology is one of the earliest and most widely used methodologies. It covers the entire ontology development process, including specification, conceptualization, formalization, integration, implementation, and maintenance, while emphasizing the importance of documentation throughout the development process. OD101 then is a methodology designed to guide ontology engineering processes. Focusing on iterative development, OD101 provides step-by-step guidelines and best practices for constructing ontologies, ensuring a systematic and efficient approach to ontology creation. The NeON Methodology next is designed to create ontology networks, rather than singular ontologies. This approach emphasizes the reuse and re-engineering of existing ontologies, fostering collaboration between ontology engineers and domain experts to facilitate a more comprehensive and efficient development process. UPON is a methodology that incorporates concepts from the Unified Modeling Language (UML) and the Rational Unified Process (RUP). This approach adheres to an iterative and incremental process, encompassing four development phases: inception, elaboration, construction, and transition. It provides a structured framework for creating and refining ontologies, ensuring consistent and systematic ontology development. SAMOD is a methodology that integrates agile software development principles into ontology engineering. It provides a flexible, iterative, and adaptable framework, fostering collaboration among ontology engineers, domain experts, and end-users. Employing iterative development cycles, SAMOD encompasses planning, design, implementation, and evaluation stages, facilitating ongoing feedback and refinements throughout the development process. Finally, MoMo is a methodology that creates and manages modular ontologies, allowing for independent development, maintenance, and reuse of smaller, more manageable modules. It facilitates collaboration and focuses on specific parts of the ontology. MoMo involves module identification, design, integration, and evaluation, resulting in a more efficient and maintainable ontology development process. In general, the use of an ontology development methodology can lead to more efficient ontology development, better ontology reuse, easier ontology maintenance, and more effective collaboration between ontology engineers and domain experts. Furthermore, the adoption of a standardized ontology development methodology can improve the interoperability and compatibility of ontologies, enabling more effective data integration and knowledge sharing among different systems and applications.

The selection of an appropriate ontology development methodology depends on several factors, such as the complexity of the ontology, the domain of application, the availability of resources, and the expertise of the ontology engineers. In this work, we have opted for SAMOD to build a user contextual profile ontology due to its flexibility, adaptability, and emphasis on collaboration among ontology engineers, domain experts, and end-users. The iterative and incremental process of SAMOD, combined with its ongoing feedback and refinements, aligns with our goal of creating a comprehensive and adaptable ontology that can evolve over time. Hence, we will describe the development process of our user contextual profile ontology in the next section.

3 Methodology for Developing a User Contextual Profile Ontology

The development of a User Context Profile Ontology is a complex process that requires a structured and systematic approach. SAMOD methodology is based on agile software development principles, enabling the efficient development of a comprehensive and adaptable UCPO. Therefore, the UCPO can be made dynamic through continuous updates and modifications to the ontology based on changes in user profiles and contextual information. This can be achieved by incorporating an agile approach to ontology development, where the ontology is built and adapted incrementally, with regular feedback and updates from users. In this section, we describe the phases of SAMOD, including kickoff, design and implementation, and test and evaluation, which serve as a framework for designing the UCPO.

3.1 Kickoff Phase

The main objective of the UCPO is to establish a standardized representation of user information and contextual information for each user profile. This ontological user modeling can facilitate the development of personalized systems and services, including personalized information retrieval, adaptive user interfaces, personalized recommendations, and other applications. Defining a standardized format for user information, context, and profile can improve the interoperability of the UCPO and enable it to adapt to the changing needs and preferences of the users. The scope of the UCPO is broad, covering a wide range of user characteristics and environmental factors that can influence user behavior and preferences. This includes, but is not limited to, demographics, psychographics, device and network context, physical context, social context, and temporal context. The ontology will be designed to be flexible and adaptable, allowing for the incorporation of new or changing user characteristics and environmental factors.

To develop the UCPO, we consult a variety of information sources and references such as academic articles, conference proceedings, textbooks, and online resources related to ontology engineering, user modeling, and personalization. In selecting the information sources, it is important to consider their relevance to the scope and goals of the UCPO, as well as the quality and credibility of the sources. We manually reviewed and analyzed the relevant literature to identify the key concepts and terms related to user context and personalization. Additionally, consulting domain experts, ontology engineers, and end-users can provide valuable insights into the specific user characteristics and contextual factors that should be incorporated into the ontology.

Table 1. Examples of informal competency questions tables.

ID	Questions
CQ1	What is demographic information of the user ?
CQ2	What is the user's preferred vehicle type?
CQ3	What is the user's budget for a vehicle purchase?
CQ4	Which particular vehicle models are favored by the user?
CQ5	What is the user's driving environment?
CQ6	What is the user's preferred vehicle brand?
CQ7	What are the primary use cases for a particular vehicle model?
CQ8	What is the user's preferred vehicle transmission type

The ontology requirements can be formulated by using informal competency questions. The set of competency questions helps to identify and define the key concepts, relationships, and attributes that should be included in the ontology. Competency questions serve as a guide for ontology development and help to ensure that the resulting ontology accurately represents the user context and profile information required for personalization. As in the Table 1, we illustrate some informal competency questions related to define the key characteristics of the user contextual profile in the vehicle sales domain, which can then be used to inform the development of the UCPO.

Defining the goal, scope, information sources, and competency questions helps to establish a shared understanding of the primary requirements of the UCPO between domain experts and ontology engineers. In the next section, we will focus on the main phases of the development process, where the key components of the UCPO are designed and implemented to obtain a comprehensive and adaptable version.

3.2 Design and Implementation Phase

Considered as the core of the ontology engineering process, the design and implementation phase consists of several iterations of ontology design, implementation, and testing until a comprehensive and adaptable ontology is achieved. The design phase begins by identifying the key concepts and relationships that should be included in the ontology based on the competency questions formulated in the previous phase. SAMOD prescribes an iterative process that aims to build the final model through a series of small steps. One key component of this process is the use of concept modelets, which are standalone models that describe particular aspects of the domain under consideration. These modelets are used to provide an initial conceptualization without being constrained by the current model available after the previous iteration of the process.

We have organized the development of the UCPO into two separate modelets. The first modelet focuses on representing mostly static and permanent user characteristics. The second modelet presents both static and dynamic information about the user profile and contextual factors that affect the user and their profile. As shown in Fig. 1, the first modelet is responsible for organizing user demographic and social information, such as *Gender*, *Age*, *address*, *occupation*, *Income*, *Language*, *NumberOfChildren*, *MaritalStatus*, *Education*, and more. *Gender* refers to a person's biological identity as male or female. *Age* refers to the user's age, typically measured in years. *Address* refers to a person's physical location, which can be used to gather contextual information about the user's environment. *Occupation* refers to a person's job or profession, which can provide insights into their daily routine and transportation needs. *Income* refers to a person's financial status, which can affect their purchasing power and preferences. *Language* refers to a person's preferred language for communication, which can influence the type of content and recommendations that they receive. *NumberOfChildren* refers to the number of children a person has, which can impact their vehicle preferences and needs. *MaritalStatus* refers to a person's current marital status, which can provide insights into their family structure and needs. *Education* level refers to the highest level of education a person has completed, which can be a factor in determining their interests and preferences. In general, these types of information have been well-organized into classes in many previous works, allowing for the description of user information in a categorical manner. In our research, we utilized the work of Bermudez et al. [15] for the basic classes of user information. Moreover, we have created a class named *PersonalProfile*, which acts as a superclass for the classes related to the demographic and social information. The *PersonalProfile* class can contains attributes such as first name, last name of the user.

The second modelet focuses on the representation of the structure and relationships between a user, their user profile, and their user context. User context refers to the circumstances, environment, and situation in which a user interacts with a system or application. It includes a wide range of factors such as the user's location, time, device, preferences, goals, and social context, among others. User context is crucial because it can have a significant impact on the user's behavior, preferences, and needs. As shown in Fig. 2, we investigated four particular contextual information about the user, including *Time*, *Location*, *Activity*, and *Device* classes. In our design, we created two subclasses of *Context* class: the *UserContext* class that describes general contextual information about the

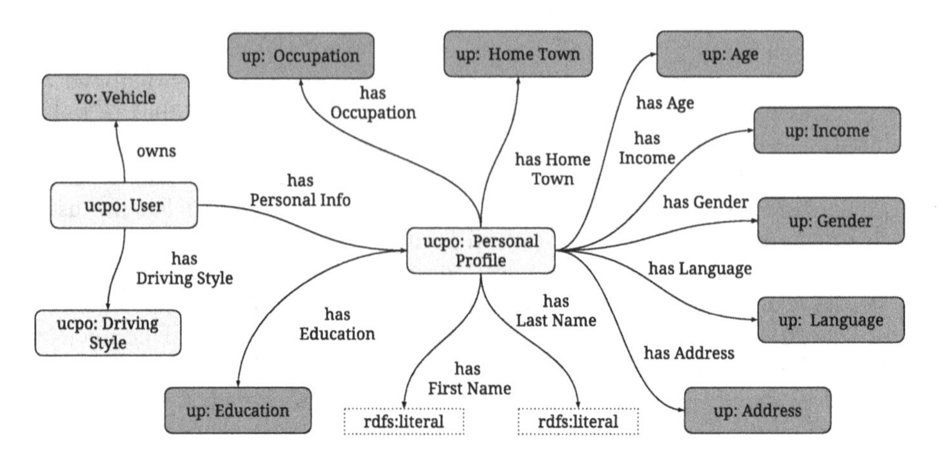

Fig. 1. Dedicated ontology section for demographic and social information of users, composed of the User Profile Ontology (highlighted in red boxes with *up* prefix), User Contextual Profile Ontology (highlighted in blue boxes with *ucpo* prefix), and the vehicle domain ontology (highlighted in green boxes with *vo* prefix). (Color figure online)

user and the *ProfileContext* class, which describes contextual information for each user profile. Furthermore, we have provided the ability for each user to declare their preferences and create various types of profiles that align with their intended purpose while interacting with the system or application. To structure this type of information, we have created two classes: *Preference* and *UserProfile*.

In order to address the needs and preferences of users in the vehicle sales domain, we have created a set of preferences that relate to the user's desired vehicle. This set of preferences includes several classes such as *VehicleType*, *RouteType*, *Mileage*, *Color*, *NumberOfPlaces*, *State*, *Budget*, and *Brand*. The *VehicleType* class refers to the type of vehicle the user is interested in, such as sedan, SUV, or truck. The *RouteType* class describes the user's preferred driving route, such as highway or city streets. The *Mileage* class indicates the user's desired mileage range. The *Color* class refers to the user's preferred color for their vehicle. The *NumberOfPlaces* class indicates the number of seats or passengers the user wants in their vehicle. The *State* class represents the vehicle state where the user intends to purchase. The *Budget* class refers to the user's budget for the vehicle purchase. The *Brand* class represents the user's preferred vehicle brand.

It is important to note that depending on the specific domain and application, additional preferences may be necessary to fully capture the user's needs and preferences. Therefore, sub-classes of *Preference* class can be created and customized based on the specific requirements of the vehicle sales domain and the user base. By incorporating these preferences into the user contextual profile ontology, we can provide a more personalized and tailored user experience.

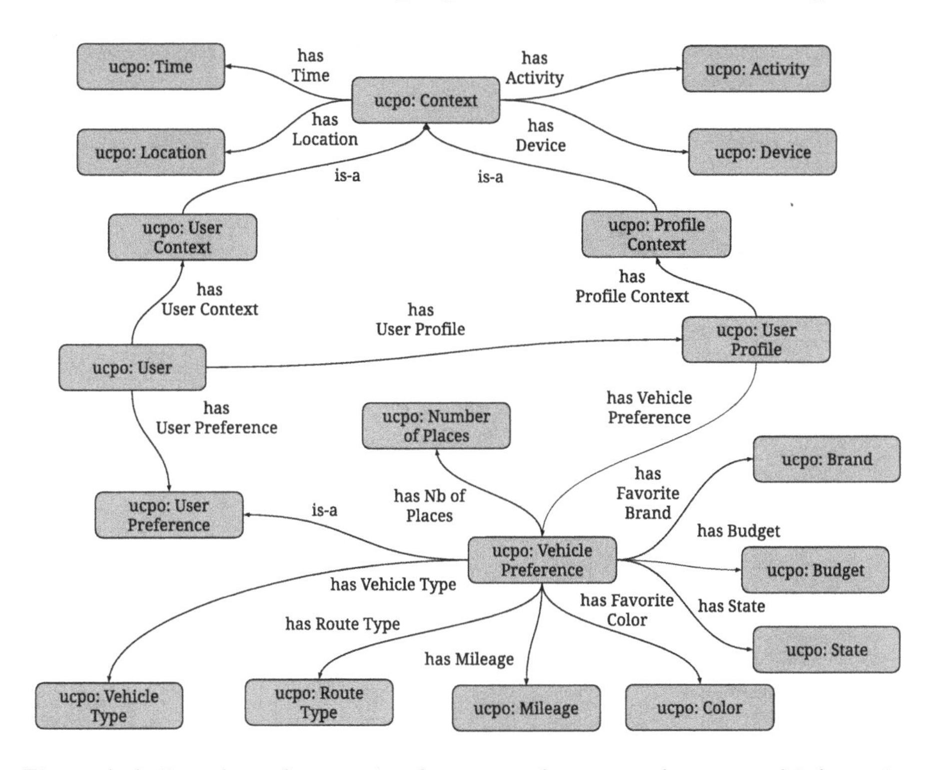

Fig. 2. A dedicated ontology section for user preferences and contextual information of user and each particular profile.

Using the conceptual model, we can outline the major classes, attributes, and relationships between them (as illustrated in Figs. 1 and 2). This model is then refined through multiple iterations, taking feedback from stakeholders and end-users into consideration to ensure that it accurately represents the domain knowledge and requirements.

The implementation of the UCPO involves the process of actually creating and deploying the ontology in a system or application. This requires converting the conceptual model of the ontology into a computer-readable format using ontology languages such as OWL or RDF. As shown in Fig. 3, we have implemented the UCPO using the Web Ontology Language (OWL) with the support of the Protégé-OWL editor [18].

After the design and implementation phase, the ontology undergoes further refinement through multiple rounds of testing and feedback until a final version is achieved. The ontology is then evaluated to ensure it meets the intended goals and requirements and can be integrated with existing systems and applications. In the next section, we present the test and evaluation phase of the ontology.

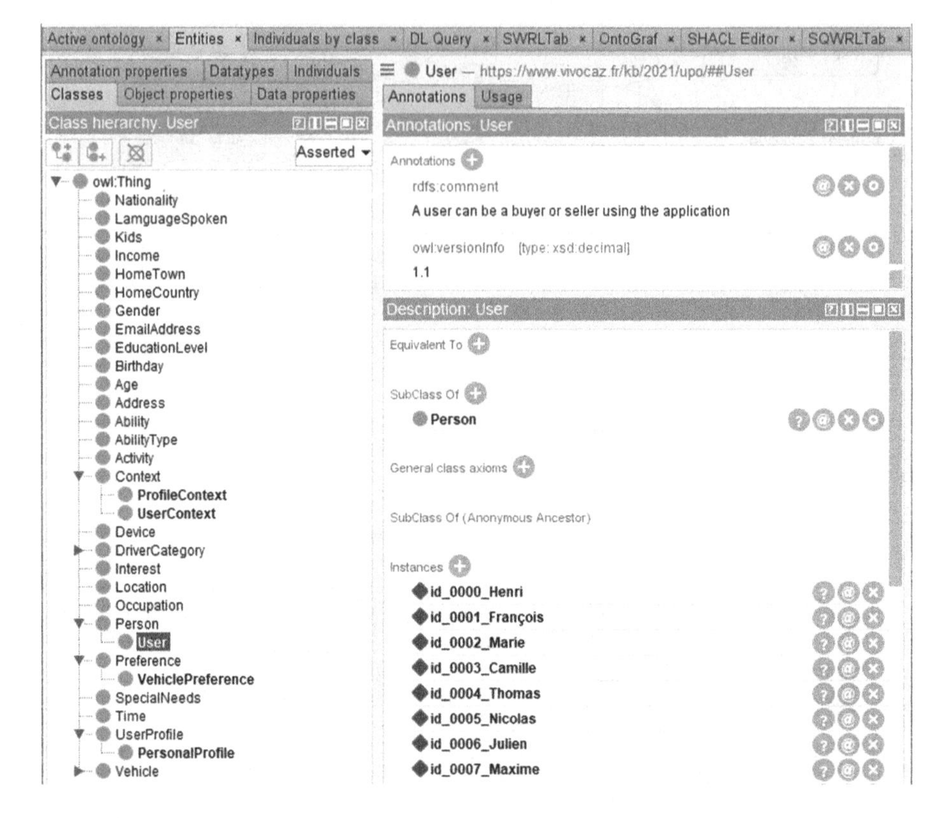

Fig. 3. A snapshot of the User Contextual Profile Ontology in the Protégé-OWL editor

3.3 Test and Evaluation Phase

In the final phase of the UCPO development, we focus on testing and evaluating the produced ontology. The tests can be categorized into a model test, a data test, and query tests. To perform the model test, we can employ evaluation metrics such as OOPS! [21], OntoQA [31], and OntoMetrics [7] to verify the overall consistency of the developed ontology. The data test consists of checking the validity of the model after populating it with instance triplets. Finally, for query tests, informal competency questions must be transformed into SPARQL queries to ensure that the expected answers are obtained. The ontology model must be adjusted until all tests are successful. This section presents our method for conducting these types of tests to ensure the ontology's adaptation.

In our work, we evaluated the quality of our ontology by examining its structure. Specifically, we observed several metrics using the OntoMetrics framework. Table 2 presents the results from the analysis of OntoMetrics with base metrics and schema metrics. The base metrics consist of simple metrics that count the number of axioms, classes, object properties, data properties, and individual instances. These metrics provide statistics about the ontology with regard to the quantity of elements present within it. Based on these statistics, we can conclude that our ontology is a lightweight knowledge graph that can be adopted in different application architectures. Moreover, the Description Logic (DL) expressivity of the ontology is $\mathcal{ALH}(D)$, which indicates the DL variant that the ontology belongs to. $\mathcal{ALH}(D)$ refers to attribute language $\mathcal{AL}$ with role hierarchy ($\mathcal{H}$).

Table 2. Results achieved from applying various metrics to UCPO.

Base Metrics	
Class count	38
Object property count	27
Data property count	16
Properties count	43
Individual count	159
SubClassOf axioms count	22
Object property domain axioms count	28
Object property range axioms count	33
DL expressivity	$\mathcal{ALH}(D)$
Schema Metrics	
Attribute richness (AR)	0.421053
Inheritance richness (IR)	0.578947
Relationship richness (RR)	0.55102
Axiom/class ratio	30.552632
Class/relation ration	0.77551

In schema metrics, the evaluation of the ontology focuses on the measure of attribute, inheritance, and relationship richness [31]. Particularly, the attribute richness (AR) is considered as the average number of attributes per class and is computed as follows:

$$AR = \frac{|NA|}{|C|} \tag{1}$$

where NA is the number attributes for all classes and C denotes the number of classes. The high score of the attribute richness indicates the high quality of ontology design and the amount of information belong to instances. The inheritance richness (IR) is defined as the average number of sub-classes per class and is computed as follows:

$$IR = \frac{|H|}{|C|} \tag{2}$$

where H is the sum of the number of inheritance relationships. The inheritance richness allows describing the distribution of information on different levels of the ontology. The relationship richness (RR) is defined as the percentage of the relationships between classes and is computed as follows:

$$RR = \frac{|P|}{|H| + |P|} \tag{3}$$

where P the sum of the number of non-inheritance relationships. The inheritance richness is used to measure the diversity of the relationship's types in the ontology. Moreover, the other metrics such as axiom/class ratio and class/relation ratio demonstrate the ratio between axioms and classes, and classes-relations. Based on the explanation of metrics, these measured results in Table 2 indicate that our proposed ontology is balanced between a horizontal (or shallow) ontology and vertical (or deep) ontology.

To conduct a data test for the UCPO, we need to check the validity of the model after populating it with instance triplets. For example, let's assume we have created instances for Henri's two profiles: his professional profile and his family profile. We can populate these instances with data such as his occupation and work location for his professional profile, and his number of children and preferred family activities for his family profile. This ensures that the UCPO can capture and express this information in a structured and consistent manner.

In order to conduct a query test on the UCPO, we can convert the set of informal competency questions, as shown in Table 1, into SPARQL queries to ensure that the expected answers are obtained. The queries should be constructed in a way that enables the retrieval of relevant information from the ontology, utilizing the appropriate classes, properties, and instances. For instance, if the competency question is "What is the user's preferred vehicle brand?", a suitable SPARQL query can be formulated as follows:

```
PREFIX vo: <http://vivocaz.fr/vo/ns#>
PREFIX ucpo: <http://vivocaz.fr/ucpo/ns#>
SELECT ?user ?brand
WHERE {
    ?user ucpo:hasUserProfile ?userProfile.
    ?userProfile ucpo:hasVehiclePreference ?userVehiclePreference .
    ?userVehiclePreference upo:hasFavoriteBrand ?brand .
} ORDER BY ?user LIMIT 10
```

Listing 1.1. A SPARQL query expression used to search for the first 10 users and their favorite brands.

The test and evaluation phase ensures that the UCPO is consistent, valid, and effective in achieving its intended purpose. By conducting model tests, data tests, and query tests, we can evaluate the overall structure of the ontology, the validity of its populated instances, and its ability to accurately answer competency questions. In the next section, we will explore an application of the UCPO through a case study focused on generating personalized recommendations.

4 Case Study - Personalized Recommendations

To demonstrate the application of the UCPO, we conducted a case study on personalized recommendations in the vehicle sales domain. Using the ontology, we were able to capture the user's contextual information, preferences, and profiles to provide tailored recommendations for vehicle purchases. Specifically, scenarios were considered where a user created a profile specifying their preferred *vehicle type*, *route type*, *mileage*, *color*, *number of seats*, *location*, *brand*, and *budget*. Additionally, the user's profile context included their *activity*, *time of day*, and *device*. This information was used to provide personalized recommendations for vehicle purchases that catered to the user's individual requirements and preferences.

Louis and Pierre are two potential buyers of used vehicles who have registered on a vehicle sales application and set up their profiles. Louis has specified his preferences for *vehicle type*, *route type*, *mileage*, *color*, *number of seats*, *location*, *brand*, and *budget*, while expressing a particular interest in sedan models. On the other hand, Pierre has emphasized the importance of safety features as a top priority due to frequent travel with his family.

The application integrates the UCPO to capture their contextual information and preferences. By leveraging Louis's contextual information and preferences, including his activity or location such as the Peugeot models he has previously liked, the application generates personalized recommendations for vehicle purchases that cater to his specific needs and preferences. The application recommends models such as *Peugeot* 206, *Peugeot* 207, or *Peugeot* 208 that Louis may like, based on his stated preferences and his profile context.

On the other hand, the application captures Pierre's contextual information and preferences, including his activity or location such as the fact that he previously owned a *Toyota* and expressed interest in hybrid vehicles, to provide tailored recommendations. By leveraging this information, the application's recommendation engine suggests several vehicles that fit Pierre's criteria, including a *Toyota RAV*4 *Hybrid*, a $Honda\ CR-V\ Hybrid$, and a *Lexus UX Hybrid*. The system takes into account factors such as safety ratings, fuel efficiency, and cost to make the best relevant recommendations for Pierre.

Consider the example of Henri, who was mentioned in the introduction section, and has two distinct profiles: one as a professional profile and the other as a family profile. It is essential to provide personalized recommendations for Henri that meet his specific needs and preferences in each profile. By leveraging the UCPO, which captures user profiles and contextual information of each profile, relevant recommendations can be provided that cater to each corresponding profile. Specifically, for Henri's professional profile who has shown interest in Renault models, vehicles such as *Renault Megane* and *Renault Talisman*, which have spacious trunks, advanced safety features, and good fuel efficiency, could be recommended, as Henri may require frequent business trips and the recommendations match the contextual information of his activities on the application. For Henri's family profile, who has indicated interest in SUV vehicle types, vehicles such as *Koleos SUV* and *Renault Scenic*, which have ample space and storage and rear-seat airbags for passengers, could be recommended as Henri has shown a preference for these models in the family profile on the applications.

These scenarios illustrate how the UCPO ontology can be utilized to capture contextual information and preferences to generate personalized recommendations that cater to the specific needs and preferences of each user. By analyzing and understanding user information, systems or applications can provide more accurate and effective recommendations, as demonstrated in various works [9,13]. By leveraging the UCPO in this way, it is fully potential to develop a system or application that provides more personalized recommendations tailored to the user's unique needs, preferences, and contextual information.

5 Conclusion and Perspectives

In this paper, we have presented the design and development process of the User Contextual Profile Ontology with a focus on the vehicle sales domain. This work established the bridge for the gap in the representation of user profile and user context by designing a user contextual profile ontology that considers both user profiles and contextual information on each profile. Thereby, the UCPO serves as a structural framework for standardizing the representation of user profiles and contextual information. It is worth emphasizing that the development of the UCPO follows the main phases of ontology engineering, ensuring its consistency, validity, and effectiveness in achieving its intended purpose. The iterative development process of the UCPO involved testing and evaluation phases, which allowed for assessing the ontology's overall structure, validity of populated instances, and ability to answer competency questions accurately. The case study presented illustrates the use of the UCPO in generating personalized recommendations in the vehicle sales domain. By leveraging the ontology's ability to capture user preferences and requirements accurately, the system can provide tailored recommendations that cater to the specific needs, preferences, and context of each user. In future work, we plan to extend the ontology to include additional information sources, such as user reviews and social media feeds, to enhance the accuracy of personalized recommendations. Finally, we aim to collaborate with industry partners to integrate the UCPO into commercial applications and evaluate its effectiveness in real-world settings.

Acknowledgment. This work was funded by the French Research Agency (ANR) and by the company Vivocaz under the project France Relance - preservation of R&D employment (ANR-21-PRRD-0072-01).

References

1. Adewoyin, O., Wesson, J., Vogts, D.: User modelling to support behavioural modelling in smart environments. In: 2022 3rd International Conference on Next Generation Computing Applications (NextComp), pp. 1–6. IEEE (2022)
2. De Nicola, A., Missikoff, M., Navigli, R.: A proposal for a unified process for ontology building: UPON. In: Andersen, K.V., Debenham, J., Wagner, R. (eds.) DEXA 2005. LNCS, vol. 3588, pp. 655–664. Springer, Heidelberg (2005). https://doi.org/10.1007/11546924_64

3. Fernández-López, M., Gómez-Pérez, A., Juristo, N.: Methontology: from ontological art towards ontological engineering. In: Proceedings of the Ontological Engineering AAAI 1997 Spring Symposium Series. American Association for Artificial Intelligence (1997)
4. He, S., Fang, M.: Ontological user profiling on personalized recommendation in e-commerce. In: 2008 IEEE International Conference on e-Business Engineering, pp. 585–589. IEEE (2008)
5. Kourtiche, A., mohamed Benslimane, S., Hacene, S.B.: OUPIP: ontology based user profile for impairment person in dynamic situation aware social networks. Int. J. Knowl.-Based Organ. (IJKBO) **10**(2), 12–34 (2020)
6. Kritikou, Y., Demestichas, P., Adamopoulou, E., Demestichas, K., Theologou, M., Paradia, M.: User profile modeling in the context of web-based learning management systems. J. Netw. Comput. Appl. **31**(4), 603–627 (2008)
7. Lantow, B.: Ontometrics: putting metrics into use for ontology evaluation. In: Proceedings of the International Joint Conference on Knowledge Discovery, Knowledge Engineering and Knowledge Management, IC3K 2016, pp. 186–191. SCITEPRESS - Science and Technology Publications, Lda, Setubal, PRT (2016). https://doi.org/10.5220/0006084601860191
8. Le, N.L., Abel, M.H., Gouspillou, P.: Apport des ontologies pour le calcul de la similarité sémantique au sein d'un système de recommandation. In: 33ème Journées Francophones d'Ingénierie des Connaissances (IC 2022), pp. 189–198 (2022)
9. Le, N.L., Abel, M.-H., Gouspillou, P.: Towards an ontology-based recommender system for the vehicle sales area. In: Troiano, L., Vaccaro, A., Kesswani, N., Díaz Rodriguez, I., Brigui, I. (eds.) ICDLAIR 2021. LNNS, vol. 441, pp. 126–136. Springer, Cham (2022). https://doi.org/10.1007/978-3-030-98531-8_13
10. Le, N.L., Abel, M.H., Gouspillou, P.: A personalized recommender system based-on knowledge graph embeddings. In: Hassanien, A.E., et al. (eds.) AICV2023. LNDECT, vol. 164, pp. 368–378. Springer, Cham (2023). https://doi.org/10.1007/978-3-031-27762-7_35
11. Le, N.L., Abel, M.H., Gouspillou, P.: Improving semantic similarity measure within a recommender system based-on RDF graphs. In: Rocha, Á., Ferrás, C., Ibarra, W. (eds.) ICITS 2023. LNNS, vol. 691, pp. 463–474. Springer, Cham (2023). https://doi.org/10.1007/978-3-031-33258-6_42
12. Le, N.L., Tireau, A., Venkatesan, A., Neveu, P., Larmande, P.: Development of a knowledge system for big data: case study to plant phenotyping data. In: Proceedings of the 6th International Conference on Web Intelligence, Mining and Semantics, pp. 1–9 (2016)
13. Le, N.L., Zhong, J., Negre, E., Abel, M.H.: Constraint-based recommender system for crisis management simulations. In: The 56th Hawaii International Conference on System Sciences (2023)
14. Li, S., Abel, M.-H., Negre, E.: Using user contextual profile for recommendation in collaborations. In: Visvizi, A., Lytras, M.D. (eds.) RIIFORUM 2019. SPC, pp. 199–209. Springer, Cham (2019). https://doi.org/10.1007/978-3-030-30809-4_19
15. Maria, B., Payam, B., Sefki, K.: User profile ontology (2015). http://iot.ee.surrey.ac.uk/citypulse/ontologies/up
16. Maria, G., Akrivi, K., Costas, V., George, L., Constantin, H.: Creating an ontology for the user profile: method and applications. In: Proceedings AI* AI Workshop RCIS, pp. 407–412 (2007)
17. Muñoz, H.J.M., Cardinale, Y.: Gente: an ontology to represent users in the tourism context. In: 2021 XLVII Latin American Computing Conference (CLEI), pp. 1–10. IEEE (2021)

18. Musen, M.A.: The protégé project: a look back and a look forward. AI Matters **1**(4), 4–12 (2015)
19. Noy, N.F., McGuinness, D.L., et al.: Ontology development 101: a guide to creating your first ontology (2001)
20. Peroni, S.: A simplified agile methodology for ontology development. In: Dragoni, M., Poveda-Villalón, M., Jimenez-Ruiz, E. (eds.) OWLED/ORE -2016. LNCS, vol. 10161, pp. 55–69. Springer, Cham (2017). https://doi.org/10.1007/978-3-319-54627-8_5
21. Poveda-Villalón, M., Gómez-Pérez, A., Suárez-Figueroa, M.C.: Oops!(ontology pitfall scanner!): an on-line tool for ontology evaluation. Int. J. Semant. Web Inf. Syst. (IJSWIS) **10**(2), 7–34 (2014)
22. Rimitha, S., Abburu, V., Kiranmai, A., Marimuthu, C., Chandrasekaran, K.: Improving job recommendation using ontological modeling and user profiles. In: 2019 Fifteenth International Conference on Information Processing (ICINPRO), pp. 1–8. IEEE (2019)
23. Shimizu, C., Hammar, K., Hitzler, P.: Modular ontology modeling. Semant. Web 1–31 (2021)
24. Sieg, A., Mobasher, B., Burke, R.: Web search personalization with ontological user profiles. In: Proceedings of the Sixteenth ACM Conference on Information and Knowledge Management, pp. 525–534 (2007)
25. Skillen, K.-L., Chen, L., Nugent, C.D., Donnelly, M.P., Burns, W., Solheim, I.: Ontological user profile modeling for context-aware application personalization. In: Bravo, J., López-de-Ipiña, D., Moya, F. (eds.) UCAmI 2012. LNCS, vol. 7656, pp. 261–268. Springer, Heidelberg (2012). https://doi.org/10.1007/978-3-642-35377-2_36
26. Sosnovsky, S., Dicheva, D.: Ontological technologies for user modelling. Int. J. Metadata Semant. Ontol. **5**(1), 32–71 (2010)
27. Suárez-Figueroa, M.C., Gómez-Pérez, A., Fernández-López, M.: The NeOn methodology for ontology engineering. In: Suárez-Figueroa, M.C., Gómez-Pérez, A., Motta, E., Gangemi, A. (eds.) Ontology Engineering in a Networked World, pp. 9–34. Springer, Heidelberg (2012). https://doi.org/10.1007/978-3-642-24794-1_2
28. Sutterer, M., Droegehorn, O., David, K.: UPOS: user profile ontology with situation-dependent preferences support. In: First International Conference on Advances in Computer-Human Interaction, pp. 230–235. IEEE (2008)
29. Sutterer, M., Droegehorn, O., David, K.: User profile selection by means of ontology reasoning. In: 2008 Fourth Advanced International Conference on Telecommunications, pp. 299–304. IEEE (2008)
30. Tamine-Lechani, L., Boughanem, M., Daoud, M.: Evaluation of contextual information retrieval effectiveness: overview of issues and research. Knowl. Inf. Syst. **24**, 1–34 (2010)
31. Tartir, S., Arpinar, I.B., Sheth, A.P.: Ontological evaluation and validation. In: Poli, R., Healy, M., Kameas, A. (eds.) Theory and Applications of Ontology: Computer Applications, pp. 115–130. Springer, Dordrecht (2010). https://doi.org/10.1007/978-90-481-8847-5_5
32. Wang, X.H., Zhang, D.Q., Gu, T., Pung, H.K.: Ontology based context modeling and reasoning using OWL. In: Proceedings of the Second IEEE Annual Conference on Pervasive Computing and Communications Workshops, pp. 18–22. IEEE (2004)

Management of Implicit Ontology Changes Generated by Non-conservative JSON Instance Updates in the τJOWL Environment

Safa Brahmia[1], Zouhaier Brahmia[1](✉), Fabio Grandi[2], and Rafik Bouaziz[1]

[1] MIRACL Laboratory, Faculty of Economics and Management, University of Sfax, Sfax, Tunisia
safa.brahmia@gmail.com, zouhaier.brahmia@fsegs.rnu.tn, rafik.bouaziz@usf.tn

[2] Department of Computer Science and Engineering (DISI), University of Bologna, Bologna, Italy
fabio.grandi@unibo.it

Abstract. τJOWL (Temporal OWL 2 from Temporal JSON) is a framework we proposed to allow users of Big Data projects to automatically create, with the closed world assumption (CWA), a temporal OWL 2 ontology from time-varying JSON-based Big Data. Such an ontology, providing the semantics to the data, facilitates complex tasks like Big Data querying, analytics and reasoning, in an environment that supports temporal versioning at both data instance and schema levels. Update operations on temporal JSON Big Data may not respect the JSON Schema associated to them, like renaming a JSON object or a JSON object member, or replacing the current value of an object member with a new one that is not conformant to its type as specified in the corresponding JSON Schema. These update operations are called "non-conservative updates" and require implicit JSON Schema changes to be performed so that they could be correctly executed. Such JSON Schema changes need then to be propagated to the OWL 2 ontology in order to maintain semantic alignment. To this purpose, in this paper, we propose an approach for automatic management of implicit OWL 2 ontology schema change operations that are generated by non-conservative updates to temporal JSON Big Data instances, in the τJOWL framework.

Keywords: Big Data · JSON · JSON Schema · NoSQL · Ontology · OWL 2 · Instance update · Schema change · Implicit schema change · τJOWL

1 Introduction

Nowadays, big data [1, 2] are being widely used in several application domains like e-business, e-health, online social networks, cloud computing, Internet of Things, smart grids, and smart cities. JSON [3] is among the lightweight and semi-structured data formats, including also XML and YAML, which are broadly used to represent, exchange and store Big Data. Although JSON Big Data are in general schemaless [4], schemas for them, specified using the JSON Schema language [5, 6] recommended by the Internet Engineering Task Force (IETF), can be discovered/extracted through the use of

I. Saad et al. (Eds.): ICIKS 2023, LNBIP 486, pp. 213–226, 2024.
https://doi.org/10.1007/978-3-031-51664-1_15

some appropriate tool (like json-schema-inferrer [7], Schema Guru [8], or Clojure JSON Schema Validator & Generator [9]). Moreover, Big Data can also be accompanied by an ontology [10], which can be very useful to provide the semantics of these data [11], from one hand, and to make easy and more efficient some complex tasks t like Big Data querying, Big Data analytics, and reasoning over Big Data, from the other hand.

The adoption of an ontology allows the formal specification of the data semantics in an information system or of the knowledge of an application domain. It is widely used in Semantic Web-based software, knowledge base management systems, and artificial intelligence-based systems. It facilitates data access, query answering, and knowledge inference. The OWL 2 Web Ontology Language [12], informally known as OWL 2, is the World Wide Web Consortium (W3C) recommendation for ontology specification in the Semantic Web. In the literature, two viewpoints have been proposed for the management of inferences and deductions from ontology statements: the Open World Assumption (OWA) [13] and the Closed World Assumption (CWA) [14]. The OWA, which means that "what is not known to be true or false is unknown", is in general applied in environments where data are considered to be incomplete (e.g., Knowledge Bases, Semantic Web repositories), whereas the CWA, which means that "what is not known to be true must be false", is in general applied in environments where data are considered to be complete (e.g., traditional databases). It is worth mentioning that the CWA can be used for ontology management via the notion of Data Box (DBox) [15]. Furthermore, with the CWA, an ontology definition (concepts, relations between concepts, axioms, …) acts as a database schema and, therefore, can be named an "ontology schema"; the ontology individuals are instances of this ontology schema.

Besides, Big Data are rapidly evolving over time, due to their "velocity" feature. In addition to this fact, several JSON-based Big Data applications require keeping track of the whole history of updates that are executed on these data. Consequently, JSON-based Big Data representation and storage need to become temporal: Big Data values are time-dependent and multi-version, so that each of them can be made of several consecutive temporal versions. Accordingly, it becomes difficult not only to manage (query, update, …) and analyze JSON Big Data, but also to formally define their semantics. Hence, it is of great help to have systems and tools that facilitate the definition of a temporal and a multi-versions ontology for temporal and multi-versions JSON-based Big Data. For that reason, and after studying the state of the art of ontologies for Big Data [16–30] and noticing that there is no work that have dealt with an automatic building of a temporal ontology from some temporal and multi-version Big Data, we have proposed in [31] a framework, called τJOWL (Temporal OWL 2 from Temporal JSON), which allows users of a Big Data project (i) to automatically create a temporal OWL 2 ontology of data, with the CWA viewpoint, from time-varying JSON-based Big Data via the use of an intermediary JSON Schema, and (ii) to manage the incremental maintenance of such a temporal ontology schema following the evolution of the underlying temporal JSON Big Data, in a multi-version setting.

As far as instance updates in τJOWL are concerned, we distinguished in [31] between conservative updates, which have no effect on the (JSON) schema of the updated JSON instance document, and non-conservative updates, which have an impact on such a schema since they require that some schema changes must be performed before executing the instance updates. Hence, the versioning of temporal ontology schemas is driven by non-conservative (temporal) updates to JSON Big Data instances. Moreover, since in [31] we have focused on the automatic building of a new temporal OWL 2 ontology schema for a temporal JSON instance document, in the present paper we complete our framework by focusing on the ontology maintenance issue. More precisely, we deal with non-conservative (temporal) JSON instance updates and show how they trigger ontology schema changes, which produce a new (temporal) ontology schema version.

The rest of the paper is organized as follows. Section 2 briefly recalls the τJOWL environment. Section 3 proposes our approach for implicit ontology schema changes that result from JSON-based big data instance updates, in τJOWL which supports temporal versioning at both instance and schema levels. Section 4 illustrates our proposal via an application example. Section 5 concludes the paper.

2 The τJOWL Environment

This section provides an overview of our τJOWL framework (as shown in Fig. 1).

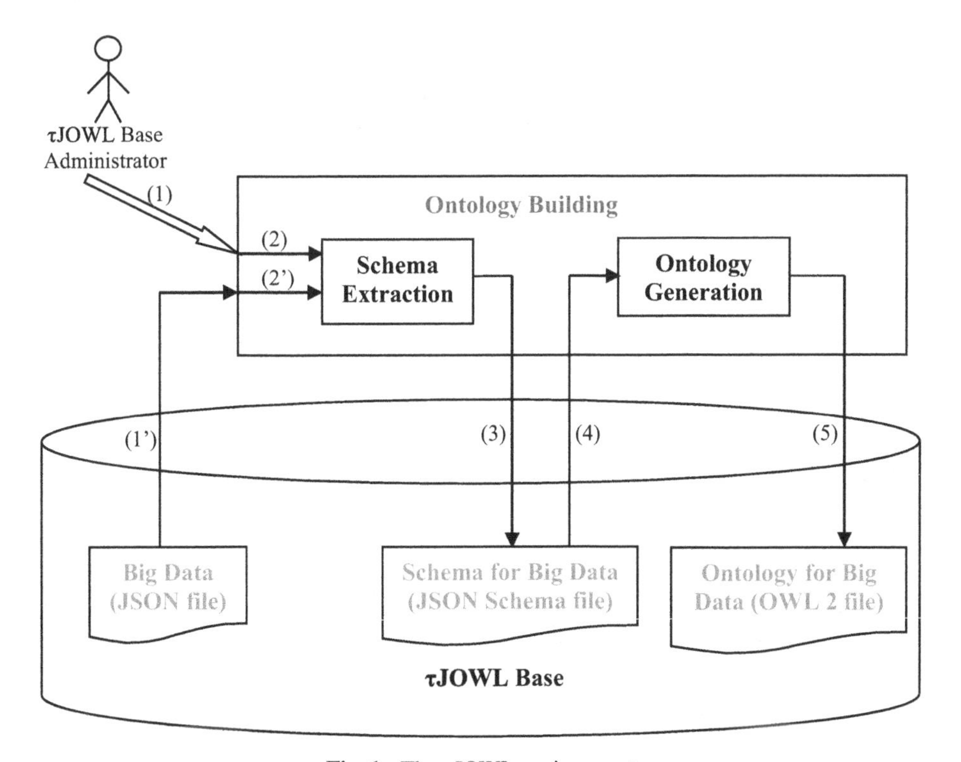

Fig. 1. The τJOWL environment

To build a temporal OWL 2 ontology of data, according to the CWA philosophy, for some temporal JSON-based Big Data instances, the τJOWL Base Administrator (arrows 1 and 2) starts by supplying a JSON instance document [3] (arrows 1' and 2'), which stores temporal big data, as input to the "Ontology Building" module. This latter first generates a JSON Schema [5, 6] (arrows 3 and 4) for the supplied JSON instance document, by calling the sub-module "Schema Extraction", and then converts this JSON Schema file into an OWL 2 ontology schema (arrow 5), by calling the sub-module "Ontology Generation".

More details on the architecture of τJOWL and the implementation of the two sub-modules "Schema Extraction" and "Ontology Generation" can be found in [31, Sec. 2]. Notice that the τJOWL Base is designed to store both big data and their ontologies. For that reason, to support the storage, at the same time, of JSON files, JSON Schema files, and OWL 2 files, the τJOWL Base could be either a (document-oriented) NoSQL database [32, 33], or a multi-model database [34].

Notice also that τJOWL considers OWL 2 ontologies with an RDF/XML syntax [35] that is the only syntax that must be supported by all OWL 2 tools, as required in the OWL 2 specification document [36].

3 Management of Implicit Ontology Schema Changes Triggered by Non-conservative JSON Instance Updates

In this section, we present our approach for managing implicit ontology schema changes generated by non-conservative JSON instance updates, in τJOWL.

As mentioned above, τJOWL is temporal and supports versioning at both instance and schema levels. Consequently, each instance update IU, when applied on some JSON instance document version D_V_i ($i >= 1$) that is conformant to some JSON Schema version S_V_j ($j >= 1$) and some OWL 2 ontology schema version O_V_j associated to (and derived from) S_V_j, leads to a new JSON instance document version D_V_{i+1}. If D_V_{i+1} is still conformant to S_V_j, then IU is conservative; otherwise, IU is non-conservative and, as a consequence, the execution of two tasks is implicitly triggered (within the same update transaction): (1) a new JSON Schema version S_V_{j+1}, to which D_V_{i+1} is conformant, is generated, and (2) a new OWL 2 ontology schema version O_V_{j+1} is automatically derived from S_V_{j+1}.

Notice that the instance update IU is a sequence of JSON instance update operations, specified in some JSON update language like JUpdate [37] or τJUpdate [38]. Therefore, IU must include at least one instance update operation that has an effect on the schema version of the updated JSON document version, like renaming a JSON object or a JSON object member, or replacing the current value of an object member with a new value that is not conformant to the type of this object member (specified in the corresponding JSON Schema version).

Non-conservative instance updates require then implicit JSON Schema changes to be executed, at schema level, so that they could be correctly performed. Moreover, the execution of such implicit JSON Schema changes has an immediate impact on the ontology side of τJOWL: the (temporal) OWL 2 ontology schema, which is related to the implicitly changed (temporal) JSON Schema, is consequently changed in an implicit and automatic way. Notice that, for changing ontology schema, we are re-using the set of ontology schema change operations that we have defined in our previous work [39, 40].

In these previous papers, we have shown that each ontology schema change operation is sound, that is, its application to a (consistent) ontology leads to a new consistent ontology (and does not give rise to any inconsistency).

The functioning of our approach is formalized via the algorithm shown in Fig. 2 and named "UpdateJSONInstancesWithImplicitOntologySchemaVersioning".

```
ALGORITHM UpdateJSONInstancesWithImplicitOntologySchemaVersioning
  Inputs: JIU, JID_Vn, JSD_Vm, OSD_Vm
  Outputs: JID_Vn+1, IJSC, JSD_Vm+1, OSC, OSD_Vm+1
BEGIN
  UpdateJSONInstanceDocument(JIU, JID_Vn, JID_Vn+1, IJSC);
  Display("Execution of instance updates.");
  If (empty(IJSC)==false) Then
    GenerateNewJSONSchemaVersion(IJSC, JSD_Vm, JSD_Vm+1);
    PropagateSchemaChangesToJSONInstances(IJSC, JSD_Vm);
    DeriveOntologySchemaChanges(IJSC, JSD_Vm, OSD_Vm, OSC);
    GenerateNewOntologySchemaVersion(OSC, OSD_Vm, OSD_Vm+1);
    Display("With execution of implicit schema changes.");
  Else
    Display("No implicit schema changes are required.");
  EndIf
END
```

Fig. 2. The algorithm for applying updates to a JSON instance document version possibly with implicit ontology schema versioning

This algorithm uses a set of input and output variables and calls a function and a set of procedures. They are all described below.

- Input variables
  - JIU: a (valid) sequence of JSON instance updates specified by the τJOWL base administrator using the JUpdate or the τJUpdate language.
  - JID_V_n: a (valid) JSON file that represents the current JSON instance document version (number n, with $n >= 1$).
  - JSD_V_m: a (valid) JSON Schema file that represents the current JSON Schema document version (number m, with $m >= 1$) to which JID_V_n is conformant.
  - OSD_V_m: a (valid) OWL 2 file that represents the current OWL 2 ontology schema version (number m) which results from the conversion of JSD_V_m to an OWL 2 file.
- Output variables:

- JID_V_{n+1}: a (valid) JSON file that represents the new JSON instance document version (number n + 1, with n > = 1).
- IJSC: a (valid) sequence of implicit JSON Schema changes automatically generated by the system during the execution of the sequence of JSON instance updates JIU.
- JSD_V_{m+1}: a (valid) JSON Schema file that represents the new JSON Schema document version (number m + 1, with m > = 1) to which JID_V_{n+1} is conformant.
- OSC: a (valid) sequence of OWL 2 ontology schema changes automatically generated by the system based on the sequence of implicit JSON Schema changes IJSC.
- OSD_V_{m+1}: a (valid) OWL 2 file that represents the new OWL 2 ontology schema version (number m + 1) which results from the conversion of JSD_V_{m+1} to an OWL 2 file.

- Function:
  - empty(IJSC): it returns either true if the sequence of implicit JSON Schema changes IJSC (passed as argument) is empty, or false otherwise.

- Procedures:
  - UpdateJSONInstanceDocument(JIU, JID_V_n, JID_V_{n+1}, IJSC): it applies the sequence of JSON instance updates JIU on the JSON instance document version JID_V_n (n > = 1) to produce the new JSON instance document version JID_V_{n+1} and possibly a sequence of implicit JSON Schema changes IJSC that could require interaction with the τJOWL base administrator.
  - Display(message): it displays the message (a string) passed as argument.
  - GenerateNewJSONSchemaVersion(IJSC, JSD_V_m, JSD_V_{m+1}): it creates a new JSON Schema version JSD_V_{m+1} (m > = 1) by applying the sequence of implicit JSON Schema changes IJSC on the current JSON Schema version JSD_V_m.
  - PropagateSchemaChangesToJSONInstances(IJSC, JSD_V_m): it propagates the effects of the sequence of implicit JSON Schema changes IJSC to all JSON instance documents that are conformant to the previous JSON Schema version JSD_V_m (m > = 1). Hence, this procedure generates a new version of each one of these JSON instance documents, which must be conformant to the new JSON Schema version JSD_V_{m+1}.
  - DeriveOntologySchemaChanges(IJSC, JSD_V_m, OSD_V_m, OSC): it derives, from the sequence of implicit JSON Schema changes IJSC that act on the current JSON Schema version JSD_V_m, the sequence of OWL 2 ontology schema changes OSC that act on the current OWL 2 ontology schema version OSD_V_m.
  - GenerateNewOntologySchemaVersion(OSC, OSD_V_m, OSD_V_{m+1}): it creates a new OWL 2 ontology schema version OSD_V_{m+1} (m > = 1) by applying the sequence of OWL 2 ontology schema changes OSC on the current OWL 2 ontology schema version OSD_V_m.

4 Application Example

To exemplify the functioning of our approach, let us consider the case of a JSON-based NoSQL data store used by a scientific research laboratory for the management of researchers' data. Let us assume that, in order to make such a management more "intelligent" as far as access, query and analysis of researchers' data are concerned, the director of the laboratory decides to deploy an ontology of data, with the CWA, within a τJOWL framework.

Assume that on March 01, 2023, the τJOWL Base Administrator created (or imported) a JSON instance document, named "ResearchersJSONInstances_V1.json" (as shown in Fig. 3), which stores data on researchers (the ORCID, the name, the discipline, characterized by an id and a description, and the h-index of each researcher). Due to space limitations, this example presents only one researcher named "Tarak Ziad", his ORCID is "0000-1111-2222-3333", his discipline is "Computer Science" (description) with an id equal to 1, and his h-index is 9. After that, he/she invoked the "Ontology Building" module (as shown in Fig. 1) of τJOWL to automatically build an OWL 2 ontology for these JSON data. In fact, this module starts by extracting the JSON Schema file, named "ResearchersJSONSchema_V1.json" (as shown in Fig. 4), of the provided JSON instance document (of Fig. 3), by calling the "Schema Extraction" sub-module, and then it generates the OWL 2 ontology file, named "ResearchersOntologySchema_V1.owl" (as shown in Fig. 5), corresponding to the extracted JSON schema, by calling the "Ontology Generation" sub-module.

```
{ "researcher":{
    "ORCID": "0000-1111-2222-3333",
    "name": "Tarak Ziad",
    "discipline": { "id": 1,
                    "description": "Computer Science"},
    "h-index": 9 } }
```

Fig. 3. The JSON instance document of researchers ("ResearchersJSONInstances_V1.json") on March 01, 2023

Furthermore, assume that on April 01, 2023, the τJOWL Base Administrator updated the current JSON instance document version (as shown in Fig. 6), by applying the following JSON instance updates on the researcher whose ORCID is "0000-1111-2222-3333":

- update the value of the object member "name" from "Tarak Ziad" to "Tarak Ben Ziad";
- replace the old value of the object member "discipline" (having an "object" type) with a new value "Computer Science" with string type;
- add to the "researcher" object two new object members: the first "citations" with the integer value 222 and the second "publications" with the integer value 39;
- rename the object members "name" and "h-index" to "fullName" and "scopusH-index", respectively.

```
{ "type": "object",
  "properties":
    { "researcher":
        { "type": "object",
          "properties":
            { "ORCID": { "type": "string" },
              "name": { "type": "string" },
              "discipline":
                { "type": "object",
                  "properties":
                    { "id": { "type": "integer" },
                      "description": { "type": "string" } } },
              "h-index": { "type": "integer" } } } } }
```

Fig. 4. The generated JSON Schema of researchers ("ResearchersJSONSchema_V1.json") on March 01, 2023

In a more technical way, we suppose that the τJOWL Base Administrator has specified the above instance updates using the JUpdate language [37]. The sequence of JSON instance update operations could be as follows:

```
UPDATE ResearchersJSONInstances_V1.json
OBJECT $.researcher[?(@.ORCID=='0000-1111-2222-3333')]
SET name="Tarak Ben Ziad", discipline="Computer Science";
ALTER DOCUMENT ResearchersJSONInstances_V1.json
OBJECT $.researcher[?(@.ORCID=='0000-1111-2222-3333')]
ADD MEMBER citations VALUE 222;
ALTER DOCUMENT ResearchersJSONInstances_V1.json
OBJECT $.researcher[?(@.ORCID=='0000-1111-2222-3333')]
ADD MEMBER publications VALUE 39;
ALTER DOCUMENT ResearchersJSONInstances_V1.json
OBJECT $
RENAME MEMBER $.researcher.name TO fullName;
ALTER DOCUMENT ResearchersJSONInstances_V1.json
OBJECT $
RENAME MEMBER $.researcher.h-index TO scopusH-index;
```

Notice that, in the above sequence, except the first instance update operation that is conservative (i.e., it respects the JSON Schema version, as shown in Fig. 4, associated to the JSON instance document version which is under update), the other operations are all non-conservative.

The execution of the above sequence of instance update operations on the first JSON instance document version of researchers ("ResearchersJSONInstances_V1.json", shown in Fig. 3) generates a new version named "ResearchersJSONInstances_V2.json" (as shown in Fig. 6). Furthermore, and since the above sequence includes non-conservative operations, such an execution also produces, in a transparent way, a sequence of implicit JSON Schema change operations. This latter is automatically applied on the first JSON Schema version ("ResearchersJSONSchema_V1.json",

```
<rdf:RDF>
 <owl:Ontology rdf:about="http://my_onto/researcher_ontology#">
  <owl:Class rdf:about="researcher"/>
  <owl:DatatypeProperty rdf:about="ORCID">
   <rdfs:domain rdf:resource="researcher"/>
   <rdfs:range rdf:resource="http://www.w3.org/2001/XMLSchema#string"/>
  </owl:DatatypeProperty>
  <owl:DatatypeProperty rdf:about="name">
   <rdfs:domain rdf:resource="researcher"/>
   <rdfs:range rdf:resource="http://www.w3.org/2001/XMLSchema#string"/>
  </owl:DatatypeProperty>
  <owl:Class rdf:about="discipline"/>
  <owl:DatatypeProperty rdf:about="id">
   <rdfs:domain rdf:resource="discipline"/>
   <rdfs:range rdf:resource="http://www.w3.org/2001/XMLSchema#int"/>
  </owl:DatatypeProperty>
  <owl:DatatypeProperty rdf:about="description">
   <rdfs:domain rdf:resource="discipline"/>
   <rdfs:range rdf:resource="http://www.w3.org/2001/XMLSchema#string"/>
  </owl:DatatypeProperty>
  <owl:ObjectProperty rdf:about="researcher_Has_discipline">
   <rdfs:domain rdf:resource="researcher"/>
   <rdfs:range rdf:resource="discipline"/>
  </owl:ObjectProperty>
  <owl:DatatypeProperty rdf:about="h-index">
   <rdfs:domain rdf:resource="researcher"/>
   <rdfs:range rdf:resource="http://www.w3.org/2001/XMLSchema#int"/>
  </owl:DatatypeProperty>
 </owl:Ontology>
</rdf:RDF>
```

Fig. 5. The generated OWL 2 ontology schema of researchers ("ResearchersOntologySchema_V1.owl") on March 01, 2023.

shown in Fig. 4) to obtain a new JSON Schema version, named "ResearchersJSON-Schema_V2.json" (as shown in Fig. 7), for the new JSON instance document version (shown in Fig. 6) Then, and within the same update transaction, the system also derives, from the generated sequence of implicit JSON Schema change operations, an equivalent sequence of OWL 2 ontology schema change operations, and applies it on the first OWL 2 ontology schema version ("ResearchersOntologySchema_V1.owl", shown in Fig. 5) to obtain a new OWL 2 ontology schema version, named "ResearchersOntologySchema_V2.owl" (as shown in Fig. 8). Notice that "ResearchersOntologySchema_V2.owl" is the second OWL 2 ontology schema version corresponding to the updated JSON Big Data instances stored in "ResearchersJSONInstances_V2.json" of Fig. 6. Changes are evidenced in red bold typeface.

Notice that we have defined all necessary JSON Schema change operations in our previous work on schema versioning in a temporal JSON-based NoSQL setting [41,

42], and all useful OWL 2 ontology schema change operations in our previous work on schema versioning in a temporal OWL 2 Semantic Web context [39, 40].

As for the implicit relation between non-conservative JSON instance update operations, JSON Schema change operations, and OWL 2 ontology schema change operations, we illustrate it below via the simple case of rename something. For example, for the following JSON instance update operation:

```
ALTER DOCUMENT ResearchersJSONInstances_V1.json
OBJECT $
RENAME MEMBER $.researcher.name TO fullName;
```

the following JSON Schema change operation will be implicitly generated:

```
RenameProperty(ResearchersJSONSchema_V1.json,
$.properties.researcher.properties.name, fullName)
```

and the following OWL 2 ontology schema change operation will be also derived:

```
RenameDataTypeProperty(ResearchersOntologySchema_V1.owl,
name, fullName)
```

```
{ "researcher":{
    "ORCID": "0000-1111-2222-3333",
    "fullName": "Tarak Ben Ziad",
    "discipline": "Computer Science",
    "scopusH-index": 10,
    "citations": 222,
    "publications": 39 } }
```

Fig. 6. The JSON instance document of researchers (“ResearchersJSONInstances_V2.json”) on April 01, 2023

```
{ "type": "object",
  "properties":
    { "researcher":
        { "type": "object",
          "properties":
            { "ORCID": { "type": "string" },
              "fullName": { "type": "string" },
              "discipline": { "type": "string" },
              "scopusH-index": { "type": "integer" },
              "citations": { "type": "integer" },
              "publications": { "type": "integer" } } } } }
```

Fig. 7. The generated JSON Schema of researchers (“ResearchersJSONSchema_V2.json”) on April 01, 2023

```
<rdf:RDF>
  <owl:Ontology rdf:about="http://my_onto/researcher_ontology#">
    <owl:Class rdf:about="researcher"/>
    <owl:DatatypeProperty rdf:about="ORCID">
      <rdfs:domain rdf:resource="researcher"/>
      <rdfs:range rdf:resource="http://www.w3.org/2001/XMLSchema#string"/>
    </owl:DatatypeProperty>
    <owl:DatatypeProperty rdf:about="fullName">
      <rdfs:domain rdf:resource="researcher"/>
      <rdfs:range rdf:resource="http://www.w3.org/2001/XMLSchema#string"/>
    </owl:DatatypeProperty>
    <owl:DatatypeProperty rdf:about="discipline">
      <rdfs:domain rdf:resource="researcher"/>
      <rdfs:range rdf:resource="http://www.w3.org/2001/XMLSchema#string"/>
    </owl:DatatypeProperty>
    <owl:DatatypeProperty rdf:about="scopusH-index">
      <rdfs:domain rdf:resource="researcher"/>
      <rdfs:range rdf:resource="http://www.w3.org/2001/XMLSchema#int"/>
    </owl:DatatypeProperty>
    <owl:DatatypeProperty rdf:about="citations">
      <rdfs:domain rdf:resource="researcher"/>
      <rdfs:range rdf:resource="http://www.w3.org/2001/XMLSchema#int"/>
    </owl:DatatypeProperty>
    <owl:DatatypeProperty rdf:about="publications">
      <rdfs:domain rdf:resource="researcher"/>
      <rdfs:range rdf:resource="http://www.w3.org/2001/XMLSchema#int"/>
    </owl:DatatypeProperty>
  </owl:Ontology>
</rdf:RDF>
```

Fig. 8. The derived OWL 2 ontology of researchers ("ResearchersOntologySchema_V2.owl") on April 01, 2023

5 Conclusion

In this paper, we have proposed an approach for an automatic management of implicit ontology schema changes that are triggered by non-conservative updates to JSON-based Big Data instances, in the τJOWL environment.

We think that our approach makes the management of ontology schema versioning more flexible and more user-friendly, as it allows to apply JSON instance updates that do not "respect" the current implicit JSON schema version and consequently the current ontology schema version. Moreover, our proposal facilitates the specification, in a progressive manner and driven by instance updates, of the correct ontology schema of some JSON-based Big Data instance document whose complete and precise structure is not fully known at its creation time. We think also that our approach could easily be applied in a context that is different from τJOWL since it is designed in an environment that is independent from any technology (i.e., from any DBMS, API, software, JSON tool, programming language, ...). It depends only on (i) the JSON format, which is the most widely used model for big data storage and management, (ii) the JSON Schema language, which is recommended by the IETF for defining structures of JSON documents, and (iii) the OWL 2 language, which is recommended by the W3C for Semantic Web ontologies.

In order to show the feasibility of our proposal, we plan in the near future to implement our approach as a new module within the τJOWL-Manager tool that is being developed to support the whole τJOWL framework. Such a module will allow us to experimentally evaluate the scalability and performances of our proposal.

Last but not least, in the future we plan to deal with the effects of implicit ontology schema changes on the programs which are using these ontologies. In fact, when an ontology schema evolves, all ontology instance update statements (specified e.g. in SPARQL 1.1 Update [43]) and ontology queries (specified e.g. in SPARQL [44], T-SPARQL [45], SWRL [46], SQWRL [47], or τSQWRL [48]), which are written against such an ontology schema, should be revised in order to take into account the new ontology schema version. These revisions could be automatically performed on the program source codes, in parallel with the execution of implicit ontology schema changes.

References

1. Rao, T.R., Mitra, P., Bhatt, R., Goswami, A.: The big data system, components, tools, and technologies: a survey. Knowl. Inf. Syst. **60**(3), 1165–1245 (2019)
2. Davoudian, A., Liu, M.: Big data systems: a software engineering perspective. ACM Comput. Surv. (CSUR), 53(5), 1–39 (2020). Article 110
3. IETF. The JavaScript Object Notation (JSON) Data Interchange Format. Internet Standards Track document, December 2017 (2017). https://tools.ietf.org/html/rfc8259. Accessed 22 May 2023
4. Banerjee, S., Shaw, R., Sarkar, A., Debnath, N.C.: Towards logical level design of big data. In: Proceedings of the IEEE 13th International Conference on Industrial Informatics (INDIN 2015), Cambridge, UK, 22–24 July 2015, pp. 1665–1671. IEEE (2015)
5. Pezoa, F., Reutter, J.L., Suarez, F., Ugarte, M., Vrgoč, D.: Foundations of JSON schema. In: Proceedings of the 25th International Conference on World Wide Web (WWW'2016), Montréal, Québec, Canada, 11–15 April 2016, pp. 263–273 (2016)
6. IETF. JSON Schema: A Media Type for Describing JSON Documents. Internet-Draft, 19 March 2018 (2018). https://json-schema.org/latest/json-schema-core.html. Accessed: 22 May 2023
7. json-schema-inferrer: java library for inferreing JSON schema from sample JSONs. https://github.com/saasquatch/json-schema-inferrer. Accessed 22 May 2023
8. Schema Guru. https://github.com/snowplow/schema-guru.Accessed 22 May 2023
9. Clojure JSON Schema Validator & Generator. https://github.com/luposlip/json-schema. Accessed 22 May 2023
10. Guarino, N. (ed.): Formal Ontology in Information Systems. IOS Press, Amsterdam, Netherlands (1998)
11. Ceravolo, P., et al.: Big data semantics. J. Data Semant. **7**, 65–85 (2018)
12. W3C. OWL 2 Web Ontology Language – Primer (Second Edition). W3C Recommendation, 11 December 2012 (2012). http://www.w3.org/TR/owl2-primer/. Accessed 22 May 2023
13. Patel-Schneider, P.F., Horrocks, I.: A comparison of two modelling paradigms in the semantic web. J. Web Semant. **5**(4), 240–250 (2007)

14. Etzioni, O., Golden, K., Weld, D.S.: Sound and efficient closed-world reasoning for planning. Artif. Intell. **89**(1–2), 113–148 (1997)
15. Seylan, İ., Franconi, E., De Bruijn, J.: Effective query rewriting with ontologies over DBoxes. In: Proceedings of the 21st International Joint Conference on Artificial Intelligence (IJCAI 2009), Pasadena, CA, USA, 11–17 July 2009, pp. 923–929 (2009)
16. Hoppe, A., Nicolle, C., Roxin, A.: Automatic ontology-based user profile learning from heterogeneous web resources in a big data context. Proc. VLDB Endowment **6**(12), 1428–1433 (2013)
17. Soylu, A., Giese, M., Jimenez-Ruiz, E., Kharlamov, E., Zheleznyakov, D., Horrocks, I.: OptiqueVQS: towards an ontology-based visual query system for big data. In: Proceedings of the 5th International Conference on Management of Emergent Digital EcoSystems (MEDES'2013), Luxembourg, Luxembourg, 29–31 October 2013, pp. 119–126 (2013)
18. Jayapandian, C., Chen, C.H., Dabir, A., Lhatoo, S., Zhang, G.Q., Sahoo, S.S.: Domain ontology as conceptual model for big data management: application in biomedical informatics. In: Yu, E., Dobbie, G., Jarke, M., Purao, S. (eds.) Conceptual Modeling. ER 2014. LNCS, vol. 8824, pp. 144–157. Springer, Cham (2014).https://doi.org/10.1007/978-3-319-12206-9_12
19. Shah, T., Rabhi, F., Ray, P.: Investigating an ontology-based approach for big data analysis of inter-dependent medical and oral health conditions. Clust. Comput. **18**(1), 351–367 (2015)
20. Verhoosel, J.P., Spek, J.: Applying ontologies in the dairy farming domain for big data analysis. In: Joint Proceedings of the 3rd Stream Reasoning (SR 2016) and the 1st Semantic Web Technologies for the Internet of Things (SWIT 2016) Workshops Co-located with 15th International Semantic Web Conference (ISWC 2016), Kobe, Japan, 17–18 October 2016, pp. 91–100 (2016)
21. Kim, A.R., Park, H.A., Song, T.M.: Development and evaluation of an obesity ontology for social big data analysis. Healthc. Inform. Res. **23**(3), 159–168 (2017)
22. Abbes, H., Gargouri, F.: MongoDB-based modular ontology building for big data integration. J. Data Semant. **7**(1), 1–27 (2018)
23. Globa, L.S., Novogrudska, R.L., Koval, A.V.: Ontology model of telecom operator big data. In: Proceedings of the 2018 IEEE International Black Sea Conference on Communications and Networking (BlackSeaCom 2018), Batumi, Georgia, 4–7 June 2018, pp. 1–5. IEEE (2018)
24. Wongthongtham, P., Salih, B.A.: Ontology-based approach for identifying the credibility domain in social big data. J. Organ. Comput. Electron. Commer. **28**(4), 354–377 (2018)
25. Nadal, S., Romero, O., Abelló, A., Vassiliadis, P., Vansummeren, S.: An integration-oriented ontology to govern evolution in big data ecosystems. Inf. Syst. **79**, 3–19 (2019)
26. Rani, P.S., Suresh, R.M., Sethukarasi, R.: Multi-level semantic annotation and unified data integration using semantic web ontology in big data processing. Clust. Comput. **22**(5), 10401–10413 (2019)
27. Djebouri, D., Keskes, N.: Exploitation of ontological approaches in big data: a state of the art. In: Proceedings of the 10th International Conference on Information Systems and Technologies (ICIST'2020), Lecce, Italy, 4–5 June 2020, Article no. 45, pp. 1–6 (2020)
28. Aghdam, M.Y., Tabbakh, S.R.K., Chabok, S.J.M.: Ontology generation for flight safety messages in air traffic management. J. Big Data **8**(1), 1–21 (2021)
29. Mhammedi, S., El Massari, H., Gherabi, N.: Cb2Onto: OWL ontology learning approach from couchbase. In: Gherabi, N., Kacprzyk, J. (eds.) Intelligent Systems in Big Data, Semantic Web and Machine Learning. AISC, vol. 1344, pp. 95–110. Springer, Cham (2021). https://doi.org/10.1007/978-3-030-72588-4_7
30. Mountasser, I., Ouhbi, B., Hdioud, F., Frikh, B.: Semantic-based big data integration framework using scalable distributed ontology matching strategy. Distrib. Parallel Databases **39**(4), 891–937 (2021)

31. Brahmia, Z., Grandi, F., Bouaziz, R.: τJOWL: a systematic approach to build and evolve a temporal OWL 2 ontology based on temporal JSON big data. Big Data Mining Analytics **5**(4), 271–281 (2022)
32. Davoudian, A., Chen, L., Liu, M.: A survey on NoSQL stores. ACM Comput. Surv. (CSUR) **51**(2), 1–43 (2018)
33. NoSQL Databases List by Hosting Data – Updated 2023. https://hostingdata.co.uk/nosql-database/. Accessed 22 May 2023
34. Lu, J., Holubová, I.: Multi-model databases: a new journey to handle the variety of data. ACM Comput. Surv. (CSUR) **52**(3), 1–38 (2019)
35. W3C. RDF/XML Syntax Specification (Revised). W3C Recommendation, 10 February 2004 (2004). http://www.w3.org/TR/2004/REC-rdf-syntax-grammar-20040210/. Accessed 22 May 2023
36. W3C. OWL 2 Web Ontology Language – Document Overview (Second Edition). W3C Recommendation, 11 December 2012 (2012). http://www.w3.org/TR/owl2-overview/. Accessed 22 May 2023
37. Brahmia, Z., Brahmia, S., Grandi, F., Bouaziz, R.: JUpdate: a JSON update language. Electronics **11**(4), 508 (2022)
38. Brahmia, Z., Grandi, F., Brahmia, S., Bouaziz, R.: τJUpdate: A temporal update language for JSON data. In: Fournier-Viger, P., Hassan, A., Bellatreche, L. (eds.) Model and Data Engineering. MEDI 2022. LNCS, vol. 13761, pp. 250–263. Springer, Cham (2022)https://doi.org/10.1007/978-3-031-21595-7_18
39. Zekri, A., Brahmia, Z., Grandi, F., Bouaziz, R.: τOWL: A systematic approach to temporal versioning of semantic web ontologies. J. Data Seman. **5**(3), 141–163 (2016)
40. Zekri, A., Brahmia, Z., Grandi, F., Bouaziz, R.: Temporal schema versioning in τOWL: a systematic approach for the management of time-varying knowledge. J. Decis. Syst. **26**(2), 113–137 (2017)
41. Brahmia, Z., Brahmia, S., Grandi, F., Bouaziz, R.: Versioning schemas of JSON-based conventional and temporal big data through high-level operations in the τJSchema framework. Int. J. Cloud Comput. **10**(5–6), 442–479 (2021)
42. Brahmia, S., Brahmia, Z., Grandi, F., Bouaziz, R.: Temporal JSON schema versioning in the τJSchema framework. J. Digit. Inf. Manag. **15**(4), 179–202 (2017)
43. W3C. SPARQL 1.1 Update. *W3C Recommendation*, 21 March 2013 (2013). https://www.w3.org/TR/sparql11-update/.Accessed 22 May 2023
44. W3C. SPARQL Query Language for RDF. *W3C Recommendation*, 15 January 2008 (2008). https://www.w3.org/TR/rdf-sparql-query/. Accessed 22 May 2023
45. Grandi, F.: T-SPARQL: a TSQL2-like temporal query language for RDF. In: Local Proceedings of the 14th East-European Conference on Advances in Databases and Information Systems (ADBIS'2010), Novi Sad, Serbia, 20–24 September 2010. CEUR Workshop Proceedings (CEUR-WS.org), vol. 639, pp. 21–30 (2010)
46. W3C. SWRL: A Semantic Web Rule Language Combining OWL and RuleML. W3C Member Submission 21 May 2004 (2004). https://www.w3.org/Submission/SWRL/. Accessed 22 May 2023
47. O'Connor, M., Das, A.: SQWRL: a query language for OWL. In *Proceedings of the 6th International Workshop on OWL: Experiences and Directions (OWLED 2009)*, Chantilly, VA, USA, 23–24 October 2009. CEUR Workshop Proceedings (CEUR-WS.org), vol. 529 (2009). https://ceur-ws.org/Vol-529/owled2009_submission_42.pdf. Accessed 22 May 2023
48. Brahmia, Z., Grandi, F., Bouaziz, R.: τSQWRL: a TSQL2-like query language for temporal ontologies generated from JSON big data. Big Data Mining Analytics **6**(3), 288–300 (2023)

A Model Driven Architecture Approach for Implementing Sensitive Business Processes

Molka Keskes(✉)

ISIMS, MIRACL Laboratory, University of Sfax, B.P. 242, 3021 Sfax, Tunisia
keskesmolka92@gmail.com

Abstract. This paper presents the application of the MDA (Model Driven Architecture) approach to implement Sensitive Business Processes (SBPs), in order to identify and localize the crucial knowledge that is mobilized and created by these processes. Therefore, we start with a conceptual specification of SBP as a CIM level, then transform it to a PIM so that SBPs can be specified and represented using extended BPMN 2.0. This extension was designed in a previous research project based on core ontologies in order to integrate all relevant aspects related to the knowledge dimension in SBP models. As well as from PIM to PSM for the implementation of SBPs. For this, we propose mapping rules for model-to-model transformation and use the standard MOF 2.0 QVT as a transformation language.

Keywords: Knowledge Management · Sensitive Business Process Modeling · BPMN4KM · Model-Driven Architecture · Model Transformation · MOF 2.0 QVT

1 Introduction

Modeling sensitive business processes (SBPs) is an effective way to manage and develop an organization's knowledge which needs to be capitalized. These processes are characterized by a high complexity and dynamism in their execution, high number of critical activities with intensive acquisition, sharing, storage and (re)use of very specific crucial knowledge, diversity of knowledge sources, and high degree of collaboration among experts [1].

In order to enrich and improve the SBP modeling, the authors [1–3] have proposed a rigorous conceptual specification of SBP organized in new multi-perspective meta-model, entitled «BPM4KI: Business Process Meta-Model for Knowledge Identification». This meta-model covers several perspectives of SBPs, namely: the functional, the organizational, the behavioral, the informational, the intentional and the knowledge perspectives. In this research work, we focus more on the representation of «Knowledge Dimension» which represents the most relevant aspect of SBP modeling. This dimension (i.e., differentiation between tacit and explicit knowledge, the different types of knowledge conversion, the

I. Saad et al. (Eds.): ICIKS 2023, LNBIP 486, pp. 227–242, 2024.
https://doi.org/10.1007/978-3-031-51664-1_16

dynamic aspects of knowledge, the different sources of knowledge, etc.) is not yet fully supported, integrated and implemented within BPs models and BPM approaches and formalisms [2,4].

In this context, Ben Hassen et al. [4] proposed BPMN4KM, an extension of the most suitable business process modeling formalism BPMN 2.0 for modeling knowledge dimension in SBPs. A specific Eclipse plug-in entitled «K4BPMN: Knowledge for Business Process Modeling Notation »developed in keskes et al. [5] to include and represent all relevant aspects related to the Knowledge dimension in SBP modelling. This plug-in is an extension of the already existing Eclipse BPMN2 Modeler plug-in[1]: it completes this later by integrating new attributes, properties, elements and specific icons for introduce new semantics.

In this paper, we are interested in the implementation of these processes modelled in accordance with BPMN4KM. To do this, we used the model-driven approach. We adopt an MDA framework, which allows the modeling and generation of an executable specification of SBPs, through the definition of models belonging to different levels of abstraction and the transformations between these models. It is then necessary to conduct a detailed study of the MDA approach, in order to delineate the models of the proposed framework and to identify the transformation rules to ensure the passage between these models.

This research work is devoted to the presentation of MDA. It first explains the MDA approach and then presents a framework designed according to the MDA approach allowing the consideration of SBPs from modelling to the generation of an executable specification.

This paper is organized as follows. Section 2 gives a short presentation of MDA approach and Sect. 3 describes our model driven framework for the consideration of SBPs from modeling to the execution. In Sect. 4, we introduce the CIM to PIM transformation and finally, Sect. 5 concludes the paper, discusses it according related works, and gives some directions for future works.

2 Model Driven Architecture

The model-driven architecture (MDA) is an initiative proposed by the object management group (OMG) in 2001, it is a new way of designing IT applications, it aims to separate the business logic of the organization, from any technical platform. Indeed, the technical architecture is unstable and undergoes many changes over time, contrary to business logic. This makes it easy to separate the two to cope with the high costs of technology migration and provides a high level of responsiveness when IT systems need to evolve to use another technology [6]. Thus, it allows the reuse of models. Indeed, it makes it possible to reuse the same "business" model for different platforms or technologies [7], and supports reengineering and reverse engineering which makes it possible to recover business logic from source codes or implementation environments [8].

[1] http://www.eclipse.org/bpmn2-modeler/.

2.1 MDA Models

MDA is based on a model-based development approach and a set of OMG standards. This approach separates the functional specifications of a system from the details of its implementation, it consists of three main models (see Fig. 1): The Computation Independent Model (CIM), in which no IT consideration appears. Thus, a CIM can be used to build a Platform Independent Model (PIM). A PIM focuses on the operation of the system while hiding details specific to a particular execution platform or technology. It remains platform-agnostic to enable the same model to be used across different platforms. The transformation of a PIM into a Platform Specific Model (PSM) involves model transformation mechanisms (see Fig. 2) and a platform description model (PDM); therefore, the PIM must be refined by the details of one or more particular architectures to obtain a PSM [9]. A PSM is a system model for a specific platform, it combines PIM specifications with the details needed to specify how a system uses a particular type of platform [10]. Essentially it is used to generate the corresponding executable code.

Figure 1 below gives an overview of an MDA process that is organized according to a "Y" development cycle by showing the different levels of abstraction associated with the models.

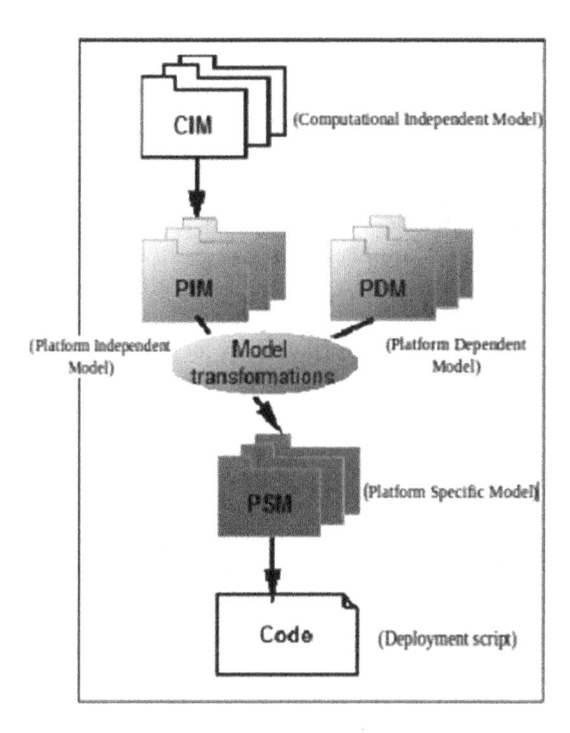

Fig. 1. Principles of the MDA Process [11]

2.2 Model Transformation

A model transformation is a process that allows the transition from a source model to a target model by matching these elements. Each of these models must conform to a meta-model which in turn conform to the meta-meta-model (MOF meta-model) (see Fig. 2).

One can distinguish two kinds of model transformations: model-to-model transformations which are used to translate source models to target models (e.g. the transition from CIM to PIM or from PIM to PSM), and model-to-code transformations concerning the generation of code from the PSM to a specific programming language as a target. Model-to-model transformations are classified as endogenous or exogenous based on the reference metamodel used for expressing source and target models. Exogenous transformations occur between models expressed by different metamodels, whereas endogenous transformations happen between models that conform to the same metamodel [12]. It is also possible to classify transformations as horizontal or vertical [13]. Source and target models reside at different abstraction levels in horizontal transformations, whereas vertical transformations are at the same abstraction level. These transformations are composed of rules that define the correspondences between the source and target metamodels, their execution makes it possible to automatically generate the target model from the source model. The standard MOF QVT (Query/View/Transformation) [14] transformation language developed by the OMG group was developed to define a metamodel that allows model transformation development. Actually, QVT supports three transformation languages: QVT-Relations and QVT-Core are declarative languages at two different levels of abstraction. QVT-Relations is a user-oriented language for defining transformations at a high level of abstraction. It has a textual and graphical syntax. QVT-Core is a low-level technical language that is defined by textual syntax.

The QVT Operational Mappings language is an imperative language, that extends the two declarative languages of QVT by adding imperative constructs (sequence, selection, repetition, etc.) as well as OCL constructs. QVT also offers a second extension mechanism called Black Box for specifying transformations, it is about invoking implemented transformations features in an external language. In this paper, we focus on QVT Relations, the high-level declarative language.

3 MDA Framework Architecture for SBPs Specification

The proposed framework is based on the MDA approach. It defines three levels of abstraction covering the CIM, PIM and PSM models respectively (see Fig. 4). Thus, these latter models support all stages of transformation of SBPs, from modeling to the automatic generation of an executable specification [15] and are described as follows.

3.1 The Framework's CIM Model

A CIM model must be independent of any computer system [16]. Therefore, the CIM of the proposed framework corresponds to the modeling of SBPs according

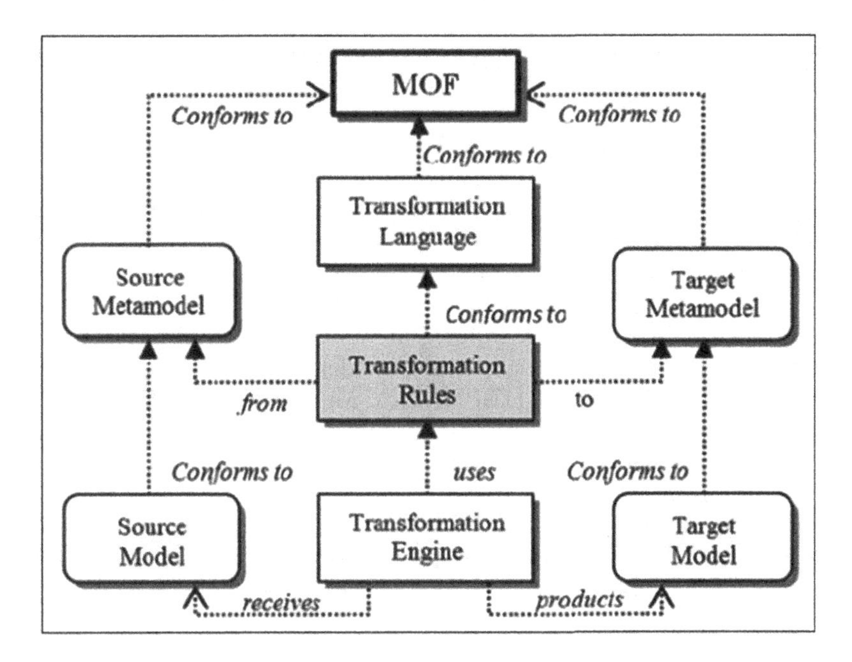

Fig. 2. Transformation Process in the MDA approach [13]

to BPM4KI meta-model. BPM4KI is a generic multi-perspective meta-model, that explicit and organize the key concepts and relations characterizing an SBP. It integrates all relevant perspectives/dimensions relating to BPM-KM, i.e., the functional, the organizational, the behavioral, the informational, the intentional and the knowledge perspectives (see Fig. 5). Basic concepts for BPM4KI have been defined in [1,2]. Indeed, the workflow application designer can create, by instantiation of BPM4KI, SBPs models. The BPM4KI instances thus form the CIM model which can be represented through a UML Object diagram containing the SBP instances to be able to visualize and locate knowledge.

3.2 The Framework's PIM Model

A PIM is a model independent of any technical platform [16]. It represents a partial view of CIM. Consequently, the PIM model of the proposed framework gives a graphical representation of SBPs from the CIM model, that is to say SBPs modelled in accordance with BPM4KI. More specifically, this model presents process designers with a wide range of languages for the graphical representation of processes. In this work we have chosen the BPMN 2.0 specification.

Such a representation is used to approve the modeling step, allowing the process designer to verify and validate a SBP model according to extended BPMN (BPMN4KM Meta-Model). BPMN4KM Meta-model [4] is the extended BPMN 2.0 meta-model for modelling the knowledge dimension in SBP. This meta-model integrates, on the one hand, the knowledge dimension, and on the other hand, the key concepts of BPM4KI meta-model [1,2] (see Fig. 6).

Figure 6 below presents the resulting extended BPMN meta-model (BPMN4KM). In this figure only the relevant standard BPMN classes are shown in white. The BPMN4KM concepts are shown in grey. The semantics and the abstract syntax of the BPMN4KM elements are based on the specification of the BPMN extension mechanism[2].

Therefore, to represent graphically SBPs according to BPMN4KM, we developed a specific Eclipse plug-in, entitled «K4BPMN: Knowledge for Business Process Modeling Notation »[5], to integrate and represent all relevant aspects related to the knowledge dimension in SBP models. Let us first remind that the BPMN 2.0 extended meta-model (BPMN4KM Meta-model) presents the BPMN elements allowing the modeling of SBPs from a perspective of knowledge localization. This meta-model contains the original elements of the notation as well as the extensions made. Thus, K4BPMN Modeler is an extension of the already existing Eclipse BPMN2 Modeler plug-in, dedicated to the definition of different BPMN diagrams: it completes this later by integrating new elements with their own properties and forms, attributes and, specific icons for representing a new semantics, to represent graphically SBPs according to BPMN4KM and consider these extensions. Therefore, K4BPMN Modeler plug-in defines a set of extension elements, allowing the new and specific graphical representation for the new concepts of BPMN4KM as illustrated in Fig. 7. Concretely, we have incorporated new notational elements with specific properties for *knowledge* typologies and *information*, *knowledge conversion modes*, *Knowledge Flows* (between *knowledge*, *activities* and *agentives entities*), *Physical Knowledge Supports*, *agentive entities* and *critical activities*, as shown in Fig. 7.

Illustrative Example of the Use of K4BPMN Plug-In. We illustrate here a SBP model using our extended plug-in. This experiment was carried out within the context of the Association of Protection of the Motor-disabled of Sfax-Tunisia (ASHMS). A depth description of the case study has been presented in [17].

We details the instantiation of BPMN4KM Meta-model related to the neuropediatric care process of a child with CP. We are interested in the graphical representation of the various modeling elements according to BPMN4KM Meta-model, in order to improve the identification and characterization of the different types and modalities of medical knowledge mobilized and produced by this process.

Figure 3 shows a screenshot of the plugin. On the right-hand side of the figure, there is the palette that contains the extension elements. The central part features the canvas where an extract of the model of representation of the SBP relating to the neuropediatric care of a child with CP is modeled. This model is enriched by the different extensions we have made to the BPMN 2.0 notation, notably the Knowledge dimension. The development of this BPMN model is based on the testimony and validation of both the Neuro-pediatrician.

[2] http://www.omg.org/spec/BPMN/2.0.

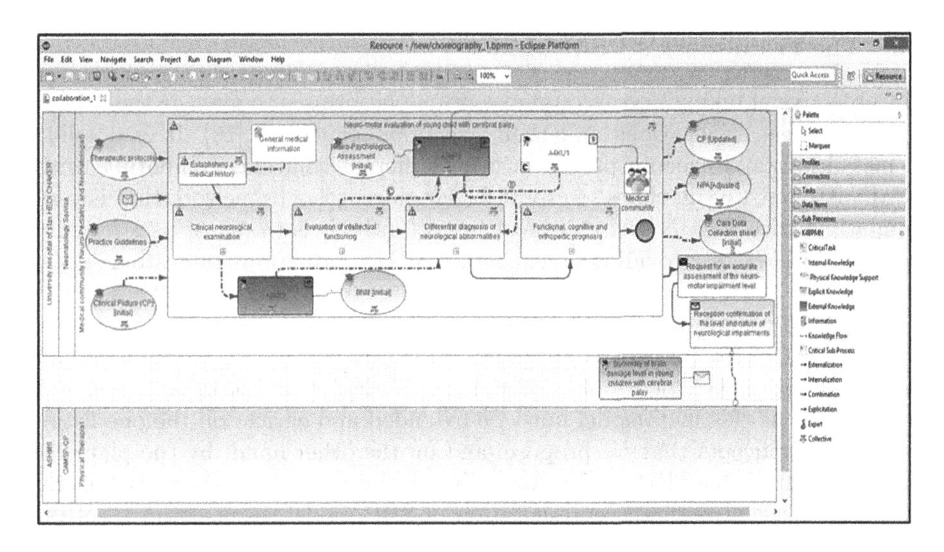

Fig. 3. Instance of the extended BPMN demonstrating a part of a (simplified) SBP model related to the neuro-pediatric consultation of a child with CP using Extended BPMN2 Modeler

During our experimentation, we identified different types of medical knowledge mobilized by each critical activity (individual and collective) related to this SBP. For example, knowledge A3Kp1 "Synthesis of neuro and psycho-cognitive, neuro-sensory and praxognosic development evaluation of young children at-risk and their disorders" is produced by critical activity A3 "Evaluation of intellectual functioning". It should be noted that this explicit/outsourced knowledge is created as a result of the performance of the A3 activity by the Neuropediatrician, during which it interacts with information (Knowledge source information) about the child with CP, interpreting them (according to their field of interest, beliefs and previous practical experiences) to generate and communicate their own knowledge.

A3Kp1 is stored in the following physical media: the neuro-psychological assessment (BNP) and the care data collection sheet. These record are recorded in the personal medical record and in the clinical picture (CT) of the young child with CP. These physical supports of knowledge are located internally in the neonatology department within the CHU Hédi Chaker, and are captured in the various drawers of the archives or in the repertoires of the patients. A3Kp1 is scientific, technical, measurement, patient-related. It represents a collective external knowledge, some of which can be represented in the form of an explicit individual knowledge recorded on the data record of care of the neuropediatrician. It is imperfect (general, incomplete and uncertain). A3Kp1 is mobilized by activity A4 "Differential diagnosis of neurological abnormalities".

3.3 The Framework's PSM Model

The objective of taking into account execution platforms is to manage the dependency between applications and their execution platforms [16]. It is possible to instantiate and execute processes using different solutions compatible with BPMN 2.0, but we will choose Activiti BPM[3] since it is a framework that is very compatible with BPMN 2.0, and most importantly, it is the most user-friendly for Java developers, providing a rich set of BPMN extensions to customize the process.

A SBP can be executed within the Activiti engine if we define a meta-model for Activiti BPMN 2.0 that is an extension of the BPMN 2.0 metamodel and includes additional elements and attributes that are specific to the Activiti BPM platform. Thus this metamodel must be extended and enrich, on the one hand, by the new concepts that we propose, and on the other hand, by the platform specific details of the SBP model. According to this extended Activiti BPMN2.0 metamodel, the structured representation of SBP model results from the PIM level that contained in BPMN XML file in XML format, is used to define the SBP model in extended BPMN 2.0 notation with platform-specific details. It contains the BPMN4SBP elements, as well as Activiti-specific extension elements and attributes that define the behavior of the SBP on the Activiti BPM platform, using the extension mechanism provided by Activiti.

Once the BPMN XML definition file is adapted, it can be used as input to a code generator that transforms the XML into platform-specific code, such as Java classes or configuration files. The generated code will contain the necessary details and logic to run the SBP on the Activiti BPM platform. Figure 4 illustrates the proposed framework.

4 Transformation from CIM to PIM

In order to specify SBPs using BPMN 2.0, we propose in this section the transformation of the CIM model to the PIM model of the proposed framework. This transformation allows the transition from SBP modelled according to the BPM4KI meta-model to their corresponding graphical representation designed according to the extended BPMN meta-model. To do this, we first specify the correspondence between concepts of these two meta-models (BPM4KI and BPMN4KM) and identify the Mapping rules between them. Then, we define an example of a mapping rule description using the QVT standard.

4.1 Mapping Rules Between BPM4KI and BPMN4KM Metamodel

With respect to the limited space of this paper, the mapping rules between concepts of each perspective cannot be presented. Figure 8 illustrates an example of the concept correspondence and the mapping rules between BPM4KI (source

[3] https://www.activiti.org/.

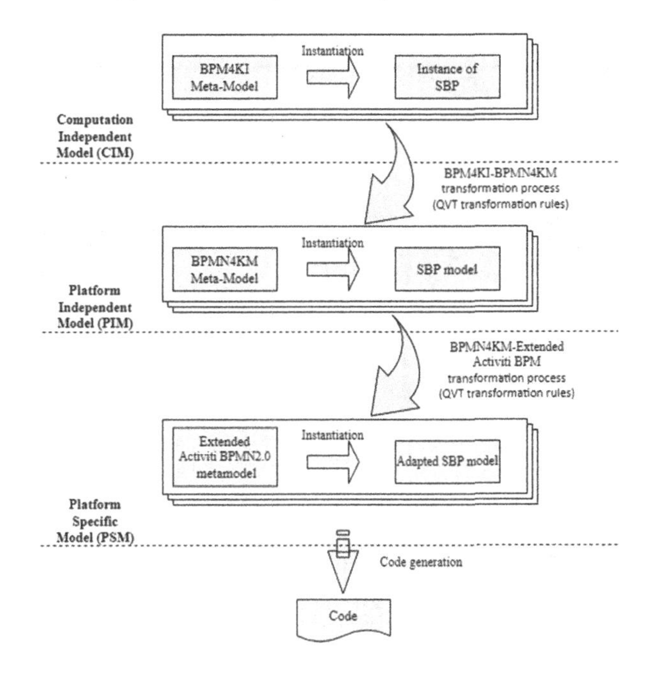

Fig. 4. MDA framework for the specification of SBPs

model) and BPMN4KM meta-model (target model) from the functional perspective. We specify in the left part of the figure concepts of the source meta-model, in the right part concepts of the target meta-model, and in the middle, we represent with red circles, the rules supporting the correspondence.

4.2 Definition of Transformation Rules with QVT

In this section, we describe an example of the transformation rules previously presented using the graphical and textual notation of the QVT language [14]. QVT defines a transformation as a list of relations. Figure 9 gives an extract of the textual representation of the transformation (called BPM4KI-BPMN4KM) allowing the transition from the bpm4ki source model to the bpmn4km target model through a set of relation.

A relation specifies the mapping between two candidate models (source and target), identifying: two or more domains, where each domain is a distinctive set of model elements. A relation domain, which describes the type of the relationship between domains and can be marked as **checkonly** (labeled with the letter C) or enforced (labeled with the letter E). A **when** clause describes the conditions under which the relation must hold. The **where** clause specifies the conditions that all model elements participating in this relation must meet.

Additionally, QVT allows for both textual and graphic notations of relations. Transformations can be declaratively defined using either notation.

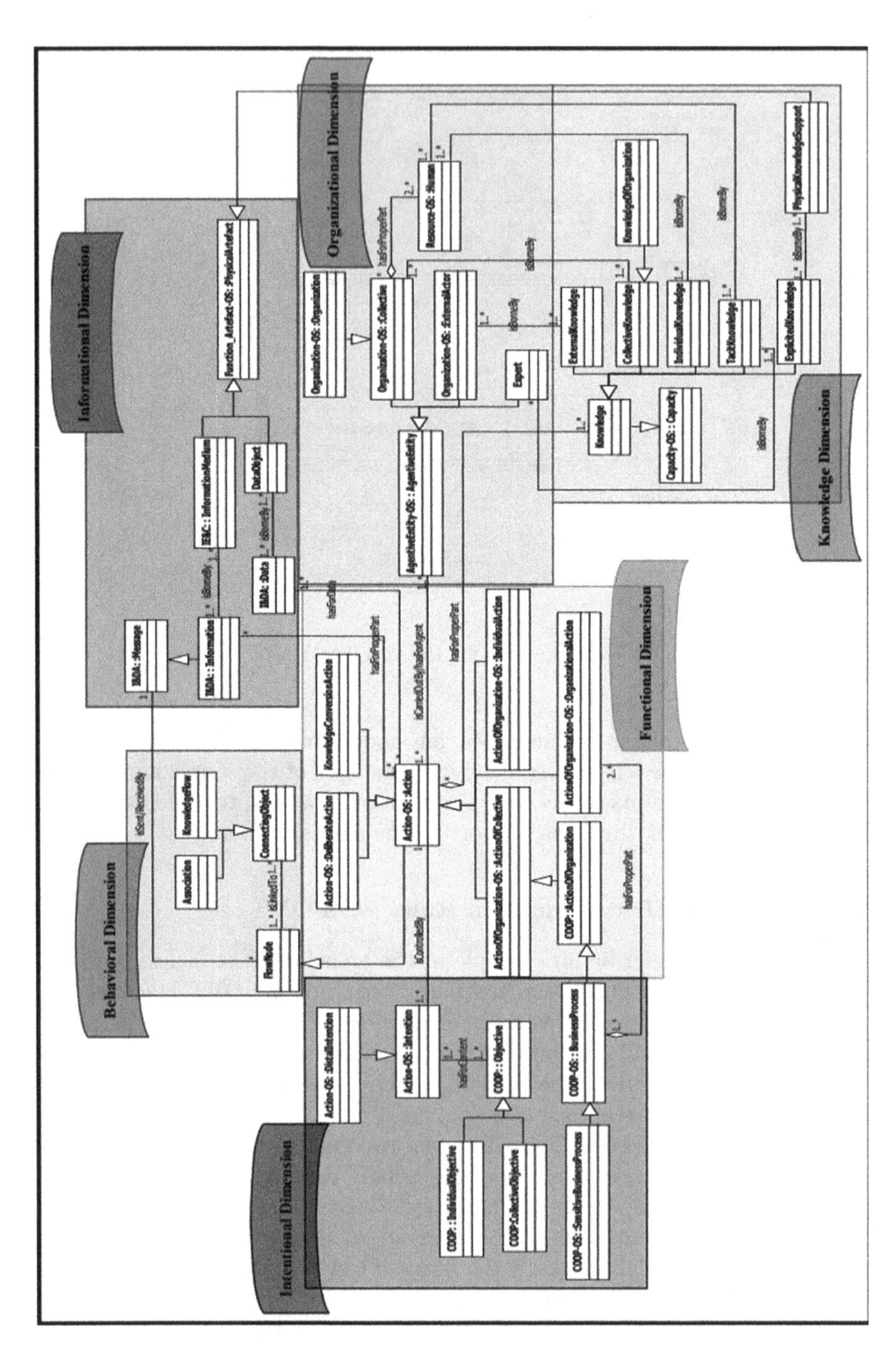

Fig. 5. A general view of BPM4KI meta-model [3]

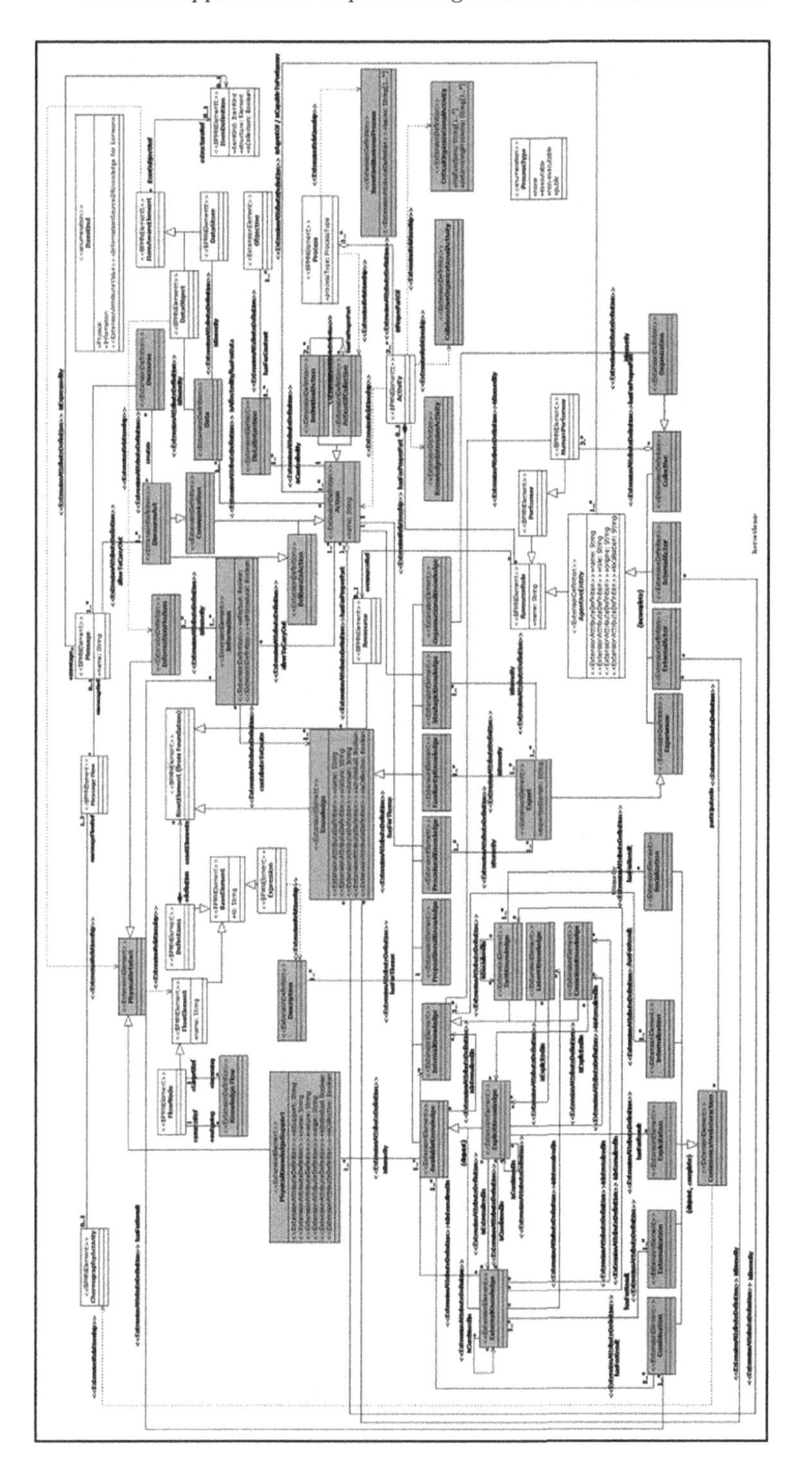

Fig. 6. BPMN4KM Meta-model for Modeling Sensitive Business Process [4]

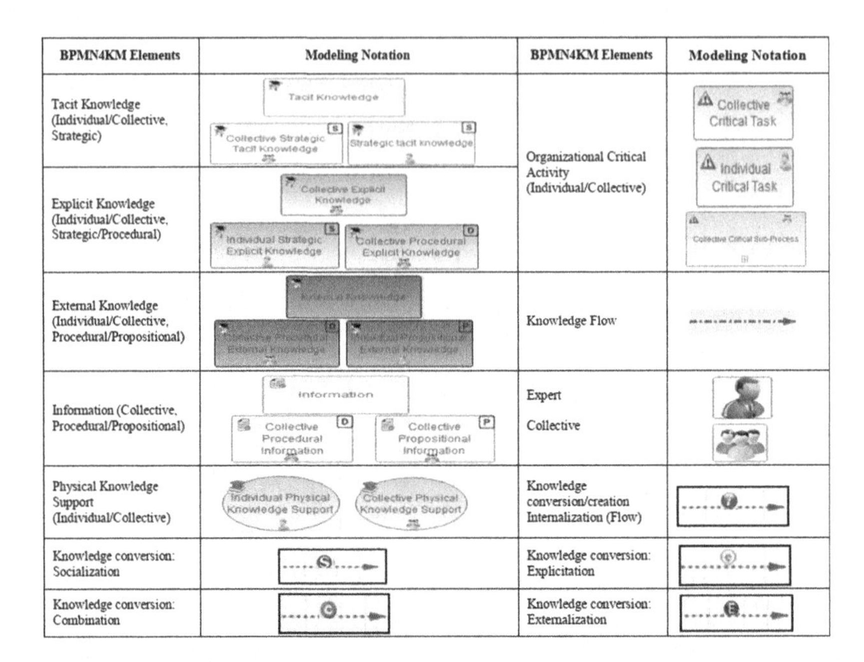

BPMN4KM Elements	Modeling Notation	BPMN4KM Elements	Modeling Notation
Tacit Knowledge (Individual/Collective, Strategic)	Tacit Knowledge Collective Strategic Tacit Knowledge Strategic tacit knowledge	Organizational Critical Activity (Individual/Collective)	Collective Critical Task Individual Critical Task Collective Critical Sub-Process
Explicit Knowledge (Individual/Collective, Strategic/Procedural)	Collective Explicit Knowledge Individual Strategic Explicit Knowledge Collective Procedural Explicit Knowledge		
External Knowledge (Individual/Collective, Procedural/Propositional)		Knowledge Flow	
Information (Collective, Procedural/Propositional)	Information Collective Procedural Information Collective Propositional Information	Expert Collective	
Physical Knowledge Support (Individual/Collective)	Individual Physical Knowledge Support Collective Physical Knowledge Support	Knowledge conversion/creation Internalization (Flow)	
Knowledge conversion: Socialization		Knowledge conversion: Explicitation	
Knowledge conversion: Combination		Knowledge conversion: Externalization	

Fig. 7. Main Objects of the K4BPMN [18]

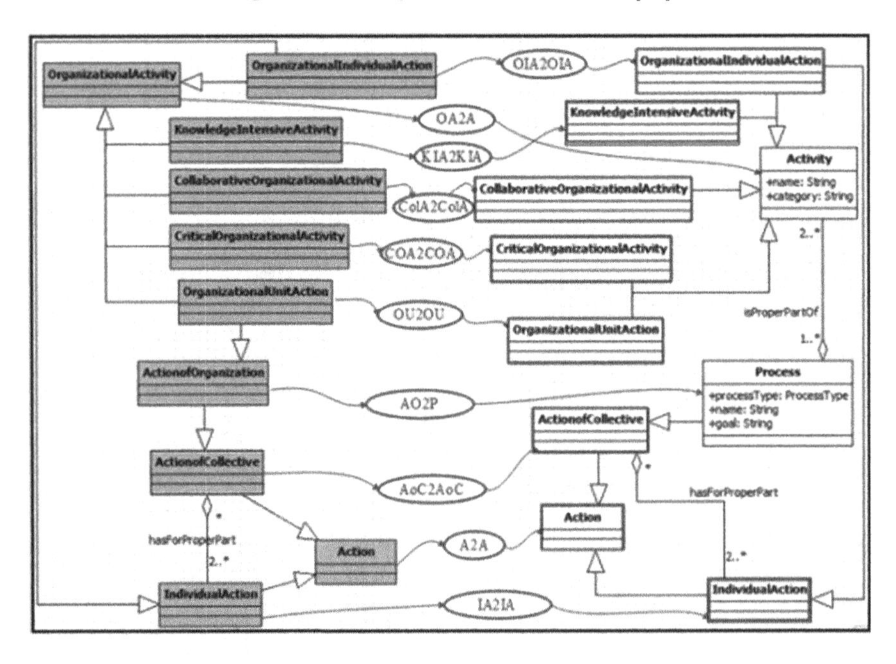

Fig. 8. Mapping between concepts according to the Functional perspective

```
Transformation BPM4KI-BPMN4KI (bpm4ki :BPM4KI, bpmn4kI : BPMN4KI)
{     top relation K2K {...}
      top relation OA2A {...}
      top relation Exp2Exp {...}
      top relation SK2SK {...}
      top relation ProcK2ProcK {...}
      top relation Prop2Prop {...}
      top relation LK2LK {...}
      top relation CsK2CsK {...}
      top relation Pks2Pks {...}
      top relation S2S {...}
      top relation COA2COA  {...}
      top relation AO2P {...}
      top relation AC2P {...}
      relation A2A {...}
      relation Ck2CK {...}
      relation INK2INK  {...}
      relation SK2SK {...}
      relation TK2TK {...}
      relation ExtK2ExtK {...}
... }
```

Fig. 9. Textual description of the QVT BPM4KI-BPMN4KM transformation

```
relation ExpK2ExpK
{ B : Boolean ;
Checkonly domain bpm4ki EpK : ExplicitKnowledge
   {
         isExternalizedIn=EK : ExternalKnowledge {},
         resultBy= Ex : Explicitation { },
         resultBy= Com : Combination { };
   };
Enforce domain bpmn4sbp EpK : ExplicitKnowledge
  {
         isExternalized= B,
         isExternalizedIn=EK : ExternalKnowledge {},
         resultBy= Ex : Explicitation { }
         resultBy= Com : Combination { };    }
When {
E2E(Ex,Ex),
Comb2Comb(Com,Com)} ;
Where {Extk2Extk(EK,EK)}
}
```

Fig. 10. Textual Description of the ExpK2ExpK relation

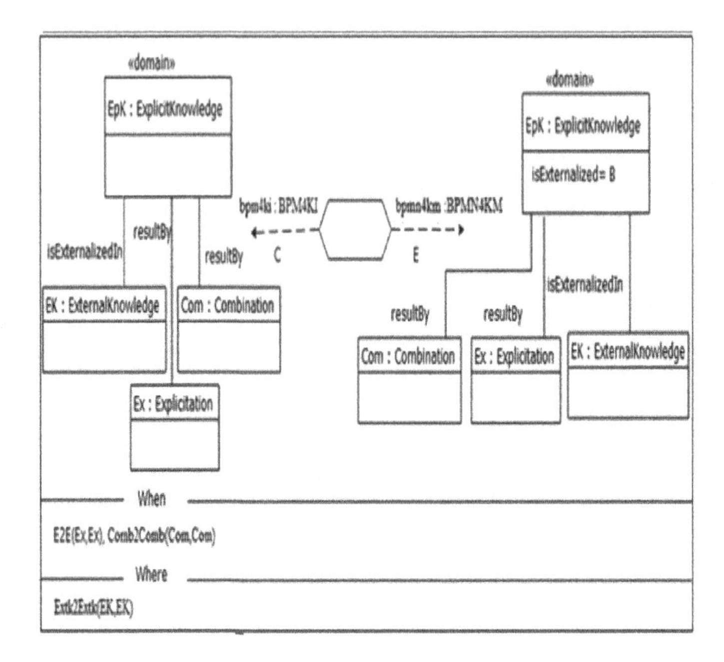

Fig. 11. Graphical Description of the ExpK2ExpK relation

An example of a relation called **ExpK2ExpK** is presented below. The relation supporting the mapping of *explicit knowledge* of the bpm4ki model to an *explicit knowledge* of the bpmn4km model through two domains. Figures 10 and 11 shows a definition of this relation. The first domain marked by **checkonly** corresponds to an *explicit knowledge* that can be *Externalized* and it results either by an *Explicitation*, or by a *Combination*. The second domain, marked by enforce, corresponds to an *explicit knowledge* having the same properties of the *explicit knowledge* of the source model and also has the *isExternalized* property (of type Boolean). In addition, the When clause of the relation indicates that the execution of the ExpK2ExpK relation is carried out following the execution of the E2E and Comb2Comb relation, while the Where clause indicates that the execution of the ExpK2ExpK relation results in the execution of the Extk2Extk relation.

5 Conclusion and Future Works

In this paper, we have presented a generic conceptual framework capable of taking into account SBPs from modelling to the generation of an executable specification designed according to the MDA approach considering (i) at the CIM level, a specific meta-model, the BPM4KI meta-model for modelling SBPs (ii) at the PIM level, an extension of the BPMN metamodel for visualizing and user validating the modelled SBP, and finally, (iii) at the PSM level, an extended Activiti BPMN2.0 meta-model for implementing SBPs. This paper mainly focuses on the automatic mapping from the CIM level onto the PIM level and to obtain a PSM model from a PIM model using a model driven approach. Specifically, we are interested in enriching the graphic representation of SBP models, previously designed according to BPM4KI, and then in their implementation.

Regarding related works, main contributions in SBPs (e.g. [1–4,17–20]) have extensively dealt with the BP/SBP identification and modeling phases. On the other hand, they did not address the automatic mapping of modelled SBP from a CIM level to a PIM level. Moreover, theses contributions do not address the crucial phase of executing these complex knowledge-intensive processes. On the other hand, there is a lack of works dealing with the implementation of extended business processes such as [21]. In addition, contributions (such as [22]) that deal with the automated transformation of BPMN specifications into executable ones do not take sensitive business processes into account.

Finally, to the best of our knowledge, it does not exist any contribution in the literature addressing the implementation of the KM notion in BPMN.

There are several open issues in this paper that we plan to address in the future to deepen the so-called problematic of implementing sensitive business processes to improve the identification and management of critical knowledge. Further work is underway to complete the application of the MDA approach we have proposed, to automate the generation of an executable specifications for SBPs.

References

1. Ben Hassen, M., Turki, M., Gargouri, F.: Functional dimension representation in sensitive business process models. In: International Conference on Knowledge Management, Information and Knowledge Systems (KMIKS 2017), Hammamet-Tunisie (2017)
2. Ben Hassen, M., Turki, M., Gargouri, F.: Towards extending business process modeling formalisms with information and knowledge dimensions. In: Benferhat, S., Tabia, K., Ali, M. (eds.) IEA/AIE 2017. LNCS (LNAI), vol. 10350, pp. 407–425. Springer, Cham (2017). https://doi.org/10.1007/978-3-319-60042-0_45
3. Ben Hassen, M., Turki, M., Gargouri, F.: Using core ontologies for extending sensitive business process modeling with the knowledge perspective. In: Proceedings of the Fifth European Conference on the Engineering of Computer-Based Systems (ECBS 2017), p. 2. ACM (2017)
4. Ben Hassen, M., Turki, M., Gargouri, F.: Extending BPMN 2.0 with the knowledge dimension. In: International Symposium on Business Modeling and Software Design, Barcelona- Spain (BMSD 2017) (2017)
5. Keskes, M., Ben Hassen, M., Turki, M.: K4BPMN modeler: an extension of BPMN2 modeler with the knowledge dimension based on core ontologies. In: Abraham, A., Muhuri, P.K., Muda, A.K., Gandhi, N. (eds.) ISDA 2017. AISC, vol. 736, pp. 755–770. Springer, Cham (2018). https://doi.org/10.1007/978-3-319-76348-4_73

6. Arrassen, I., Esbai, R., Meziane, A., Erramdani, M.: QVT transformation by modeling: from UML model to MD model. In: 2012 6th International Conference on Sciences of Electronics, Technologies of Information and Telecommunications (SETIT), pp. 86–91. IEEE (2012)
7. Gerber, A., Lawley, M., Raymond, K., Steel, J., Wood, A.: Transformation: the missing link of MDA. In: Corradini, A., Ehrig, H., Kreowski, H.-J., Rozenberg, G. (eds.) ICGT 2002. LNCS, vol. 2505, pp. 90–105. Springer, Heidelberg (2002). https://doi.org/10.1007/3-540-45832-8_9
8. Bezivin, J., Ploquin, N.: Tooling the MDA framework: a new software maintenance and evolution scheme proposal. J. Object-Oriented Program. (JOOP) 1–10 (2001)
9. Bousetta, B., El Beggar, O., Gadi, T.: A methodology for CIM modelling and its transformation to PIM. J. Inf. Eng. Appl. **3**(2), 1–21 (2013)
10. Truyen, F.: The fast guide to model driven architecture the basics of model driven architecture. Cephas Consulting Corp (2006)
11. Alili, H., Drira, R., Ghezala, H.H.B.: Model driven framework for the configuration and the deployment of applications in the cloud. Cloud Computing (2016)
12. Agner, L.T.W., Soares, I.W., Stadzisz, P.C., Simao, J.M.: Model refinement in the model driven architecture context. J. Comput. Sci. **8**(8), 1205 (2012)
13. Kriouile, A., Addamssiri, N., Gadi, T.: An MDA method for automatic transformation of models from CIM to PIM. Am. J. Softw. Eng. Appl. **4**(1), 1–14 (2015)
14. OMG. Meta object facility (MOF) 2.0 query/view transformation specification. OMG Document Number: formal/2016-06-03 (2016)
15. Hassen, M.B., Turki, M., Gargouri, F.: Choosing a sensitive business process modeling formalism for knowledge identification. Procedia Comput. Sci. **100**, 1002–1015 (2016)
16. OMG. MDA guide version. 2.0. OMG Document: ORMSC/2014-06-01, Juin (2014)
17. Ben Hassen, M., Turki, M., Gargouri, F.: Sensitive business processes representation: a multi-dimensional comparative analysis of business process modeling formalisms. In: Shishkov, B. (ed.) BMSD 2016. LNBIP, vol. 275, pp. 83–118. Springer, Cham (2017). https://doi.org/10.1007/978-3-319-57222-2_5
18. Hassen, M.B., Keskes, M., Turki, M., Gargouri, F.: BPMN4KM: design and implementation of a BPMN extension for modeling the knowledge perspective of sensitive business processes. Procedia Comput. Sci. **121**, 1119–1134 (2017)
19. Turki, M., Kassel, G., Saad, I., Gargouri, F.: A core ontology of business processes based on DOLCE. J. Data Semant. **5**, 165–177 (2016)
20. Hassen, M.B., Turki, M., Gargouri, F.: Extending BPMN models with sensitive business process aspects. Procedia Comput. Sci. **207**, 2968–2979 (2022)
21. Appel, S., Kleber, P., Frischbier, S., Freudenreich, T., Buchmann, A.: Modeling and execution of event stream processing in business processes. Inf. Syst. **46**, 140–156 (2014)
22. Zhe, L., Zacharewicz, G.: Model transformation from BPNM format to OpenERP workflow. In: The 11th International Multidisciplinary Modeling and Simulation Multiconference, Bordeaux, France (2014)

Extension of the Functional Dimension of BPMN Based on MDA Approach for Sensitive Business Processes Execution

Zohra Alyani(✉)

MIRACL Laboratory, University of Sfax, Sfax, Tunisia
Zohra.alyani@gmail.com

Abstract. In this paper of study, we propose a rigorous scientific approach to ensure the transition to Enhanced BPMN2.0 Transform the metamodel into an executable business process model. We are particularly interested in extending the functional dimension of BPMN2.0. We then position ourselves on the modeldriven MDA architecture Ensure that models are translated into executable business processes. In fact, at the CIM (Computational Independent Model) level of MDA, Corresponding to the BPM4KI model, the PIM layer (Plateform Independent Model) corresponds to the extended BPMN2.0 meta-model and The PSM level (Platform Specific Model) corresponds to the execution Sensitive business processes. Therefore, modeling and deploying This sensitive process requires the development and implementation of Specific business process management systems that support the definition expand.

Keywords: MDA · Meta-model · BPMN · Sensitive business process · Activiti · PSM · Extension

1 Introduction

Nowadays, it is essential for companies to constantly enhance their efficiency, foster innovation across all domains, minimize their production costs, and boost their adaptability. One way to achieve these objectives is to implement a knowledge management system. Consequently, organizational knowledge has become a pivotal factor that sets process-driven companies apart, as they acknowledge the value of their staff's knowledge and expertise. This competitive edge hinges largely on the abilities of employees to create a culture of continuous learning within the organization. Within the IT industry, experts in knowledge engineering offer techniques, methods, and concepts for acquiring, formalizing, modeling, and implementing knowledge utilized by an organization's business procedures. To achieve this, these processes must be identified, assessed, mapped, modeled, and executed.

To this end, previous research conducted by [1] focused on presenting an approach for identifying critical processes [1]. Subsequently, Mariem BEN HASSEN's thesis work proposed an extension of the BPMN2.0 formalism that enables

I. Saad et al. (Eds.): ICIKS 2023, LNBIP 486, pp. 243–254, 2024.
https://doi.org/10.1007/978-3-031-51664-1_17

the modeling of these significant processes [2–4]. This extension relies on the BPM4KI (Business Process Modeling for Knowledge Identification), which is a business process model. The BPM4KI model relies on six dimensions: functional, informational, organizational, intentional, behavioral, and knowledge dimensions. It is worth noting that the proposed extensions have not been integrated or supported by existing business process management systems.

Hence, the primary aim of this research paper is to present a scientific approach to transform the extended BPMN2.0 meta-model into an executable business process model. The focus is mainly on the extension of the functional dimension of BPMN2.0. The MDA (Model-Driven Architecture) is utilized to ensure the transformation of models into an executable business process. Specifically, the CIM (Computational Independent Model) level of MDA corresponds to the BPM4KI model, the PIM (Platform Independent Model) level corresponds to the extended BPMN2.0 meta-model, and the PSM (Platform-Specific Model) level corresponds to the execution of significant business processes. Consequently, modeling and implementing these sensitive processes require the development and integration of a specialized business process management system that supports the proposed extensions. This constitutes the second objective of this research paper.

The paper is structured as follows: Sect. 2 discusses the related work, while Sect. 3 presents an MDA approach to extend the functional dimension of BPMN2.0 for executing sensitive business processes. Section 4 deals with implementation and validation. Finally, Sect. 5 concludes the paper and highlights future research topics.

2 Related Work

Undoubtedly, in recent years, modeling Business Processes (BP) has emerged as a prominent research field and has significantly contributed to the development of modern information systems. Although there are various languages for BP modeling, BPMN has become the leading choice due to its expressiveness and simplicity. However, despite its common elements, BPMN often requires extensions to fully meet specific requirements. In the following discussion, we will explore the efforts made to extend BPMN and successfully execute business processes. Our proposal in Extensions to BPMN involves an extension that establishes a linkage between process models and events. To showcase the feasibility of this extension, we have applied it to a logistics scenario in a European project. Baumgraß, A., Herzberg, N., Meyer, A., and Weske, M. formally introduces the concept of complex event processing (CEP) in the context of BPM in [5] and establish a link between the two worlds. Following Stroppi, L., Chiotti, O. and P. Villarreal in [6] a model-driven architecture-based approach for developing extensions to the BPMN 2.0 metamodel. This approach allows for the creation of extensions to be modeled using UML, visualized using BPMN extension mechanisms, and transformed into XML schema documents that can be processed by BPMN tools. M. Ben Hassen, M. Turki, and F. Gargouri proposed a scientific approach in their previous research [7] to enhance BPMN2.0 for knowledge management. The resulting extension, named "BPMN4KM", is firmly grounded in the core domain ontology and contains rich semantics. It encompasses all pertinent knowledge-related aspects of SBP modeling to enhance the identification and localization of crucial knowledge that is utilized and produced throughout these processes.

2.1 Business Process Management System

The role of the workflow management system is to manage and execute workflow process definitions. The WfMC is an international organization dedicated to promoting workflow development by establishing standards for workflow management systems. Since the 90s, it has published a glossary that is considered the reference for defining terminologies in this field (Hol99). According to the WfMC, a workflow management system is a system that defines, creates, and manages the execution of workflows using tools that run on one or more workflow engines [8]. The workflow engine is responsible for initiating and tracking process instances, notifying roles to complete their tasks, routing data flows, and more. The engine interprets the process definition, interacts with workflow participants, and invokes the use of IT tools and applications to create process instances.

The BPMS consists of four main components:

- Modeler: the BPMN modeling part which can be done with Eclipse or Alfresco.
- Designer: the form modeling part.
- Engine: the launch and deployment part of the BPMN process.
- Explorer: the workflow management part.

In this section, we present a comparative study of existing workflow management systems in order to choose a system that will help us achieve our project objectives. We have chosen the most recommended tools in the field of workflow management, namely, KissFlow which is a paid tool, and Activiti, BonitaSoft, and Camunda which are open source tools.

Table 1. Comparison between existing tools.

Languages/Criteria	Bonita	Activiti	KissFlow	Camunda
Open Source	Yes	Yes	No	Yes
Distributed by	Bonitasoft	Alfresco	KissFlow	Camunda
Notoriety	Excellent	Excellent	Excellent	Medium
Dynamism	Excellent	Excellent	Excellent	Excellent
Extensibility	Low	Excellent	Low	Low
Open source API	Rest API	Rest API	No	Rest API
Process Model	BPMN	BPMN	Form	BPMN

In Table 1 we used several criteria to compare the systems and make the right choice of the to make the right choice of the system that will help us automate our processes according to our project vision:

- Open Source: It is a software engineering method that consists in developing software, or software, or software components, and letting the source code produced be freely available.
- Notoriety: The notoriety measures for a brand the fact of being known or recognized by consumers. The notoriety of a brand can be associated with a more or less positive or negative perceived image.
- Dynamism: It allows us to study the behavior of a computer program and the effects of its execution on its environment.
- Scalability: It refers to the ability of a product to adapt to a change in the order of magnitude of the demand (increase in load), in particular its capacity to maintain its functionality and performance in the event of high demand.
- Open Source API: It allows developers to access the different functionalities of software.
- Accepted file format: Allows to know the inputs of the software.

2.2 Principle of Transition from Modeling to Execution of Processes

The design of business processes has been the subject of intense work in recent years, with many years where many design tools have been developed. It is certain that the way in which business processes are structured has a great influence on the performance of the products and services on which they will be built. We present, below, the different design approaches and techniques to assist in the design of business processes in workflow systems. We also present the approaches needed to re-configure business processes, and thus to re-design workflows.

The requirements analysis and planning phase is the first step which consists of analyzing the needs and objectives of the business process to be designed. It also consists in to place it in its true context of use. Workflow users produce these specifications and transmit them to the designers in the form of input parameters. The objectives can be quantified to be used later in the last phase as measurement tools for validating the performance of the designed workflow. Thus, during the requirements planning phase, the guidelines for the different functionalities and the expected results of the process are defined. Based on the specifications from the analysis phase and the requirements planning, the second step of building workflow construction is started. During this construction step, the different process specifications are put forward in the form of a workflow model using the editing tool of the workflow system chosen for the implementation of the business process.

A workflow engine is a software application that helps organizations initiate and automate tasks. It manages and monitors activities in a workflow and determines which task to move to base on predefined processes.

3 MDA Approach for the Execution of the Functional Dimension

The MDA (Model-Driven Architecture) approach consists of three stages: Conceptual Information Model (CIM), Platform Independent Model (PIM), and Platform Specific Model (PSM), which covers the entire software development process from modeling to implementation. Through a transformation process, the MDA approach ensures traceability among the different models. The transformation of models is an important phase of the MDA process, they can be - Transformation from CIM to PIM: this transformation consists of building, partially, PIM models from the CIM. The goal is to transcribe the information contained in the CIM to the PIM models. This is what will ensure that the user's needs are conveyed and respected throughout the MDA process [9]. - Transformation PIM to PSM: this transformation consists of creating PSM models from the information provided by the PIM models, and adding technical information related to the target execution platform. This is how the link with the execution platform is formed.

In our work, we are interested in the transformation of CIM, PIM and PSM models. In order to generate SBP graphical models corresponding to BPM4KI instances, we identified correspondence rules that ensure the transformation between BPM4KI concepts and extended BPMN metamodel concepts. Figure 1 gives a graphical representation of the functional dimension transformation rules between BPM4KI (source model) and extended BPMN (target model). On the left side of the figure, we show the source metamodel, on the right side, we show the target metamodel, and in the middle, we describe the correspondence rules between the source and target metamodels with blue circles.

We ensure the transformation of the extended BPMN2.0 metamodel into an executable business process model. We are particularly interested in extending the functional dimension of BPMN2.0. To achieve this goal, we have oriented the model-driven MDA architecture to ensure the translation of models into executable business processes. In fact, the CIM (Computational Independent Model) layer of MDA corresponds to the BPM4KI model, the PIM (Platform Independent Model) layer corresponds to the BPMN2.0 extended meta-model, and the PSM (Platform Specific Model) layer corresponds to the sensitive business processes executed. It is described in the following sections.

3.1 Activiti BPMS Meta-model

Figure 1, The Platform Specific Model (PSM) level is concerned with the implementation of vital business processes. Therefore, designing and implementing these crucial processes require developing and implementing a personalized Business Process Management System (BPMS) that supports the required extensions [10]. Figure 2 displays the metamodel utilized by the Activiti BPMS, which was created through a reengineering process. The construction of the meta-model was based on the technical documentation and source code of the BPMS.

Figure 3 shows the meta-model on which the Activiti BPMS is based with the added extensions.

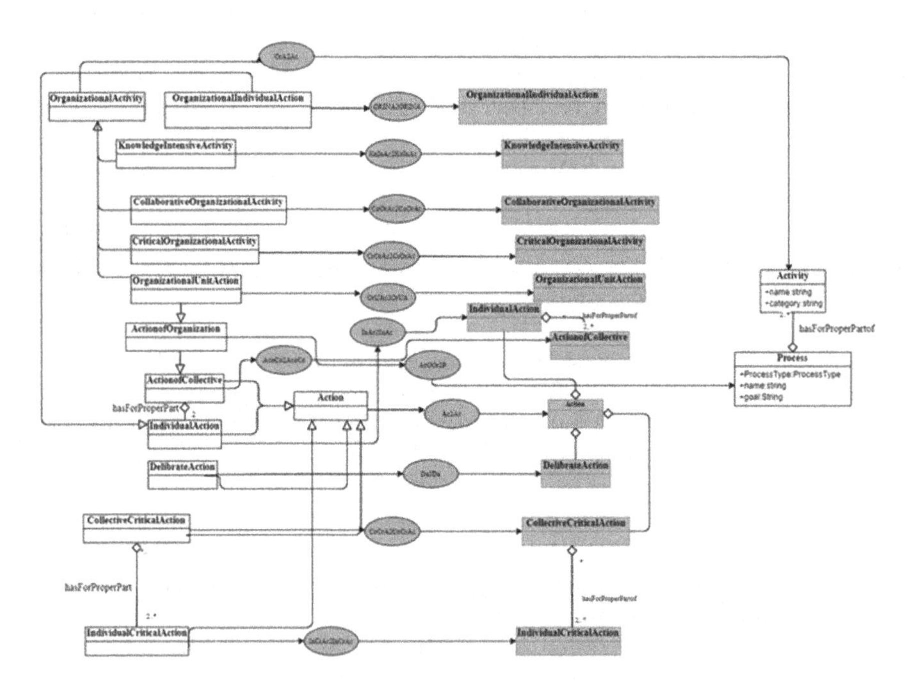

Fig. 1. Correspondences between concepts according to the Functional dimension

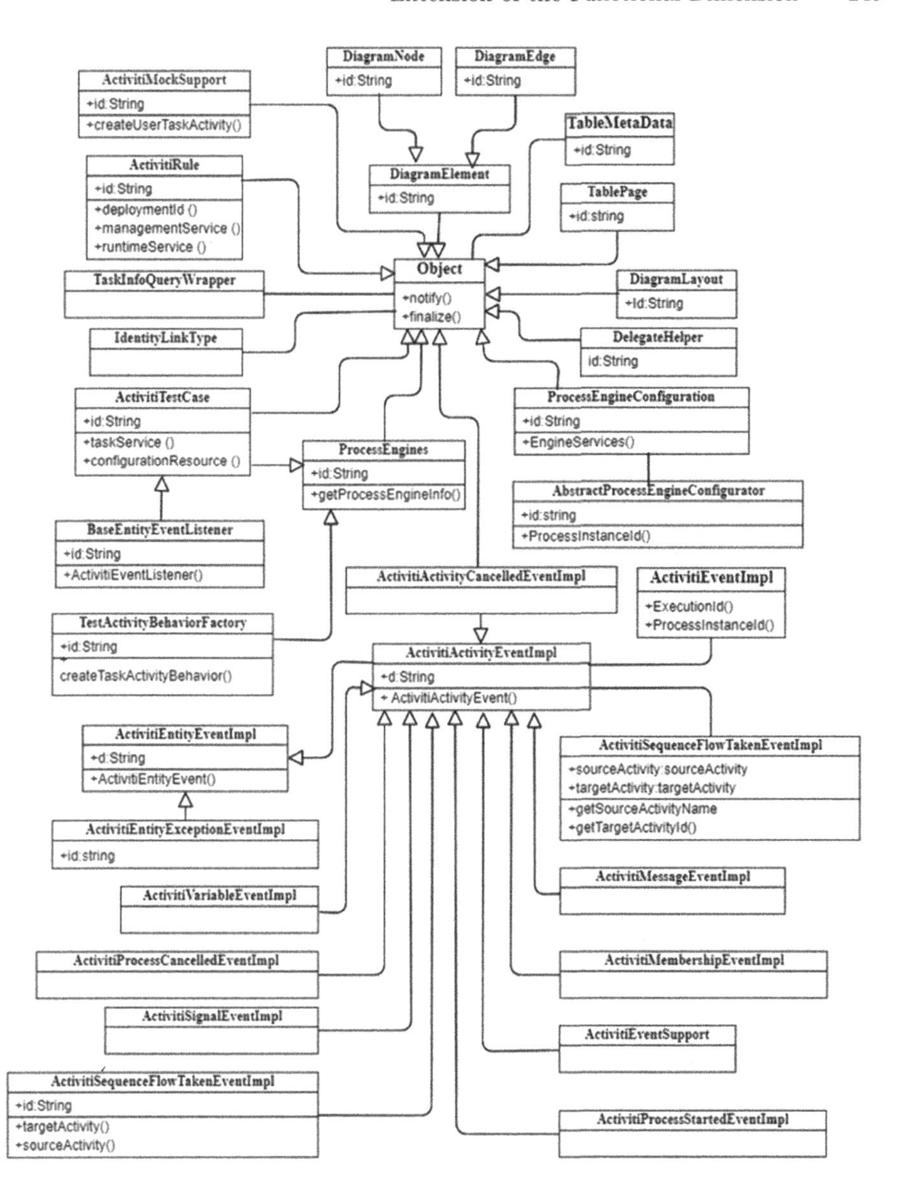

Fig. 2. Meta-model of bpms "Activiti"

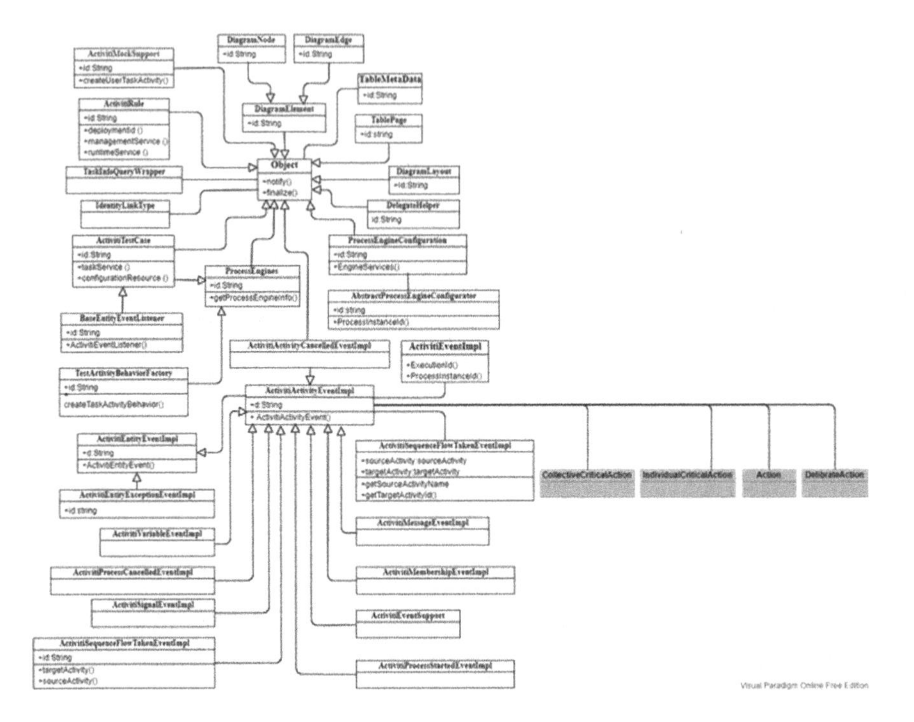

Fig. 3. Meta-model of bpms "Activiti"

3.2 Definition of the INAC2INAC Relationship with QVT

Two Figs. 4, 5 illustrate the definition of the INAC2INAC relationship using text (first image) and graphical (second image) notation for the QVT language. According to the figure, the INAC2INAC relationship allows the translation of individual actions of the bpmn4sbp model into collective actions of the bpmn4sbp model of both domains.

The first domain, represented by checkonly, represents a single action that can be turned into a collective action, supported by ActivitiRule or ActivitiTestCase. The second field, denoted "enforce", represents a single task that can be passed as a collective action as in the source model, backed by ActivitiTestCase or ActivitiRule. In addition, the when clause of the relationship specifies that the execution of the INAC2INAC relationship is executed after both the A2AR2 and ATC2ATC1 relationships have been executed.

```
top relation InCA2CCA
{B:Boolean;
Checkonly domain bpmn4sbp IA:IndividualCriticalAction
{
isTransmittedIn=a1:CollectiveCriticalAction{},
isBornBY=a:ActivitiRule{},
isBornBY=atc:ActivitiTestCase{},
Enforce domain bpmn4sbp AC:CollectiveCriticalAction
isTransmittedIn=a2:CollectiveCriticalAction{},
isBornBy=ar2:ActivitiRule{},
isBornBy=atc1:ActivitiTestCase{},
}
when
A2AR2(a,ar2);
ATC2ATC1(atc,atc1);
}
where{
A12A2(a1,a2);
}}
```

Fig. 4. Definition of the INAC2INAC relationship with the QVT text notation

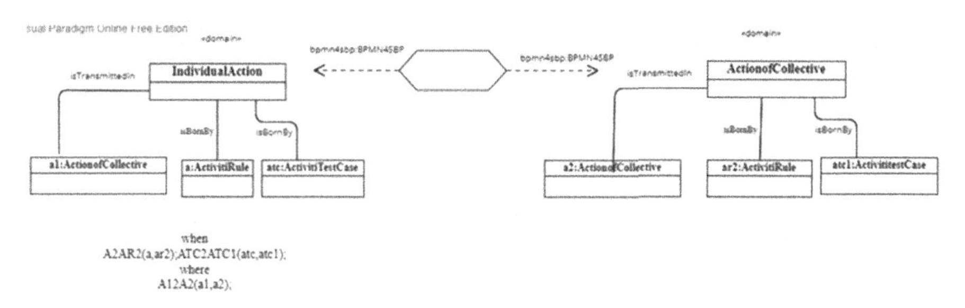

Fig. 5. Definition of the INAC2INAC relationship with the QVT graphic notation

4 Implementation and Validation

Our project centers on improving the representation of BPM4KI meta-model concepts through an enhancement of the BPMN 2.0 notation. We accomplished this by creating a plugin for Eclipse called BPMN4FM Activiti that expands the functional dimension of BPMN 2.0. With the help of this plugin, it is possible to generate BPMN diagrams that include the new extensions to the BPMN 2.0 meta-model that we introduced [11] (Figs. 6, 7, 8 and 9).

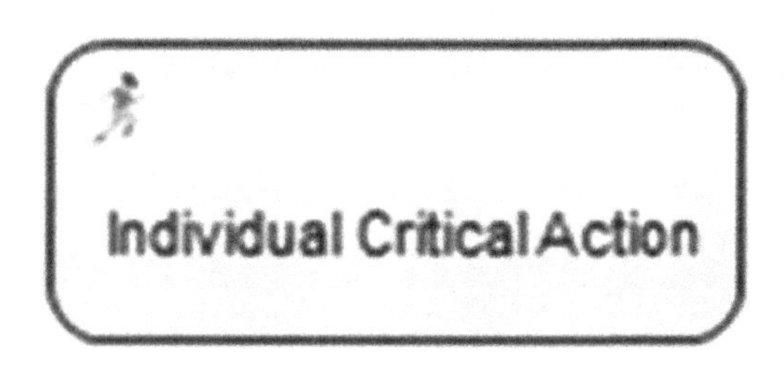

Fig. 6. The element individual critical Action

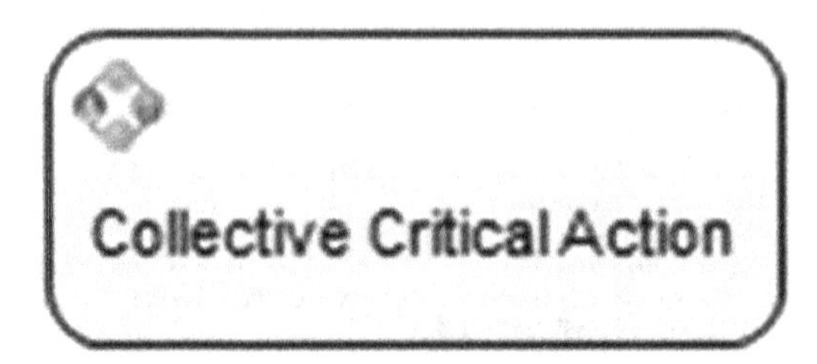

Fig. 7. The element collective critical action


```xml
<?xml version="1.0" encoding="UTF-8"?>
<definitions xmlns="http://www.omg.org/spec/BPMN/20100524/MODEL" xmlns:xsi="http://www.w3.org/2001/XMLSchema-instance" xmlns:xsd="http://www.w3.org/.
  <collaboration id="Collaboration">
    <participant id="pool1" name="Demande" processRef="process_pool1"></participant>
  </collaboration>
  <process id="process_pool1" name="process_pool1" isExecutable="true">
    <laneSet id="laneSet_process_pool1">
      <lane id="lane1" name="Emplyé">
        <flowNodeRef>startevent1</flowNodeRef>
        <flowNodeRef>servicetask1</flowNodeRef>
      </lane>
      <lane id="lane2" name="Responsable">
        <flowNodeRef>servicetask2</flowNodeRef>
        <flowNodeRef>endevent1</flowNodeRef>
      </lane>
    </laneSet>
    <startEvent id="startevent1" name="Start"></startEvent>
    <serviceTask id="servicetask1" name=" Saisir demande" activiti:expression="${echoBean.echo(execution)}" activiti:extensionId="org.bpmnfp.service
    <sequenceFlow id="flow1" sourceRef="startevent1" targetRef="servicetask1"></sequenceFlow>
    <serviceTask id="servicetask2" name="valider la demande" activiti:delegateExpression="${echoJavaDelegateExpressionBean}" activiti:extensionId="o
    <sequenceFlow id="flow2" sourceRef="servicetask1" targetRef="servicetask2"></sequenceFlow>
    <endEvent id="endevent1" name="End"></endEvent>
    <sequenceFlow id="flow3" sourceRef="servicetask2" targetRef="endevent1"></sequenceFlow>
  </process>
  <bpmndi:BPMNDiagram id="BPMNDiagram_Collaboration">
    <bpmndi:BPMNPlane bpmnElement="Collaboration" id="BPMNPlane_Collaboration">
      <bpmndi:BPMNShape bpmnElement="pool1" id="BPMNShape_pool1">
        <omgdc:Bounds height="300.0" width="500.0" x="190.0" y="140.0"></omgdc:Bounds>
      </bpmndi:BPMNShape>
      <bpmndi:BPMNShape bpmnElement="lane1" id="BPMNShape_lane1">
        <omgdc:Bounds height="150.0" width="480.0" x="210.0" y="140.0"></omgdc:Bounds>
      </bpmndi:BPMNShape>
      <bpmndi:BPMNShape bpmnElement="lane2" id="BPMNShape_lane2">
        <omgdc:Bounds height="150.0" width="480.0" x="210.0" y="290.0"></omgdc:Bounds>
      </bpmndi:BPMNShape>
      <bpmndi:BPMNShape bpmnElement="startevent1" id="BPMNShape_startevent1">
        <omgdc:Bounds height="35.0" width="35.0" x="250.0" y="190.0"></omgdc:Bounds>
      </bpmndi:BPMNShape>
```


Fig. 8. Extract from the file including the grammar of the extension point BPMN2 Activiti runtime

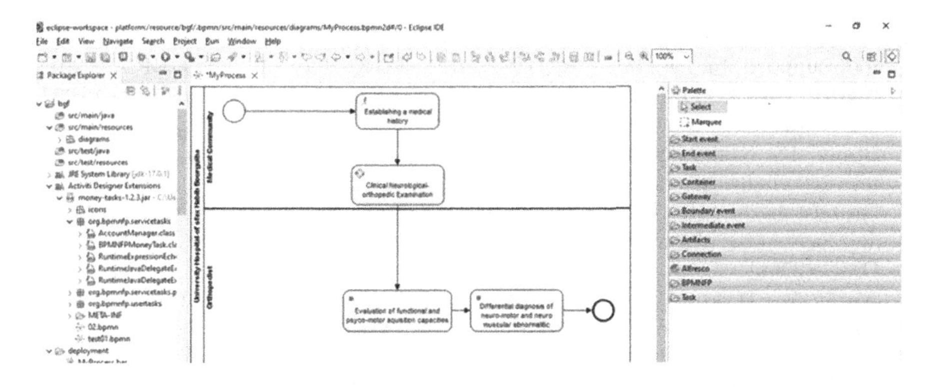

Fig. 9. Extract from the graphic representation model of the process related to the neuro-pediatric management of a child with Cerebral Palsy with extended BPMN Activiti

5 Conclusion

The aim of our research was to develop an Eclipse plugin, called Activiti, that allows for the modeling and execution of SBPs using the BPMN2.0 standard. Our plugin incorporates the functional enhancements of BPMN2.0 to improve its performance.

Although this work has already achieved notable advancements, there are still opportunities for further investigation. In the near future, it could feasible to entirely integrate the upgrades to the BPMN2.0 notation, in alignment with the BPMN4SBP metamodel.

Our objective in the medium term is to automate the conversion process between PIM and PSM models, utilizing the proposed MDA framework. This would facilitate the automatic creation of executable SBPs.

References

1. Turki, M., Saad, I., Gargouri, F., Kassel, G.: A business process evaluation methodology for knowledge management based on multi-criteria decision making approach. Inf. Syst. Knowl. Manag. 978-1848216648 (2014)
2. Hassen, M.B.: Une approche d'extension de bpmn 2.0 basée sur des ontologies noyaux pour la spécification des processus métier sensibles dans une perspective de gestion de connaissance. Thèse de doctorat en informatique (2021)
3. Hassen, M.B., Turki, M., Gargouri, F.: Sensitive business processes: characteristics, representation, and evaluation of modeling approaches. Int. J. Strateg. Inf. Technol. Appl. **9**(1), 41–77 (2018)
4. Ben Hassen, M., Turki, M., Gargouri, F.: A multi-criteria evaluation approach for selecting a sensitive business process modeling language for knowledge management. Data Semant. **72**(2), 157–202 (2019)
5. Baumgraß, A., Herzberg, N., Meyer, A., Weske, M.: BPMN extension for business process monitoring. In: Enterprise Modeling and Information Systems Architectures (2014)

6. Stroppi, L.J.R., Chiotti, O., Villarreal, P.D.: Extending BPMN 2.0: method and tool support. In: Dijkman, R., Hofstetter, J., Koehler, J. (eds.) BPMN 2011. LNBIP, vol. 95, pp. 59–73. Springer, Heidelberg (2011). https://doi.org/10.1007/978-3-642-25160-3_5
7. Braun, R.: Semantics in the context of BPMN extensions state of affairs and research challenges. Numer. Math. (2016)
8. Grundstein, M., Rosenthal-Sabroux, C.: A process modeling approach to identify and locate crucial knowledge. Numer. Math. 49–54 (2008)
9. OMG (2008)
10. Alyani, Z., Turki, M.: An MDA approach for extending functional dimension for sensitive business processes execution. In: Abraham, A., Pllana, S., Casalino, G., Ma, K., Bajaj, A. (eds.) ISDA 2022. LNNS, vol. 715, pp. 147–155. Springer, Cham (2022). https://doi.org/10.1007/978-3-031-35507-3_15
11. Turki, M., Saad, I., Gargouri, F., Kassel, G.: A business process evaluation methodology for knowledge management based on multi-criteria decision making approach (2014)

Inclusive Mobile Health System for Yoruba Race in Nigeria

Olutola Agbelusi[1](✉), Isaiah Olomi Aladesote[2], and Olumuyiwa Matthew[3]

[1] Department of Software Engineering, Federal University of Technology, Akure, Ondo State, Nigeria
tola52001@yahoo.com

[2] Department of Computer Science, Federal Polytechnic, Ile Oluji, Ondo State, Nigeria
isaaladesote@fedpolel.edu.ng

[3] School of Computing, University of Portsmouth, Portsmouth, UK
olumuyiwa.matthew@port.ac.uk

Abstract. Language differences are major barrier to the dissemination of information between patients and health workers in the Southwestern part of Nigeria. Developing a system that would assist an individual to explain their health issues in their preferred language using mobile devices is a spot for concern to effective communication. This idea will promote and achieve good health for all which is the NO 3 of the seventeen sustainable development goals (SDGs) adopted by the general assembly of the United Nations. The authors have developed an inclusive mobile health application that would translate English text to Yoruba text using a direct rule-based machine translation approach. Production rules based on English sentence structure were formulated using context-free grammar and a dictionary data set containing English words and their Yoruba language equivalent was also adopted. The system development was achieved with Java programming, Extensible Markup language (XML), and Mongo DB. The new system (RTCHAT) was developed and its translation result was compared with Google Translate. One hundred (100) translated sentences used in the new application (RTCHAT) were extracted and given to Yoruba linguistics to identify correctly, partially, and wrongly translated sentences. The result was compared with google translate of the same group of sentences and the result shows that eleven (11) sentences were correctly translated in Google translate and eighty-nine (89) in RTCHAT, ten (10) were partially translated in RTCHAT and thirteen (13) in Google translate and finally, seventy-two (72) sentences were wrongly translated in Google and three (3) in RTCHAT. Performance of the new system (RTCHAT) was tested at Rufus Giwa Polytechnic (Nigeria) community and the result shows a high percentage of performance and ratability.

Keywords: Translation · Bilingual dictionary · Direct Rule-Based machine translation · Extensible Markup Language

I. Saad et al. (Eds.): ICIKS 2023, LNBIP 486, pp. 255–264, 2024.
https://doi.org/10.1007/978-3-031-51664-1_18

1 Introduction

The significant of language as a means of dissemination of information cannot be ruled out in our day-to-day activities, especially in the area of health. The target of this work is to remove language barrier of communication between health workers and mobile users in the South-western Nigeria using the direct machine language translation technique. Translation in this context is the process of translating English text to Yoruba text and this is not always a direct substitution of a word(s). It also requires a total translation system that must have all the grammatical features of the target language.

2 Review of Related Works

This section comprises related works on Health Application and Language Translation.

2.1 Related Works on Some Health Application

Many researchers have worked on health application development, and translation of health information over the years, intending to use mobile devices to improve health challenges among citizens of the South western Nigeria. A review of some related works that motivated this research on mobile development and machine translation is hereby presented. Authors in [1] developed a mobile personal health care system to keep glucose at a healthy level for people with diabetes. The research objective is to develop a personal diabetes monitoring system which integrates wearable sensors, 3G mobile phone, smart home technologies and Google sheet to facilitate the management of chronic disease like diabetes. The system is robust and simple, but is only for the elderly with diabetics and this makes the system not inclusive.

The design of a mobile platform for proper distribution of rescue and relief materials to disaster-affected areas in the Philippines using android technology to design a mobile [2]. The genetic algorithm was adopted to determine the most optimum route for the volunteers and rescuers to use for the proper distribution within the shortest possible time during the period of this calamity. The result shows that victim without mobile phone found it difficult to locate an escape direction and the optimum route along the given geographical locations were determined. This application is a mono-tasking one because it is used for only disaster management.

The development of a Framework for Usable Operations Support in E-health Based Systems using Algebraic specifications in object constraint language (OCL) and unified modelling language (UML) for the designing of sub-system. Wireless markup language (WML) and Java programming language were used for the development. The system is not inclusive because it is limited to the health workers alone [2].

Authors in [3] developed a data encryption solution for mobile health applications in corporation environments (DESMHA). The objective of the study is to present a robust solution based on encryption algorithms that guarantee the best confidentiality, integrity, and authenticity of users' health information. A mobile health application was developed with an encryption algorithm (DE4MHA) and its performance was evaluated through a prototype using a mHealth app for obesity prevention and cares (SapoFit). The result shows that the availability of a data encryption algorithm (DE4MHA) allows users to safely obtain secured health information. The limitation of the study shows that security on mobile devices, due to low processor capacity, is not guaranteed.

[4] presented the strength of passwords used to protect personal health information in clinical trials. The objective of this study is to evaluate the security practices used to transfer and share sensitive files in clinical trials. The researchers used commercial password recovery tools on 15 password-protected files to crack their passwords and compare the result with the opinion of 20 study coordinators to understand file-sharing practices in clinical trials for files containing personal health information. The result suggested that the password technique tended to be relatively weak to protect personal health information. The limitation of the research is that the use of the password is not strong for securing data.

2.2 Review of Related Works on Language Translation

Yoruba race in Nigeria speaks two major languages (English and Yoruba) and in order to make the new system (RTCHAT) an inclusive one, there is a need to include language translation as part of this research so as to assist in the translation of language from English text to Yoruba text and vice versa.

[5] worked on English text to Yoruba text translation using a rule-based approach. The visual studio 2012, ASP.net, and C# programming language were adopted for the system implementation. A comparative analysis was made between the new system translate and that of Google and the result shows that partial translation could be achieved in the new system.

A direct machine translation approach was adopted in [6] to build a dictionary for English disease names and their Yoruba equivalent. The system implementation was done using Visual Basic 6.0 and the system has good and effective performance capability.

[7] adopted a transfer rule-based machine translation method to develop a system that can convert English text to Yoruba text. The system implementation was achieved and the new system gives a better result than Google Translate a thorough comparison between the two systems. The limitation is that this was not deployed as mobile application.

[8] worked on a computational analysis of the translation process using a rule-based approach. The system modeling was achieved using context-free grammar and re-write rules were generated. The limitation is that a model was developed in this research but the new system will develop and deployed on a mobile application.

This researcher developed a translation system in which finite state automation was adopted for the modeling process. Translated noun-phrase noun phrases to Yoruba were achieved (Adeoye, O.B. 2012). This translation system was not deployed as a mobile application.

[9] proposed Malayalam text to English text translator using a transfer-based translation approach. The system implemented was done using the python language and the system performance is good and encouraging. The limitation is that the research was hosted on a local host and not as mobile application.

[10] proposed a noun–phrases machine translation system using twenty-nine specified with context-free. The model was achieved with PHP Hypertext Processor programming language, MYSQL tested on four hundred randomly selected noun phrases and

gives an accuracy of 91% which is quite encouraging. This translation system was not deployed on a mobile app.

The works of [5, 7–9, 11] and [10] were reviewed. Translation in this research is playing a vital role of translating English text to its Yoruba equivalent and vice versa and invariably making the new developed system inclusive. The inclusivity of the new system addresses the limitation of the existing systems of not been inclusive.

3 Research Methodology

Figure 1 is the software architecture of the new system (IHS). It comprises of three layers (presentation, middle and backend layers. The presentation layer consists of authentication support, identity support and view/post request support services. The middle layer also consists of the webserver, WAP gateway, data access and update service, expert and user support service and translation module that helps in the translation of text from English to Yoruba language and vice versa. Translation model is developed and the process of translation is explained as seen in Fig. 2. Finally, the backend or data layer consist of the database.

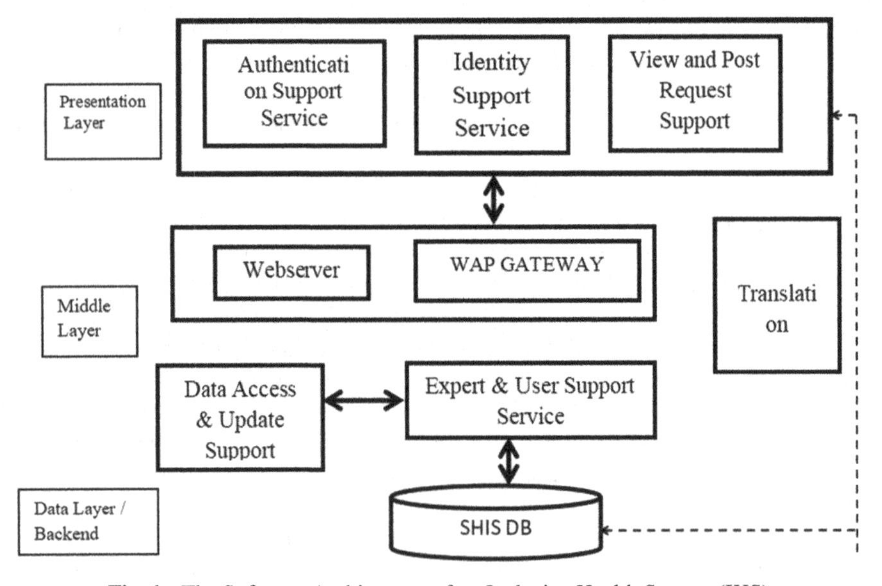

Fig. 1. The Software Architecture of an Inclusive Health System (IHS)

3.1 Translation Model

Figure 2 Describes the translation process in this research where the source language was analyzed with a Stanford parser. The bilingual dictionary for the translation process consists of Yoruba text and the English text equivalent. After the tagging process, there is a lookup of every word (English/Yoruba) from the dictionary with their equivalents. The lexicon contains different grammatical attributes which are assigned by the sentence structure and stored in the status flag which is a temporary location using queuing process. Syntactic re-ordering is optional and it is represented with dotted lines; that is, sentences in English that are translated to Yoruba directly do not need any re-arrangement. The target sentences are generated as output after the whole process of translation is carried out.

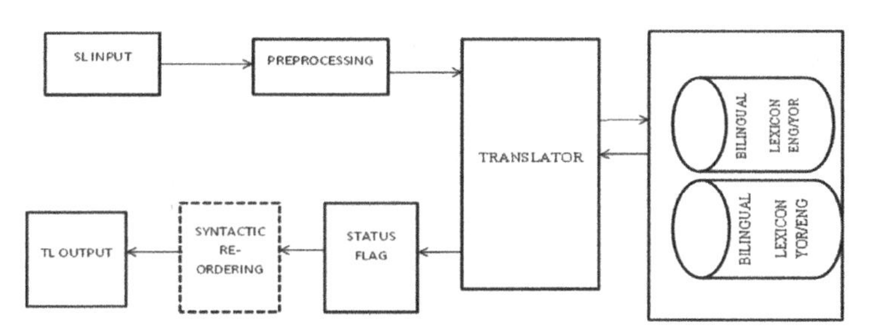

Fig. 2. Block Diagram for Translation Process

The production rules based on English sentence structure are as follows:

S→<NP><VP>

NP→<N>|<PN>|<DETN>|<ADJN>|<ε>

N→<SN>|<PN>

VP→<VNP>

V→<AVLV>|<AV>|<LV>|<LVAD>

LV→<LVC>|<LVP>|<LVS>

PP→<PREPNP>

The production rules based on Yoruba sentence structure are as follows:

S→<NP><VP>

NP→<N>|<NP>|<NDET>|<NADJ>|<ε>

N→<SN>|<PN>

VP→<VNP>

V→<AVLV>|<AV>|<LV>|<LVAD>

LV→<LVC>|<LVP>|<LVS>

PP→<PREPNP>

where: S = Sentence, NP = Noun phrase, VP = verb phrase, N = Noun, P = Pronoun, DET = Determinant, V = Verb, ADJ = Adjective, AV = Auxiliary Verb, LV = Lexical Verb, LVP = Plural lexical verb, LVS = Singular lexical verb, SN = Singular Noun, PN = Plural Noun, AD = adverb, PREP = Preposition, ε = Empty.

4 Result and Discussion of the New System

The chat in Fig. 3 is from English text to Yoruba text. The individual was able to send chat in Yoruba and also getting reply in Yoruba. Figure 4 showing a chat from English text to English text.

Fig. 3. Chatting process from Yoruba text to Yoruba text

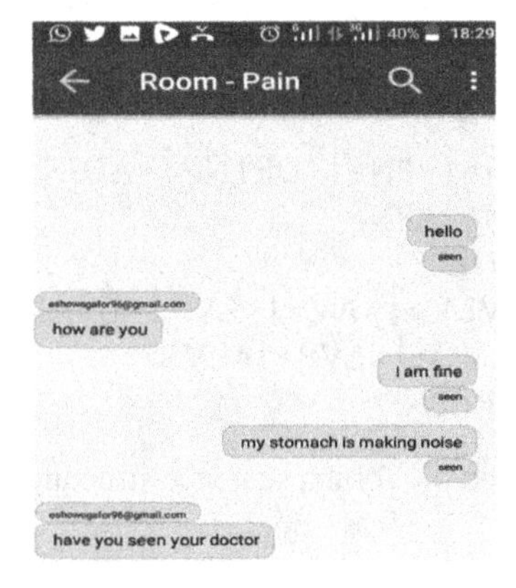

Fig. 4. Chatting process from English Text to English text

4.1 Performance Evaluation of the New System (RTCHAT) Translation Process

The new system performance evaluation was done using eight (8) attributes. A Statistical tool was adopted to determine the mean rating, standard deviation, variance values, and average mean rating. The average value of the mean rating was compared with each attribute's mean value to test for the level of significance. The decision rule is significant if the mean value of each attribute exceeds or equals the average mean as shown in Table 1. Table 3 shows the reliability test of the application in which Cronbach's Alpha value is more than70%. Table 3 shows the reliability test of the application (Table 2 and Fig. 5).

Table 1. Descriptive Statistical Analysis of Questionnaire Data.

	Mean	Std. Deviation	Variance	Decision Rule
Completion Time	3.14	.752	.566	Significant
Easy to learn	3.16	.735	.540	Significant
Effective Support	3.16	.884	.782	Significant
Error Recovery	3.14	.752	.566	Significant
Comfort ability	3.16	.735	.540	Significant
Satisfaction	3.14	.921	.849	Significant
Visual Display	3.14	.779	.606	Significant
Acceptance	3.14	.954	.909	Significant
Average Mean	**3.14**			

Table 2. Case Summary.

		N	%
Cases	Valid	100	100.0
	Excluded(a)	0	.0
	Total	100	100.0

Table 3. Reliability Statistics.

Cronbach's Alpha	No of Items
.712	8

4.2 Comparative Analysis of RTCHAT and GOOGLE Translate

Table 4 shows GOOGLE and RTCHAT translation process comparison in which (100) sentences translated from English to Yoruba were given to Yoruba linguistic to identify correctly, partially and wrongly translated sentences. Their responses show that Eleven (11) sentences were correctly translated in GOOGLE translate and eighty-nine (89) in RTCHAT, ten sentences were partially translated in RTCHAT and thirteen (13) in GOOGLE translate and finally, seventy-two (72) sentences were wrongly translated in GOOGLE translate and three in RTCHAT. The results in Fig. 6 shows that RTCHAT translate perform better than Google translate.

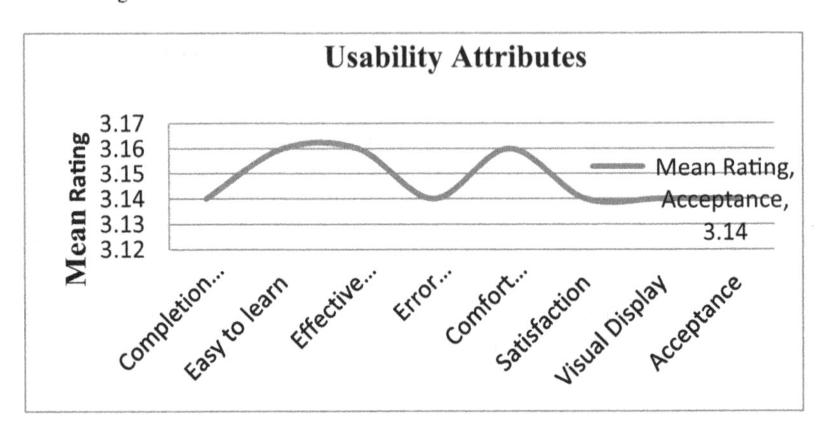

Fig. 5. Graphical Representation of the Usability Attribute Rating

Table 4. .

S/N	English Sentence	Rtchat Translate	Google Translate
1	Mother what about your leg	Ìyá bá woni esè re	Iya ohun ti o jẹ nipa ẹsẹ rẹ
2	The girl didn't sleep well at night	Omodébìrinnáà ò sùn ni òru	Ọmọbirin naa ko sùn daradara ni alẹ
3	My stomach is making noise	Inú mi ń pariwo	Iya mi n ṣe ariwo
4	Go and see your doctor	Lo rídókítà rẹ	Lọ ki o wo dokita rẹ
5	The boy is eating	Omodékùnrinnáà n jẹun	Ọmọkunrin n jẹun
6	My doctor said I should go for medical test	Dókítà mi sọwípéki n lọsẹàyèwò	Dokita mi sọ pe mo yẹ ki o lọ fun idanwo egbogi
7	My knee-cap is paining me	Orókún mi ń dùnmí	Okun ikunkun mi ti npa mi
8	I cannot hear very well with my right ear	Mi ò le gbóran dáradárapèlú ètí òtún mi	Emi ko le gbọ eti ọtun mi daradara
9	I have headache	Mo ní èfórí	Mo ni orififo
10	I have rashes on my body	Ifòn wàniara mi	Mo ni rashes lori ara mi

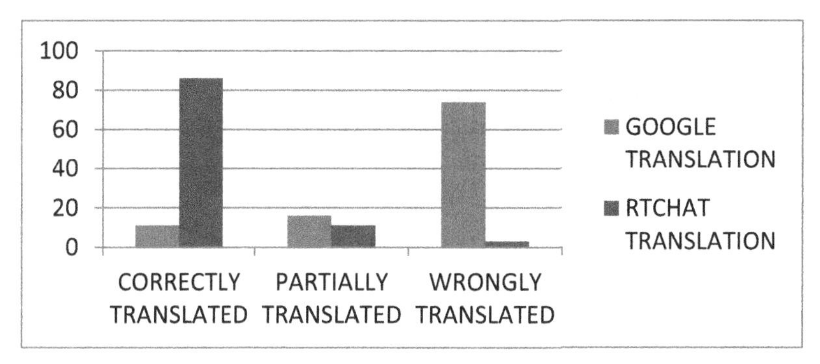

Fig. 6. Graphical Representation of Comparative Analysis of RTCHAT and GOOGLE Translate

5 Conclusion

The new system has been able to remove the barrier of language inclusivity and this will promote the "well-being for all as proposed in the SDG. Direct rule-based machine translation technique has been able to translate sentences efficiently and the result from system evaluation also shows that RTCHAT translation is better than Google translate.

References

1. Fachao, Z., Yang, H., Alamo, J., Wong, J., Chang, C.: Mobile personal health care system for patients with diabetes. In: Lee, Y., et al. (eds.) Aging Friendly Technology for Health and Independence. ICOST 2010. LNCS, vol. 6159, pp. 94–101. Springer, Heidelberg (2010). https://doi.org/10.1007/978-3-642-13778-512. Accessed 14 Nov 2020
2. Nicholas, O.: Development of a Formal Framework for Usable Operations Support in E-Health Based Systems. Unpublished Ph.D Thesis Submitted to the Department of Computer and Information Sciences to the School of Postgraduate Studies. Covenant University, Ota, Ogun State (2007)
3. Bruno, S., Rodrigues, J., Canelo, F., Lopes, I., Zhou, L.: A data encryption solution for mobile health apps in cooperation environments. J. Med. Internet Res. **15**(4), 1–11 (2015). https://doi.org/10.2196/jmir.2498
4. Khaled: How strong are passwords used to protect personal health information in clinical trials? J. Internet Res. **13**(1), e18 (2011)
5. Akinwale, O.: Web-based English to Yoruba machine translation. Int. J. Lang. Linguistics. **3**(3), 154–159 (2015). https://doi.org/10.11648/j.ijll.20150303.17
6. Abiola, O.B., Adetunbi, A.O., Oguntimilehin, A.: A review of the various approaches for text to text machine translation. Int. J. Comput. Appl. **120**(18), 7–12 (2015)
7. Agbeyangi, A., Eludiora, S.I., Adeyinka, A.O.: English to Yoruba machine translation system using rule-based approach. J. Multi. Eng. Sci. Technol. (JMEST). **2**(8), 2275–2280 (2015)
8. Eludiora, S.: Development of an English to Yoruba machine translation system. Unpublished Ph.D Thesis Submitted to Department of Computer Science and Engineering, Obafemi Awolowo University Nigeria (2013)
9. Latha, R., David, P., Renjith, P.: Design and implementation of a Malayalam to English Translator-A Transfer Based Approach. Int. J. Comput. Linguist. (IJCL). **3**(1), 1–11 (2012)

10. Abiola, O.B., Adetunbi, A.O., Fasiku, A.I., Olatunji, K.: A web-based English to Yoruba noun-phrases machine translation. Int. J. English Lit. **5**(3), 71–78 (2014)
11. Adeoye, O.: A Web-Based English to Yoruba Noun Phrase Machine Translation. Unpublished Master Thesis Submitted to Department of Computer Science. The Federal University of Technology Akure, Nigeria (2012)

Cybersecurity and Intelligent Systems

Epistemology for Cyber Security: A Controlled Natural Language Approach

Leigh Chase(✉), Alaa Mohasseb, and Benjamin Aziz

School of Computing, University of Portsmouth, Portsmouth, UK
up348663@myport.ac.uk, {alaa.mohasseb,benjamin.aziz}@port.ac.uk

Abstract. In this paper we introduce a new Controlled Natural Language (CNL) known as "Noam". It is used to express cyber security knowledge and for reasoning over it. The approach follows examples set by other domain-specific languages and constrained grammars, but is highly unusual due to its singular focus on cyber security. Like most CNLs Noam is both human-readable and machine-solvable, thus fulfilling important assurance requirements with respect to transparency and explainability. The language seeks to address a growing problem faced by security engineers and architects; namely, that their endeavours are constrained by the complexity and sheer interconnectedness of the systems they protect. This is further compounded by year-on-year vulnerability disclosure rates and diversification of the Tactics, Techniques and Procedures used by threat actors. Our approach is analogical in which the Noam CNL is used to construct a system model, instrument it with data from the real environment and apply functional programming techniques in order to 'solve-for' certain conditions of interest. The intention is to demonstrate the value of CNLs and semantic reasoning within cyber security, framed in the context of improving the information available to security engineers, architects and other decision-makers.

Keywords: cyber security · knowledge representation · controlled natural language · epistemology · machine reasoning · explainable learning

1 Introduction

Understanding the security posture of information systems is an important topic because it tells us something about how robust, reliable and trustworthy these systems are. In turn, this informs decisions about which systems we will use, what information we are prepared to share and what functions we will allow them to perform. Security Architecture is a discipline whose focus is to assure the Confidentiality-Integrity-Availability (C-I-A) of Information Technology - specifically, it is "the practice of designing computer systems to achieve security goals"

Supplementary Information The online version contains supplementary material available at https://doi.org/10.1007/978-3-031-51664-1_19.

I. Saad et al. (Eds.): ICIKS 2023, LNBIP 486, pp. 267–282, 2024.
https://doi.org/10.1007/978-3-031-51664-1_19

(NCSC 2018). Cyber security attacks are made possible by the discovery and exploitation of vulnerabilities that affect their components or are otherwise inherent in their design. This introduces important relationships between vulnerabilities, system design (architecture) and the Tactics, Techniques and Procedures (TTPs) used by adversaries (NIST-CSRC 2023). The problem for security architects is that whilst these concepts are all related, they are also quite different fields of study. Furthermore, there are only very limited mechanisms with which to evaluate these ideas within a single frame of reference. A potential solution is to create an abstraction in which one which expresses the necessary concepts such as systems, networks, vulnerabilities and adversaries to create models that express and allow us to evaluate cyber security scenarios. This akin to how Goal Modelling has been used in other domains (Ballard et al. 2022).

In this paper we introduce "Noam": a Controlled Natural Language (CNL) designed to address these problems. It is highly domain-specific and only to express cyber security concepts, construct model-based scenarios and reason over them. The language provides a specialised grammar for knowledge representation, a rule-based scenario evaluation system and a novel mechanism that accounts for the temporal components of IT security. Like most CNLs it is both human-readable and machine-solvable, fulfilling a need for transparency and "interpretability" (as the term is used in Machine Learning). This paper introduces the language's formal specification abstract of any particular implementation, but Noam has been tested using Prolog and miniKanren (discussed below). The language is named in reference to Noam Chomsky on account of his work-established work in linguistics. Chomsky's Hierarchy of Grammars is considered a seminal contribution that spans a (perhaps unlikely) combination of computer science and linguistics[1].

2 Related Work

This approach builds upon the examples set by other domain-specific CNLs, according to how they are applied to complex and semantic problems in real-world environments - such as Gellish (van Renssen 2003). Noam is unusual amongst CNLs because of its specialism, however the field of knowledge modelling and expression within cyber security is well-served by the literature. It is important to note that this work is part of a wider research project considering the use of knowledge sharing and ensemble learning algorithms to predict cyber security threats. The CNL described herein was developed as an in Intermediate Representation (IR) with which to describe the organisation of a system, conduct reasoning/inferencing over that 'space' and to explore domain-specific knowledge representation formats compatible with shallow machine learning techniques (specifically ensembles of classification algorithms).

[1] Use of this name is purely a convenient shorthand and recognises the technique's inherent overlapping of information science and language. There is no reference - intended or otherwise - to any of the namesake's specific work, writings or politics. The authors, nor the ideas herein seek any controversy and the naming is purely symbolic.

These ideas are highlighted by (Rahman et al. 2020) in the context of predicting attacks - notably, where the authors aims to predict possible future attacks using data known to contain hostile activities. These include such as Denial of Service (DoS) attacks, port scanning (survey and discovery), botnet activity, brute force authentication attempts, SQL Injection and exploitation of the Heartbleed vulnerability (MITRE 2014). Indeed, this is a hard problem and the authors show excellent performance using a J48 decision tree algorithm and predictive model, on data sets obtained from the Canadian Institute for Cyber Security. However, their work highlights some of the difficulties in generalising both effective learning algorithms and "cyber knowledge". (Rahman et al. 2020) and (Straub 2020) individually highlight the challenge of a representation framework that is genuinely a priori with respect to any given cyber security case, incident or investigation. This is also discussed in the context of threat profiling by (Irwin 2014).

Semantic computing methods within cyber security are rather more rare. Cyber Threat Intelligence (CTI) is garnering more attention in areas related to this work due to its inherently associative nature. CTI formerly defined in Special Publication 800-150 from the National Institute for Standards and Technology (C. et al. 2019):

> "By exchanging cyber threat information within a sharing community, organizations can leverage the collective knowledge, experience, and capabilities of that sharing community to gain a more complete understanding of the threats the organization may face."

The MITRE ATT&CK project (MITRE 2021b) is amongst the most well documented public CTI dataset, though this is a purely reference source. From computer and information science more widely, projects such the Dublin Core Metadata Initiative (DCMI 2022) highlight its relevance to many knowledge representation and reasoning problems. (Souza L.O. 2019) introduces and explains "Focused Experience Sharing" as a mechanism for improving the learning performance of deep methods - crucially, it is the experience sharing component of their work that yields these improvements. Similarly, (F. et al. 2020) introduces a model of shared experience using the Actor-Critic method that applies to more general types of learning agents. However, these topics do not emphasise transparency, nor do they attempt to express or work upon the semantics of the underlying environment. Separately are ideas relating to cyber security-specific knowledge models, such as (H.S. et al. 2018) in their discussion of Attack Trees. The topic is also described by (Petrica et al. 2017) and generalised further by (Ampel et al. 2021) and (Shahid and Debar 2021). However, these do not feature CNLs in their approach. The YARA standard was originally developed by VirusTotal as tool designed to help malware researchers identify and classify samples (VirusTotal 2022). It offers a lightweight and flexible solution to the problem of encapsulating the results of malware analysis, but is not a processing or reasoning environment. It is a structured data format with which to codify the results of analysis, independent of the methods that have been applied during those analyses. The Common Vulnerabilities and Exposures (CVE) and Common Vulnerability Scoring System (CVSS) are widely adopted as the basis for expressing vulnerability data, but again are data standards first-and-foremost. The US Government's Cybersecurity and Infrastructure Security Agency (CISA) maintains a list of exploited vulnerabilities (CISA 2023), which associates vulnerabilities and what they affect via pairwise mappings, but is offered as an authoritative reference rather than as the basis for scenario building/evaluation.

More generally, (Haykin 2012) sets-out four functions (or tasks) that are basic to human cognition comprising the perception-action cycle, memory, attention and intelligence. The author describes the field as "integrative" rather than a "discipline", which summarises the approach used in the Noam language design. There exists a substantial body of work dealing with security information and its adjacent topics, but comparatively little that applies these ideas. The perception-action cycle is a more inspirational model than many of the more cyber-specific approaches described in the literature.

Specifically, this CNL is intended to address the problem of combining CTI data with a generalised, abstract yet accurate description of how a system is organised. Whilst CTI informs us as to how attackers structure/effect attacks, a system's design and its implementation (including the 'running configuration' of its components) have a tangible effect on its security posture. Consequently, being able to make effective use of both sources is considerable practical use. However, at the time of writing there no single solution that allows us to combine these datasets in a straightforward fashion. Both CTI and systems configuration datasets can be complex - as such, a language that is human readable and machine parsable helps to address parallel problems of scalability and representational completeness, whilst providing a framework for machine reasoning is a helpful asset. Finally, the CNL-based approach outlined in here is also considered the foundation for the future application of machine learning classification algorithms.

3 Proposed Approach

Noam is a 'macro modelling' CNL with which to construct and evaluate cyber security scenarios. Each scenario combines components called **Objects**, **Relationships** and **Data Flows** that articulate its characteristics. This follows a system-of-systems approach in which a 'system' is not an individual computer but a collection of hosts, networks and other apparatus assembled to serve a given purpose. For example, Noam can be used to model a company's payroll system and networks, but not the internal workings of an individual device (such as a web server). The design combines Semantic Computing (IEEE 2022) with CNLs and extends them into this domain to create a rich, flexible descriptive framework for cyber security knowledge. Noam is strongly influenced by the principles of Analogical Reasoning (AR). When explaining the topic, (Luger 2005) highlights two important principles:

1. **Generalisation**: the purpose of AR is to generalise experience in some meaningful way. AR uses analogies to infer qualities of x (of which one has limited knowledge) based on general similarities it has with y (of which one has more detailed knowledge or experience) - akin to induction.
2. **Soundness**: compared to Explanation-Based Learning (EBL), AR is not logically sound. In the strictest sense, AR is not a deductive process and this must be considered when interpreting any conclusions.

Both are useful in the context of cyber security - Noam is not concerned with modelling 'perfect knowledge' of systems because it is simply unrealistic to acquire. AR shows how well-judged approximations and analogues can be used to infer properties that are likely, but of which we cannot be certain. While Noam is rules-based, it cannot be said to be truly deductive since components within its models and processes may lack precision. Noam can be said to approach conclusions "on the balance of probabilities". Like most CNLs, Noam is 'interpretable', in that it is both:

- **Human-readable** - meaning that the language is comprehensible to humans without need of an intermediate parser or transpiler; and
- **Machine-solvable** - meaning that Noam scenarios are evaluated programmatically without intermediate parsing or transpilation.

More intuitively, this means that human users and machines can "work in the same language". Similarly, Noam is not a general computer language nor is it Turing-complete. An overview of Noam's components is shown in Fig. 1.

The **Domain Model** (Sect. 3.1) combines the objects, relationships and data flows to describe a given **Scenario** of interest the user. The **Realms** are used to separate what the system *is*, from what is *was* and what it *can be*. This is somewhat more conceptual than the Domain Model but is an important part of the design, discussed below in Sect. 3.3. Finally, **Solvers** (Sect. 3.5) evaluate the Domain Models expressed in one of the realms using rules applied to a given scenario and which yield an output based on these evaluations.

3.1 Domain Model

The Domain Model is an ontological representation of an information system and contains three entity types: **Objects**, **Relationships** and **Data Flows** (each is described below). All entities have a set of sub-types specified by the CNL. Again, these are described below. The architecture of objects, relationships and data flows is illustrated in Fig. 2.

Objects. Domain Model Objects represent components of the subject system. Each object has a single type and each type has a set of corresponding properties. The objects are constrained to types meaningful in cyber security such as hosts, operating systems, vulnerabilities and networks. Noam contains 30 unique entity types in total. Object types and properties are important to the integrity of the

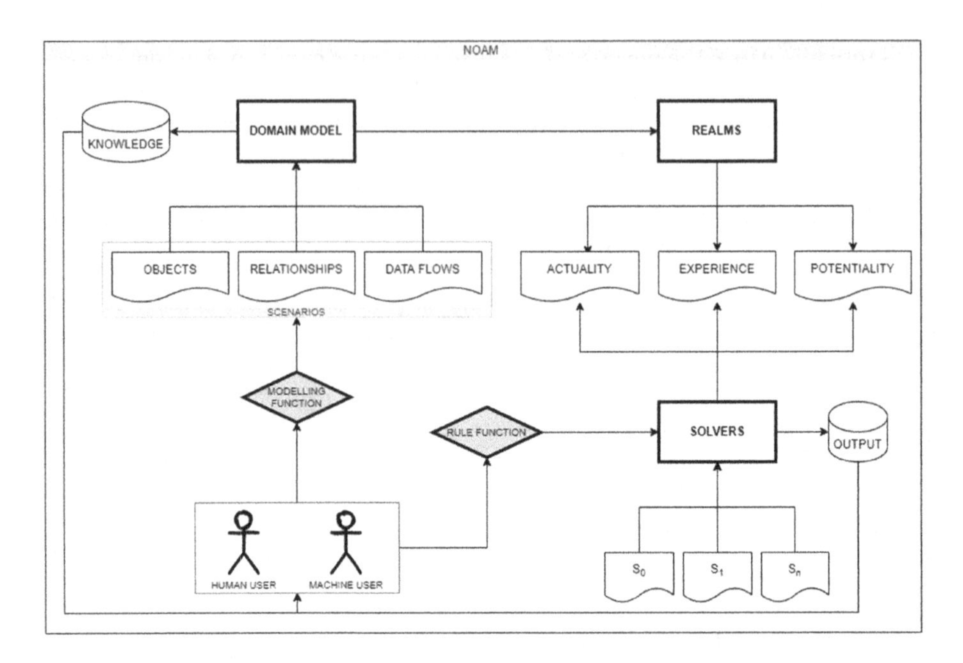

Fig. 1. NOAM High-level design (Cross-references: Domain Model, Sect. 3.1; Realms, Sect. 3.3; Solvers, Sect. 3.5; Users, Sect. 3.2)

Domain Model (i.e., correctness) and to the Solvers (see Sect. 3.5), which apply a form of functional computing to these `type:{properties:[key:value,...]}` structures. Objects are atomic and their properties may not contain other objects. All object types are given in Appendix A (supplementary material), including "SECINT" (Security Intelligence) where the language adopts the Structured Threat Information Expression (STIX) standard (OASIS 2021) and the object types are derived from the OASIS STIX specification directly. The relative layout of the objects is shown in Fig. 3, which alludes to how some of the relationships are formed (see Sect. 3.1).

Relationships. Domain Model Relationships describe how objects are linked. They are directional, have uniquely two endpoints and have an inverse that provides the 'reverse mapping' between two entities. The set of relationship types is given in Appendix A. Modelling associations is a vital part of the language - (Zhao et al. 2022) illustrates the 'connected' nature of cyber security analysis using graph computing techniques. Noam applies a different approach whereby relationships of a prescribed type, link entities together according to the semantic relationships they share. This is an important principle because it means that objects can only participate in a finite range of relationships with other objects, bounded by their type (hence, Controlled Natural Language). The 'allowed relationships' logic is vital important to the overall design and alludes to the constraint-based reasoning performed by the solvers. The group of valid

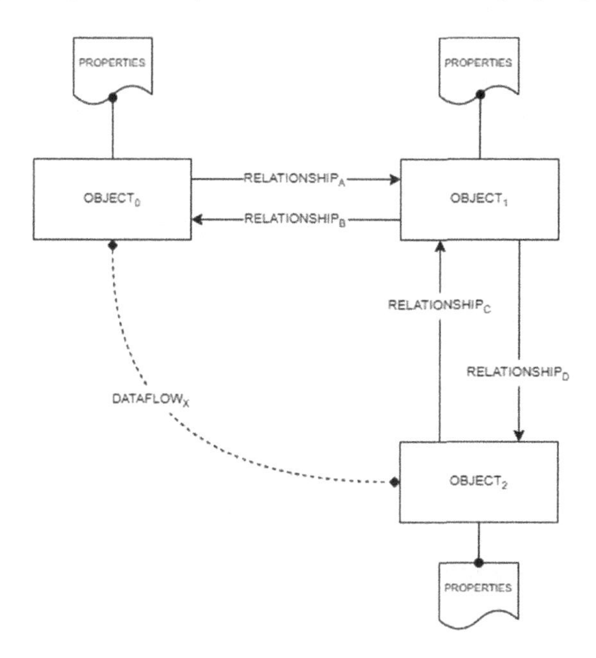

Fig. 2. Domain Model Components

relationships is described by a two-dimensional "Relationships Matrix" - similar to a Cayley Table - and describes the language's **relational algebra**. This is applied by the solvers during evaluation. The rows and columns (x, y) correspond to the various object types (one-for-one) and each element in the matrix contains the valid relationships: $RE((x, y)_m, (x, y)_n)$, where $RE()$ is the relationship assignment function. Tracing either a single row or column will yield all relationships for than particular object type. Note that some object pairings can have multiple relationship types - for instance, the `Software` object has both `IS_DRIVER_FOR` and `IS_DEPENDENT_ON` relationships with the `Hardware` object (plus the corresponding inverse relationships). The STIX types have the densest concentrations of object-to-object relationships. The `SECINT.Malware` object shares four relationship types with the `SECINT.Intrusion_set` object - namely:

1. `IS_USED_BY`;
2. `IS_ASSOCIATED_WITH`;
3. `IS_DEPENDENCY_OF`; and
4. `STIX_RO_TYPE`.

The last of these is a special type reserved for STIX objects and is used to model STIX Relationship Objects (SROs) defined by the STIX standard itself.

Data Flows. The final element of the Domain Model is Data Flows, which describe communications that took place between two objects. They signify the existence of an actual connection and contain information that summarises the

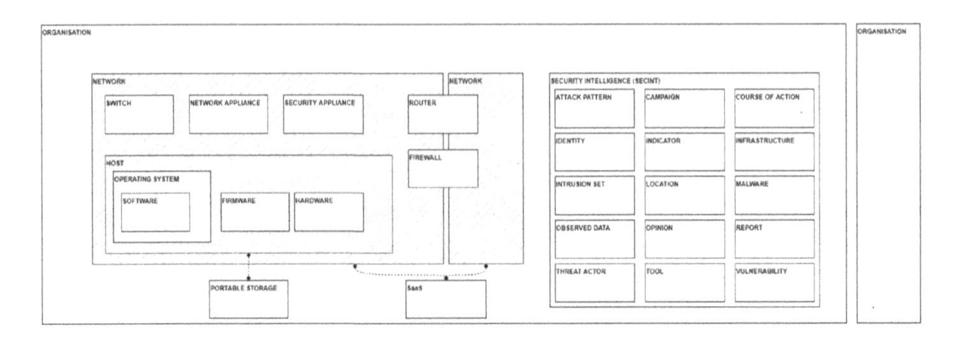

Fig. 3. Domain Model Object Types

nature of the interaction - for example a network six-tuple[2]. Like objects and relationships, data flow entities have a type and these are also prescribed by the language - see Appendix A. The difference between relationships and data flows is important because whilst they are related, they are modelling very different concepts. Relationships describe the semantics of a the environment - hosts being contained by networks, software being run by hosts, threat actors exploiting vulnerabilities, etc. These are all conceptual associations defined according to the language's rules. By contrast, data flows are instances of technical connections between objects - such as network traffic or data transfer across a Universal Serial Bus (USB) port. These are not conceptual, they are actual. Relationships are concerned with placement, whereas data flows express the dynamics of the system under review.

3.2 Users of the Systems Modelled by Noam

Figure 1 includes human and machine users of Noam scenarios, but not the users of the modelled system. This is very much intended and preserves two important tenets of the architecture:

1. A focus on the programmatic, machine-machine level organisation and interactions of the system under review.
2. Prioritise simplicity over complexity within the Domain Model by avoiding the vagaries of human-computer interaction.

However, there is value in modelling the presence of users where this is done without becoming embroiled in the details of their behaviour or overwhelmed by their numbers. In this regard the language contains a final element: the **System User**, which is a meta-object of the Domain Model. It has only a single relationship type (`USES`) that may only be applied to the `Host` object, whose reverse relationship is `IS_USED_BY`. The system user meta-object is optional in the scenario and has only two properties: `uid:"user_id", type:`

[2] (source IP address, source port, destination IP address, destination port, protocol, timestamp).

"[std||priv||fct]" (standard, privileged or functional). A special data flow type of `System Access` is used to show that a system user meta-object accessed a host object. This type takes a only a single ISO-8601 timestamp as its value. Figure 4 is a revised Fig. 2, showing the System User meta-object type, its relationship type and data flow in context.

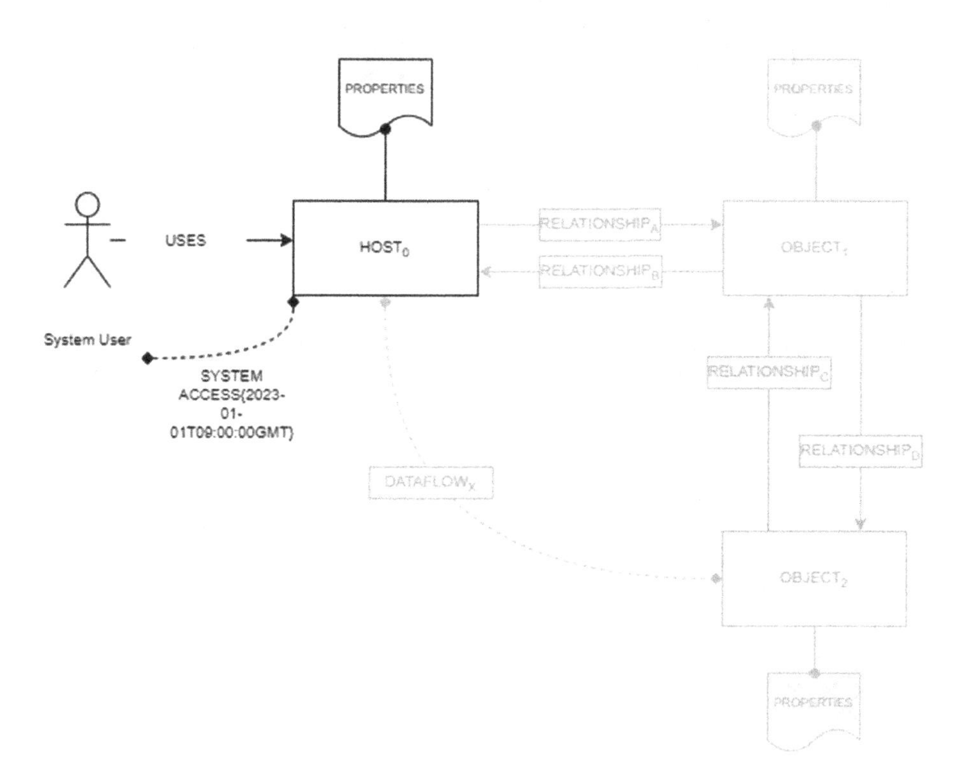

Fig. 4. Domain Model Meta-object "System User"

3.3 Realms

Realms bind the scope of a Data Model to given point in time or set of conditions. The Data Model is an observational construct, whereas the realms are temporal - providing the "when" and ordering scenarios as a time-series. There are three realms: **Actuality** (describes "what is"), **Experience** (describes "what was") and **Potentiality** (describes "what can be"). Their arrangement is shown in Fig. 5, which indicates that experience and actuality comprise facts whilst potentiality is contains with predictions.[3]

[3] There is some overlap in which actuality may also be considered predictive (i.e., due to uncertainty). Since no new measurements can be taken to verify experience, we simply assert that this is solely a domain of facts.

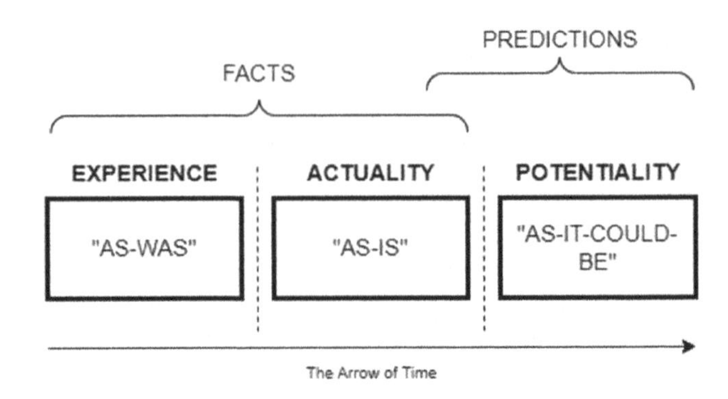

Fig. 5. Realms - order, facts, predictions and the arrow of time

Actuality is built by instantiating Domain Model objects and instrumenting them according the system's design. For instance, a `Host` type that takes the properties: `{"hostname": "dev0.local", "type": "server", "CPU": 8, "RAM": 512, "disk": 6000}`, is paired with an `Operating System` object with the properties `{"uname": "Ubuntu", "version": "20.04"}`, via a `RUNS` relationship. The same is then applied for other objects modelled in the scenario. Data flows are created when the model builder wishes to capture certain events described by the types (Appendix A). Scenario modellers need to be selective about what is included and how the potential volume of objects, relationships and data flows is constrained to avoid 'bloating' the Domain Model. This is discussed in Sect. ??. **Experience** is in effect "prior actuality" and represents knowledge of events that have already happened. Crucially, these must have been verified and so satisfies the epistemological definition for "justified true belief" (Audi 1998) to ensure the validity of the model. Experience may also be used as training data to teach learning algorithms to detect patterns of interest (outlined in Sect. 4 under "Further Work"). **Potentiality** is a projection of possible future states of the scenario. This is akin to "what-if" questions in which the modeller wishes to evaluate potential scenarios, the realism of which is provided by experience and actuality. However, potentiality doesn't allow for arbitrary speculation and only supports projections rooted in the facts contained within the other two realms.

Cyber security is a dynamic subject and if we must take account of this changeability to assure the integrity of our reasoning. Time is what bounds the accuracy of statements in an environment where circumstances change continually. When discussing truth and logic, (Winston 1992) introduces "Situation Variables" as mechanism for dealing with problems whose statements are both `TRUE` and `FALSE` at different times. This is a common occurrence in cyber security, for example:

- Statement A: `System A runs the Ubuntu 20.04 Linux distribution.`
- Statement B: `System B has the IPv4 address 10.0.2.6.`
- Statement C: `APT41 uses the Mikikatz password dumper.`

Each of these statements can be both true and false over time - System A upgrades to Ubuntu 22.10, System B changes IP address to 192.168.56.100 and APT41 develops and uses an alternative exploit to Mimikatz. Time as a "measure of change" (Rovelli et al. 2018) is reflected in Noam's design. Realms are used to address these problems and so essential to the languages usefulness.

3.4 Solvers

Solvers are the final component of the language. They are Finite State Machines (FSMs) that evaluate the Domain Model to 'solve-for' some target condition. This is performed algorithmically, akin to how the term is used in the fields of simulation and modelling (such as Computational Fluid Dynamics). For example, one may run a solver to detect all instances of hosts affected by CVE-2021-44832 (MITRE 2021a) in the scenario. The solver evaluates the entire domain model to detect any Software object running versions 2.0–2.17 of Apache's Log4j2 software. Then, the solver iterates the set of relationships to identify which Host objects are running the affected software, together with their corresponding networks. This is a powerful capability and one whose impact is increased with the size and complexity of the systems being modelled. The power of solvers is that they are exhaustive in processing of the Domain Model - something that quickly becomes intractable for human analysts as systems grow to a non-trivial size.

3.5 Worked Example

Figure 6 shows an example Domain Model comprising two hosts, one network, one piece of software, one owning organisation, one threat actor, one router, one firewall, one external IP address, one malware implant and one vulnerability (CVE-2021-44832). The scenario features a set of relationships that link the objects and two data flows. The host `dev0.local` is running a Java program affected by the Log4j vulnerability. The threat actor `APT41` is known to exploit this vulnerability and to target the organisation identified as the owner of the vulnerable host. The second host `dev1.local` is on the same network (`192.168.56.0/24`) as `dev0.local` (the victim) and is protected by the firewall at `192.168.56.0/24`. Let us suppose this firewall is able to block remote exploitation of the CVE. However, the `dev0.local` is not behind this device and so does not have the same protection. However, `dev1.local` is vulnerable through abuse of peer trust enabling lateral movement due to its location on the shared internal network. The `LokiBot` implant is dropped onto `dev1.local`, where `dev0.local` is the infection vector. We also see that `dev1.local` is attached to an outbound router `192.168.56.254`, the routing table of which permits onward connections to the external IP address `195.51.100.17`. Finally, let us suppose that this is a command and control domain used for tasking and exfiltration of the `/etc/shadow` file on `dev1.local`.

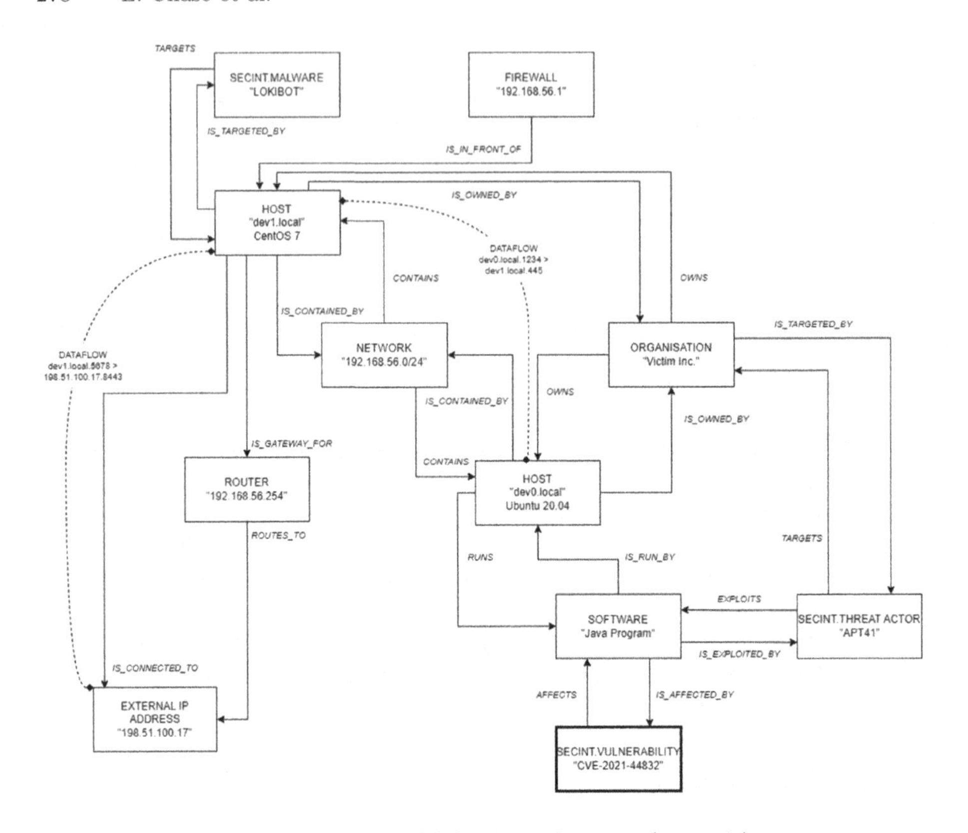

Fig. 6. Domain Model for Log4j Scenario (example)

Figure 7 illustrates a simple solver configuration used to detect the Log4j vulnerability in this scenario. The objectives of the solver are to determine whether the organisation `Victim Inc.` are affected and to identify why. In this case, the logic is straightforward and the goals identified by the major-numbered items (such as `1`), with each evaluation indicated by the minor-numbered items (such as `1.1`, `1.2`, etc.).

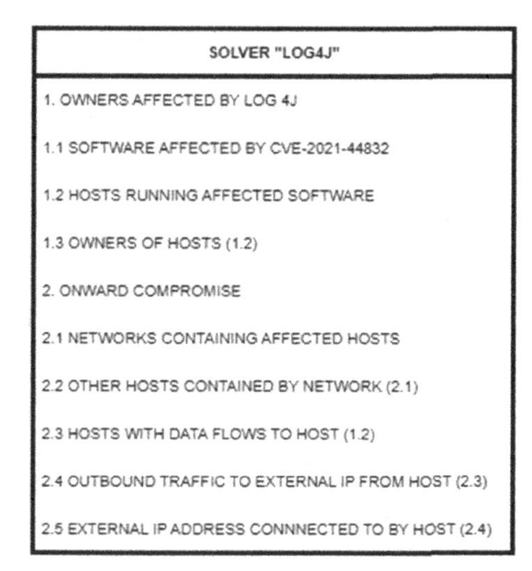

Fig. 7. Solver for Log4j Scenario (example)

4 Conclusions and Further Work

We have introduced the base specification for Noam and a simple worked example. To validate its usefulness, consider how else this problem might be solved. It would likely be based upon log data analysis, protective monitoring, network traffic inspection, configuration management databases, event correlation/deduplication, CTI processing, etc. These yield various data and in different formats that require normalisation. Consider also how are the results to be displayed and how are they consumable by humans working in multiple teams. There are practical problems too, such as how repeatable a process would this be and what skills are required to develop/maintain the tools that determine these results. By contrast, Noam offers a single representational environment in which to store and work with cyber knowledge. Of course, this does incur the cost of building Noam scenarios - notably, creating accurate Domain Models and writing the required solvers.

The literature indicates research into the use of semantic computing and CNLs within this field remains limited, though Noam offers an example of how powerful these techniques can be. The problems of environment complexity and human comprehensibility are very real and Noam looks to address these by offering an interpretable language that models the most significant, information-carrying features of the system under review. In turn programmatic solvers allow specific, well-defined questions to be asked and answered in meaningful terms. However, Noam requires considerable human input and knowledge of the system under review. The language is intentionally non-exhaustive as to be so would generate such data and internal complexity as to render the approach ineffective. Instead, the power of Noam is its capacity to express, reason with and exchange cyber knowledge to support human decision-making. There is a heuristic component to optimising Noam in any given setting, so developing and validating test-cases is an important next step. This would likely lead to tightening of the specification in areas such as object properties and the rules relating to relationships. Simplifying and in some cases automating, the object/relationship/dataflow population tasks would be highly advantageous - as would some type of central "Noam scenario builder"-like tool. Additional experimentation could be used to guide development of the language - e.g., adding new object types such as Platform-as-a-Service (PaaS) and Infrastructure-as-a-Service (IaaS), together with new relationship types that extend the interoperability of the model (e.g. software-to-network, host-to-security appliance). Serialisation in the JSON-LD format means Noam Domain Models can be converted into a matrix representation that can be used as input to statistical machine learning algorithms - similar to how co-occurrence matrices are used in Natural Language Processing (NLP) and computer vision. A formal exploration of machine learning using Noam might begin by exploring embedding techniques and matrix representations of models, which can then used as input to the training of learning agents. Threat Actor TTPs would be an interesting candidate of learning target, since this could be generalised and applied in other areas - such as detection in SIEM tools or aggregated log data. Finally, learning techniques could be used to generate the realm of potentiality by extrapolating likely future scenarios based upon historical evidence (learned from experience and actuality). This may also be suitable as a candidate for building learning ensembles that combine the most suitable arrangement of techniques to detect specific threats.

As described in earlier sections, this approach is part of a larger body of research concerned with predicting cyber threats. The primary focus of subsequent work should be to address the limitations of the approach. This should be begin with how to incorporate CVE (vulnerability) data into the scheme - including how resources like CPE could be used to extend the resolution/accuracy of the solvers. Similarly the relationships matrix could be extended to provide more comprehensive mappings between Domain Model entities, though further experimentation is likely the best way to approach this. Whilst the language goes some way towards marrying the concepts of system events (i.e., Data Flows) and organisation (the entities within the Domain Model), further work is needed to improve the representational effectiveness of this approach. Ideally, the scheme would combine CTI and vulnerability data with 'systems organisation' data (expressed using Noam) - to which further solvers could then be applied. The Realms model could be meaningfully extended to express 'security posture over time'; simply, in what direction (positive, negative) is a system's risk profile headed. Finally, to fulfil the longer-term ambition of applying ensembles of classification algorithms to this data a suitable 'space' needs to be defined using Noam. One approach is to create vectors using graph embedding-type schemes, based on the Noam Domain Model. This vector space (likely a Hilbert space) may then be used to generate input for various classification algorithms.

References

Ampel, B., Samtani, S., Ullman, S., Chen, H.: Linking common vulnerabilities and exposures to the MITRE ATT&CK framework: a self-distillation approach (2021)
Audi, R.: Epistemology: A Contemporary Introduction to the Theory of Knowledge. Routledge, Amsterdam (1998)
Ballard, T., Neal, A., Farrell, S., Lloyd, E., Lim, J., Heathcote, A.: A general architecture for modeling the dynamics of goal-directed motivation and decision-making. Psychol. Rev. **129**(1), 146–174 (2022)
Johnson, C., Badger, M., Waltermire, D., Snyder, J., Skorupka, C.: NIST Special Publication 800-150: Guide to Cyber Threat Information Sharing. National Institute for Standards and Technology (2019)
CISA: Known exploited vulnerabilities catalog (2023)
DCMI: Dublin core metadata initiative (2022). https://www.dublincore.org/
Christianos, F., Schäfer, L., Albrecht, S.: Shared experience actor-critic for multi-agent reinforcement learning. In: Advances in Neural Information Processing Systems 33, pp. 10707–10717 (2020)
Haykin, S.: Cognitive Dynamic Systems. Cambridge University Press, Cambridge (2012)
H.S., L., K., D., J., B.: Studying cyber security threats to web platforms using attack tree diagrams, vol. 13 (2018)
IEEE: International Conference on Semantic Computing (2022). https://www.ieee-icsc.org/
Irwin, S.: Creating a threat profile for your organization. Technical report, SANS Institute (2014). https://www.giac.org/paper/gcih/1772/creating-threat-profile-organization/110995
Luger, G.F.: Artificial Intelligence: Structures and Strategies for Complex Problem Solving. Addison-Wesley (2005)
MITRE: Cve-2014-0160 (2014). https://cve.mitre.org/cgi-bin/cvename.cgi?name=cve-2014-0160
MITRE: Cve-2021-44832 (2021a). https://cve.mitre.org/cgi-bin/cvename.cgi?name=CVE-2021-44832
MITRE: MITRE ATT&CK (2021b). https://cti-taxii.mitre.org/stix/collections/95ecc380-afe9-11e4-9b6c-751b66dd541e/objects/
NCSC: How the NCSC thinks about security architecture (2018). https://www.ncsc.gov.uk/pdfs/blog-post/how-ncsc-thinks-about-security-architecture.pdf
NIST-CSRC: Tactics techniques and procedures (2023)
OASIS: STIX Version 2.1 (2021). https://docs.oasis-open.org/cti/stix/v2.1/cs02/stix-v2.1-cs02.html
Petrică, G., Axinte, S.-D., Bacivarov, I.C., Firoiu, M.: Studying cyber security threats to web platforms using attack tree diagrams, vol. 9 (2017)
Rahman, M.A., Al-Saggaf, Y., Zia, T.: A data mining framework to predict cyber attack for cyber security. In: 2020 15th IEEE Conference on Industrial Electronics and Applications (ICIEA), pp. 207–212 (2020)
Rovelli, C., Segre, E., Carnell, S.: The Order of Time. Riverhead Books (2018)
Shahid, M., Debar, H.: CVSS-BERT: explainable natural language processing to determine the severity of a computer security vulnerability from its description (2021)
Souza, L.O., Ramos, G.O., Ralha, C.: Experience sharing between cooperative reinforcement learning agents. Technical report, Numenta, Universidade do Vale do Rio dos Sinos, University of Brasilia (2019)

Straub, J.: Modeling attack, defense and threat trees and the cyber kill chain, ATT&CK and STRIDE frameworks as blackboard architecture networks. In: 2020 IEEE International Conference on Smart Cloud (SmartCloud), pp. 148–153 (2020)
van Renssen, A.: Gellish: an information representation language, knowledge base and ontology. In: Proceedings of the 33rd European Solid-State Device Research - ESSDERC 2003 (IEEE Cat. No. 03EX704), Standardization and Innovation in Information Technology. The 3rd Conference on, Standardization and Innovation in Information Technology, pp. 215–228 (2003)
VirusTotal: Yara documentation (2022). https://yara.readthedocs.io/en/latest/
Winston, P.H.: Artificial Intelligence. Addison-Wesley, Boston (1992)
Zhao, J., Shao, M., Wang, H., Yu, X., Li, B., Liu, X.: Cyber threat prediction using dynamic heterogeneous graph learning. Knowl.-Based Syst. **240**, 108086 (2022)

Application of Fuzzy Decision Support Systems in IT Industry Functioning

Dudnyk Oleksii[1(✉)], Sokolovska Zoia[1], Alexander Gegov[2], and Farzad Arabikhan[2]

[1] Economic Cybernetics, Odessa National Polytechnic University, Odessa, Ukraine
frightempire@gmail.com, nadin_zs@te.net.ua

[2] School of Computing, University of Portsmouth, Portsmouth, UK
{alexander.gegov,farzad.arabikhan}@port.ac.uk

Abstract. The information technology (IT) market occupies a significant place in the world's economy and in times of rapid digitalization will only continue to increase its influence. Under conditions of environment's increased functioning entropy caused by crises or military conflicts, the importance of IT industry's stable functioning can't be overstated. This paper considers the possibility of fuzzy decision support systems (DSS) efficient usage to forecast and predict the dynamics of IT industry functioning at a scope of a single country, using Ukraine's IT industry as an example.

The software platform used for demonstration purposes is the authentically designed fuzzy DSS FuzzyKIDE. The mathematical basis of platform's functioning is fuzzy logic, which makes it possible to take into account the influences of a set of factors of different nature and obtain a reasonable assessment with potentially inaccurate information. The structure of DSS is offered, using a combined model of the semantic network and fuzzy implication rules. Fuzzy rules are based on predetermined relationships between key influence factors, the analysis of which is offered as well.

The use of the DSS is aimed at forming a holistic forecast of IT industry development. The operation of the DSS is demonstrated on the example of Ukraine's IT industry functioning in different points in time. Comparing the conclusions of the DSS with the historical statistical data of Ukraine's IT industry functioning proves the feasibility of fuzzy DSS usage.

Keywords: IT industry · development trends · fuzzy logic · decision support system · FuzzyKIDE software platform

1 Introduction

This paper starts with a short statistical analysis of recent changes in the international market and how the IT industry is part of this change. The possibilities of using the apparatus of fuzzy logic are stated to improve decision making capabilities on macro-economy scale. That is followed by a section on the DSS application design, which describes the development of the fuzzy DSS for IT industry dynamics forecasting. The

I. Saad et al. (Eds.): ICIKS 2023, LNBIP 486, pp. 283–295, 2024.
https://doi.org/10.1007/978-3-031-51664-1_20

next section focuses on the application of the fuzzy DSS, discussing data collection and fuzzy logic-based analysis. The article concludes with a section on conclusions, summarizing the research findings and highlighting the contributions and limitations of the study.

The role of digital data and technology in today's world is rapidly expanding, fueled by the exponential growth of digital data aggregation and the increasing use of big data analytics, artificial intelligence, cloud computing, and digital platforms. As more devices connect to the Internet and more people use digital services, the digital economy is evolving at an increasing pace. At the heart of much of this activity is the IT industry, which plays a crucial role in the economy by not only providing a potential source of income but also driving cross-growth and making changes in various sectors of the economy. Technologies such as cloud computing, big data analysis, the Internet of Things, artificial intelligence, and more are already transforming the way businesses design, produce, and provide services. IT industry is an essential component of the digital economy, serving as a reliable measure of its effectiveness.

Over the past decade, the growing importance of large technology companies and digital platforms has become most apparent. A significant shift is evident when comparing the sectors of the 20 largest companies in the world by market capitalization in 2009 and 2020. In 2009, the top 20 companies were dominated by oil and gas companies, which represented 39% of the total market capitalization. Financial services followed with a 14% share, while technology and household services accounted for 15%. However, by 2020, the situation had changed dramatically. The share of technology and household services had grown to 43%, while financial services' share had decreased significantly to 10%. Additionally, the telecommunications and basic materials sectors were completely absent from the list of top 20 companies in 2020 (see Table 1) [1, 2].

Given the numerous crisis phenomena in the world economy, the significance of key players in the IT industry and the IT industry itself has been amplified. Determining an appropriate mathematical framework for IT industry's operations research is a major challenge, especially when dealing with complex tasks that involve numerous qualitative factors that cannot be fully formalized or quantified. In such cases, relying solely on quantitative methods is often inadequate. Additionally, the presence of indirect influences and complex relationships further complicates the use of stochastic approaches. The conceptual problem can be addressed through the use of fuzzy logic. This involves processing fuzzy incoming information and using fuzzy mathematics to transform the fuzzy result information into a meaningful conclusion for analytical assessment and forecasting.

Thus, it is often necessary to incorporate the knowledge and expertise of many different experts, and this can be achieved through the use of intelligent information technologies, such as DSS. While the utilization of such systems is not new, their application and research over the last few decades has undergone significant fluctuations, from an active development phase to a decline, and currently, to a trend of scientific resurgence and practical adoption by experts in diverse domains. By applying this tool in combination with fuzzy logic apparatus, it is possible to assess the development directions of the IT sector, which facilitates timely identification of problem areas and decision-making in the hierarchy of IT industry management. That includes strategic planning for specific

IT clusters and product/outsourcing IT companies professional associations, such as the IT Ukraine Association, which has a strategic partnership with various IT clusters.

Table 1. Market capitalization represented by the top 20 companies in 2009 and 2020

Industry	2009			2020			Change in market capitalization 2009–2020 (%)
	Market capitalization		№ of companies	Market capitalization		№ of companies	
	Bn $	%		Bn $	%		
Oil and gas	1264	39.64	7	1,741	13.68	1	37.74
Financial services	455	14.27	3	1379	10.84	4	203.08
Telecommunications	324	10.16	2	-	-	-	−100.00
Technologies	293	9.19	2	5,528	43.45	6	1786.69
Consumer goods	267	8.37	2	291	2.29	1	8.99
Pharmaceutical	264	8.28	2	941	7.40	3	256.44
Domestic services	204	6.40	1	2,844	22.35	5	1294.12
Basic materials	118	3.70	1	-	-	-	−100.00

To be more specific, it could be used in various areas, such as the formulation of regulatory policies for the industry's growth, including tax legislation; providing a favorable business environment for cluster development, both domestically and internationally; promoting a balance between the product and IT outsourcing sectors, favoring the integrated product approach and supporting an effective investment policy towards industry entities, which involves attracting foreign and domestic investors, reinvesting profits into growth, investing in IT education, and enhancing the human capital. Thus, continuous monitoring and forecasting of global and regional IT industry trends will aid in developing viable solutions aimed at fostering the industry's balanced growth, increasing the value of software products and services, directing financial resources to the economy, and boosting the industry's share in the country's GDP.

2 DSS Application Design

We propose a novel development, namely a fuzzy DSS called FuzzyKIDE, which is an applied software implementation of the mathematical apparatus of fuzzy logic. The architecture and technology of the software platform exhibit typicality, demonstrating its potential for solving a wide range of problems under conditions, where inaccurate data are prevalent, utilizing effective mechanisms of logical inference. The general architecture of the system is shown in Fig. 1.

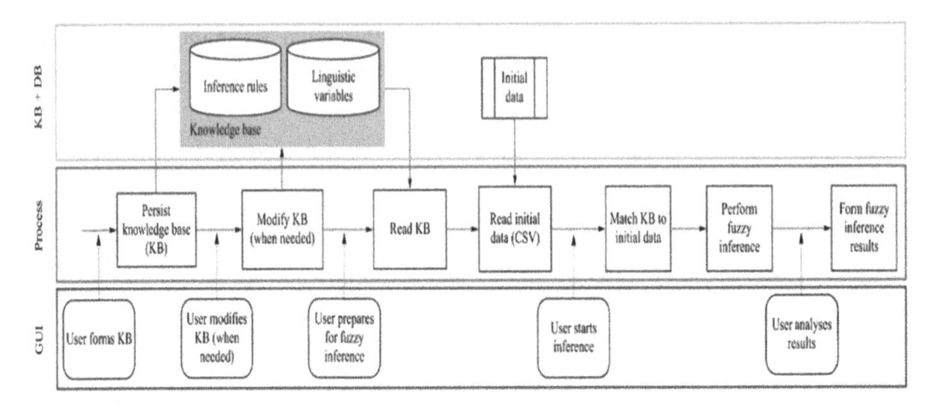

Fig. 1. FuzzyKIDE DSS application design.

The system operates based on a simplified knowledge base structure comprising easily comprehensible fuzzy rules and linguistic variables of a specific type.

IF (X = A) THEN (Y = C),
X:Initial:[A:Trapezoidal:(A1,A2,A3,A4)|B:Trapezoidal:(B1,B2,B3,B4)|…],
Y:Derivative:[C:Trapezoidal:(C1,C2,C3,C4)|D:Trapezoidal:(D1,D2,D3,D4)|…].

In mentioned format, X, Y are the linguistic variables that are specified in the database of linguistic variables; A, C – fuzzy linguistic equivalents of some clear meanings associated with the corresponding linguistic variable; A, B – sets of values for the linguistic variable X; C, D – sets of values for the linguistic variable Y; (A1–A4), (B1–B4), (C1–C4), (D1–D4) – limits of crisp values for fuzzy values of A, B, C, and D, respectively; "Trapezoidal" indicates a trapezoidal type of membership function used to describe the values of a linguistic variable.

The system's fuzzy inference core is based on a simplified SNePS semantic network, which eliminates the intermediate facts base during the fuzzy inference process [3]. This results in obtaining all necessary information for fuzzy inference without repeated queries to the database or knowledge base, potentially speeding up the process.

3 Related Work

Previous authors' research has explored the application of fuzzy expert systems in IT project management. The advantages of fuzzy expert systems in terms of their ability to handle uncertain and incomplete data were emphasized, which is a common challenge in project management. The research provided insights into the effectiveness of fuzzy expert systems in improving decision-making in IT project management [4].

The FuzzyKIDE system has already been tested by making simplified forecasts of Ukraine's IT industry functioning [5]. A methodology based on fuzzy logic for analyzing and predicting the state and dynamics of the product/outsourcing components of the IT industry in a country was proposed. The approach aimed to create a system for responding to crisis phenomena and non-standard situations in the industry's functioning, which could support relevant management decisions.

Building upon these works, the present study aims to generalize the usage of decision support systems to forecast IT industry development trends so it can be done in any country. As examples, we still focus on the use of fuzzy logic in our decision support system to analyze statistical data from Ukrainian IT industry and predict trends in industry development.

4 Application of Fuzzy DSS for IT Industry Dynamics Forecasting

As an example of application of FuzzyKIDE for IT industry processes forecasting, Ukraine IT industry is used. The IT industry in Ukraine has emerged as a significant contributor to the country's economy. Despite being a relatively young industry, it accounted for 37% of Ukrainian exports for services and generated USD 6.8 billion in revenues by the end of 2021. As of the beginning of 2022, the IT industry engaged 285,000 IT specialists and contributed USD 800 million in tax revenue. Additionally, the export of IT services accounted for approximately 2.7% of Ukraine's Gross Domestic Product (GDP) by the end of 2021 [6]. The National Bank of Ukraine (NBU) reported that the IT industry's influence continued to grow until the onset of the war, reaching a record monthly export figure of USD 839 million in February 2022. This marked a 43% increase from the same period in 2021, which was USD 480 million. However, in March 2022, the Ukrainian IT industry experienced a decline in exports, losing 35% of its volume of exports of computer services, which amounted to USD 317 million, as compared to the previous month [7].

The impact of hostilities on emigration among IT specialists was evaluated, given the humanitarian crisis resulting in a significant number of Ukrainian citizens being displaced. Since February 24, 2022, over 6 million Ukrainians have been forced to seek asylum in neighboring countries, with more than 2.5 million continuing on to other nations with about 100–150 thousand of those being IT specialists [8–10]. Consequently, approximately one-third of the Ukrainian population has been either internally or externally displaced. Taking this into account UNDP predicts that Ukraine's development will significantly regress in the medium term due to the ongoing war. If the conflict continues, up to 90% of the population may fall below the poverty line [11]. Such a forecast would mean the loss of 18 years of socio-economic progress in Ukraine and a return to poverty levels seen last in 2004.

Based on the information provided, it can be concluded that new opportunities are emerging for IT investments in regions less affected by the war, such as North America, Middle East, and Africa. Additionally, the specific characteristics of the IT industry make Central Asia and India potential alternative markets. The expected increase in unemployment in Central Asia due to the return of migrant workers from Russia creates an opportunity to accelerate the development of the region's own IT sector, providing employment opportunities and creating favorable conditions for investments from abroad. India, which is already a popular market for IT outsourcing, may see increased investment as a result of the current situation [12]. The labor market crisis in Europe is another factor that needs to be taken into account. The current statistics on the emigration of IT specialists indicate that there are favorable conditions for leveling the labor market crisis in Europe through the influx of Ukrainian refugees. In the long term, this may have a negative impact on the overall Ukraine economy [13].

Efforts are being made to promote the IT industry in Ukraine to attract more investments and to increase workforce demand closing capability of the industry. To counteract the projected decrease in the IT sector, the Ukrainian government has established the IT Generation initiative through the Ministry of Digital Transformation of Ukraine, which aims to train IT professionals with government funding [14]. The long-term success of these initiatives is uncertain, but it's reasonable to anticipate that investors may prefer to invest in less risky regions. Due to the global economic impact of the Russian-Ukrainian war, Central Asia has the potential to emerge as a new IT outsourcing center, while India is likely to solidify its position as an existing IT outsourcing hub. Upon analysis of the present state of Ukraine's IT industry, key patterns and interrelations between factors that impact its development were identified (refer to Fig. 2).

Using a conceptual model of these factors' interrelationships, a collection of fuzzy inference rules was established for investigating industry growth trends. It is important to note that this rule set is not comprehensive and may be extended as necessary, and another fuzzy inference may be conducted to assess additional factors.

Based on current and prior research on the Ukrainian IT industry [3], a set of initial data was compiled to forecast industry performance for the end of 2021 and 2022 under three different scenarios (refer to Fig. 3). Table 2 presents partial results from the experiments for first scenario, which tries to forecast the behavior of Ukraine IT industry by the end of 2021. Tables 3 and 4 present partial results for the second and third scenario respectively. Second scenario tries to forecast the behavior of IT industry by the end of 2022 if Russian-Ukrainian war didn't occur and using expert prediction for 2022 as a base for comparison. Thirds scenario is similar, but the impact of Russian-Ukrainian war is considered and actual statistical data from the end of 2022 is used as a base for comparison. The results show a reasonable level of correlation between the fuzzy inference results and actual statistical data [6, 15–17].

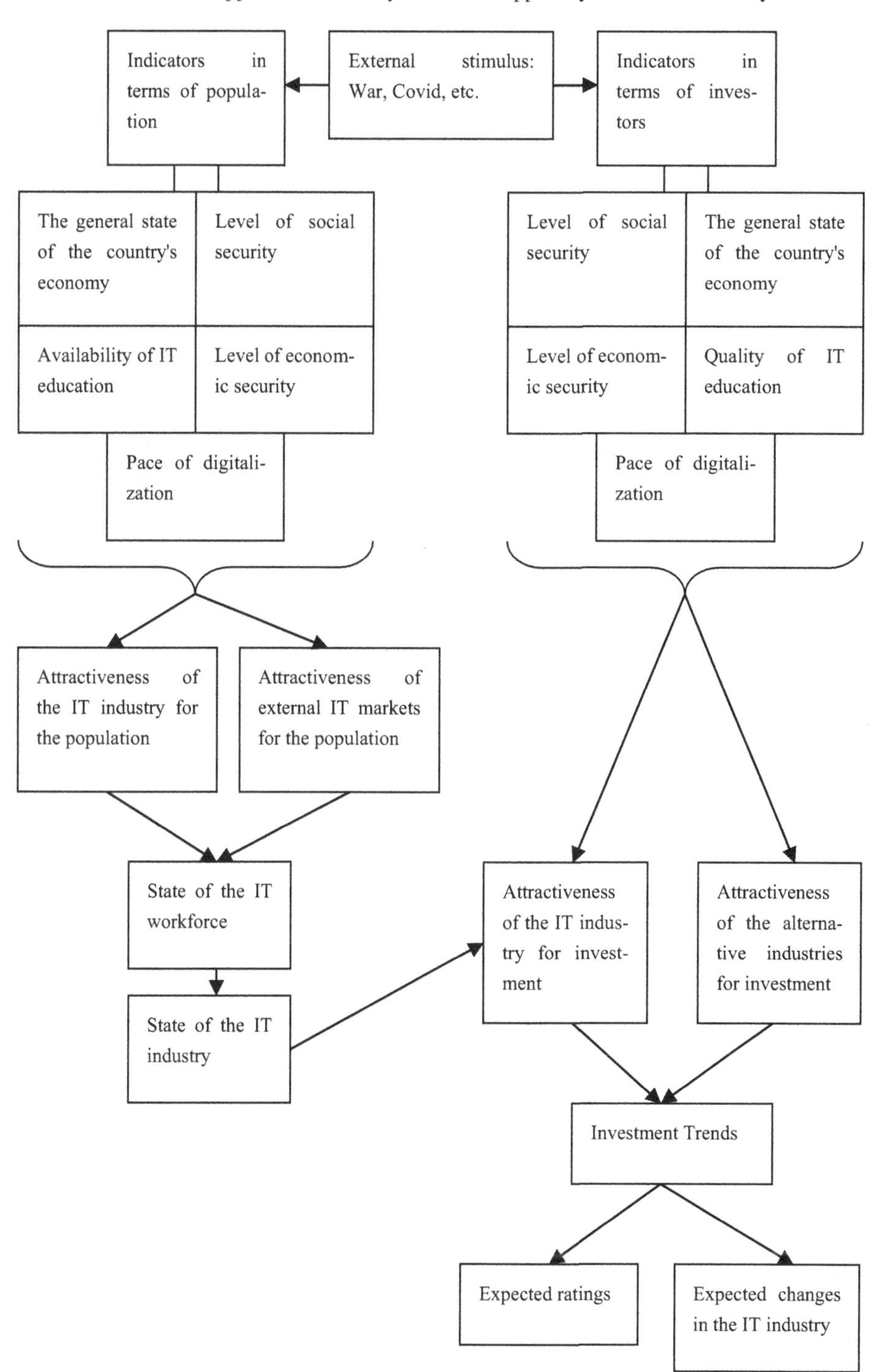

Fig. 2. External and internal factors influence graph.

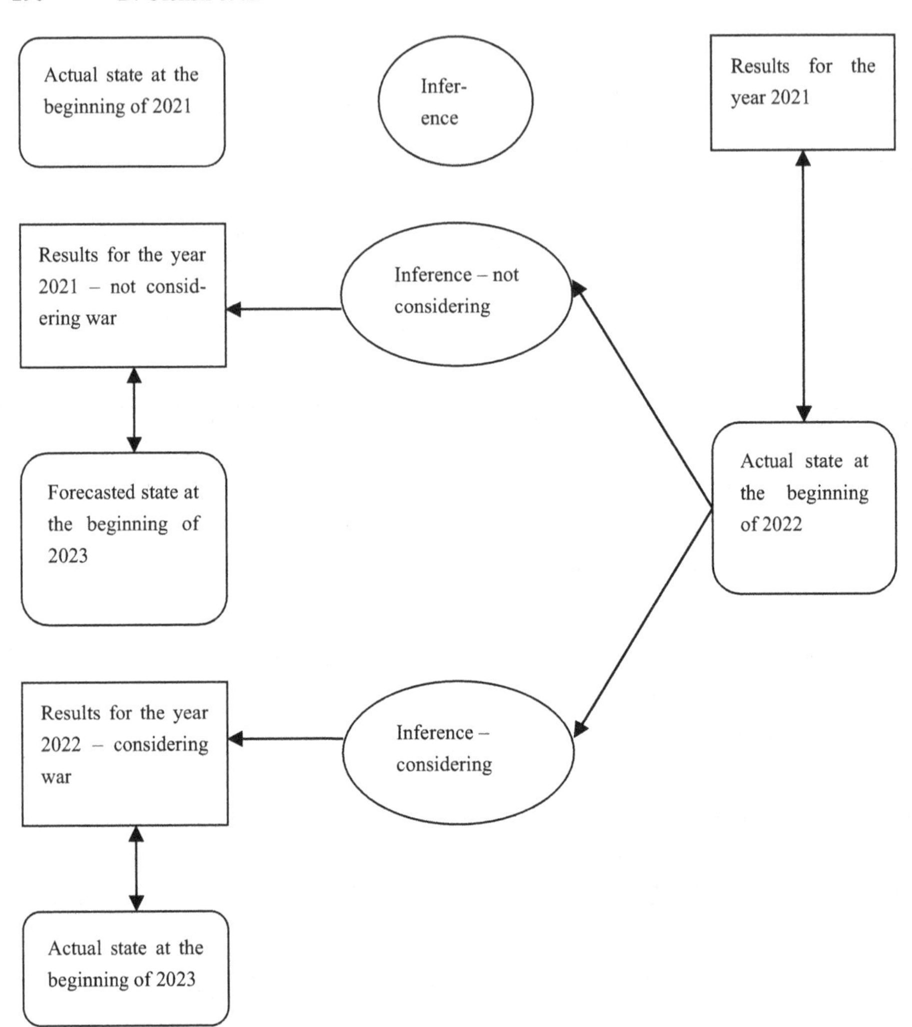

Fig. 3. Experiment scenarios overview.

Table 2. Partial experiment results for 2021 forecast.

Indicator	Result analysis		
	Obtained result	Statistical data	Clarification
General population and IT specialists emigration trend	BELOW MODERATE	5% emigration rate	DSS forecasts no change in emigration trend, which is supported by statistical data available by the end of 2021
IT workforce growth trend	FAST GROWTH	20% of IT workforce growth	DSS forecasts the IT workforce will continue to grow rapidly due to the popularity of IT as a workplace, driven by the availability of IT education and persistent pay inequality compared to other industries
IT lindustry subindustry shares	NO CHANGE	51% - outsourcing companies; 49% - product and start-up companies	DSS forecasts no changes in the companies' shares, which is supported by statistical data indicating a lack of stimuli for any significant shifts
IT industry investments trend	FAST GROWTH	35% private investments growth	DSS forecasts continued fast growth of private investments into Ukraine IT industry. Growing size of IT industry and relatively cheap and highly educated IT workforce keep it attractive for investments
Ukraine IT industry ratings	GROWTH	45th place globally	DSS forecasts continued growth of ratings. Statistical data supports that as Ukraine IT industry moved from 50th place to 45th in one year

Table 3. Experiment results for 2022 forecast (not accounting for Russo-Ukrainian war).

Indicator	Result analysis		
	Obtained result	Experts' forecasts	Clarification
General population and IT specialists emigration trend	BELOW MODERATE	No change compared to 2021	DSS forecasts no change in emigration trend, which is in line with experts' forecasts for the end of 2022
IT workforce growth trend	FAST GROWTH	Continued workforce growth	DSS forecasts that IT workforce will continue to grow as there are were no significant economic changes to make other industries more significantly attractive
IT industry subindustry shares	NO CHANGE	-	DSS forecasts no changes in the companies' shares, which is in line with experts' forecasts, which indicate a lack of stimuli for any significant shifts
IT industry investments trend	FAST GROWTH	Continued investments growth	DSS forecasts continued fast growth of private investments into Ukraine IT industry. Expected growing size of IT industry keeps it attractive for investments
Ukraine IT industry ratings	GROWTH	No change compared to 2021	DSS forecasts continued increase in ratings for Ukraine's IT industry. Doesn't entirely match to experts' forecasts as they expected coming recession to be a counteracting factor

Table 4. Experiment results for 2022 forecast (accounting for Russo-Ukrainian war).

Indicator	Result analysis		
	Obtained result	Statistical data	Clarification
General population and IT specialists emigration trend	MODERATE	13% emigration rate	DSS forecasts increasing emigration trend of IT specialists due to the worsening economic situation in a war-torn country, despite the industry's growing popularity compared to other sectors
IT workforce growth trend	STABLE	<1% of IT workforce growth	DSS forecasts no change in workforce growth, which is supported by statistical data by the end of 2022
IT industry subindustry shares	SHIFT TO PRODUCT	41% - outsourcing companies; 59% - product and start-up companies	DSS forecasts an increasing share of product companies due to preference for more financially stable workplaces. Statistical data supports this trend as outsource IT companies become less popular
IT industry investments trend	DECLINE	50% drop in private investments	DSS forecasts a decline in Ukraine's IT industry private investments, which is supported by statistical data available by the end of 2022
Ukraine IT industry ratings	DECLINE	60th place globally	DSS forecasts a decline in global ratings for Ukraine's IT industry, as indicated by statistical data showing a drop from 45th place to 60th

5 Conclusions

Based on available statistics, an analytical review of global IT market trends has shown that the industry is experiencing high growth rates with potential for long-term sustainable development. However, the sector's heterogeneity, intensifying crisis phenomena, and market uncertainties, including those related to Russian-Ukrainian war and COVID-19, are inhibitory factors that require careful consideration. Therefore, effective IT industry management is crucial to achieving strategic goals and maintaining financial stability. To support decision-making under conditions of uncertainty and risk, decision support systems are becoming increasingly important. Fuzzy decision support systems, which use methods and models of fuzzy logic, are proving to be effective and gaining popularity as an instrumental base for decision support in these conditions.

To improve the operation of IT industry, the use of fuzzy DSS FuzzyKIDE has been proposed. This application offers a modified mechanism of fuzzy inference that accelerates the process and reduces the time needed for expert consultations. The effectiveness of the fuzzy DSS has been demonstrated by performing forecasts of Ukraine IT industry functioning dynamics under various circumstances. The study also highlights the benefits of updating the knowledge base frequently and the ability to consult with minimal dependence on precise data. While the proposed approach has demonstrated promising results, there are some limitations that should be considered, such as the subjectivity of knowledgebase filling, which can lead to conflicting inputs from different users. Additionally, the method currently does not support the combination of fuzzy and clear computations, which may impact its applicability in certain scenarios.

Decision support systems are considered a reusable tool, and their effectiveness in management improves with the enrichment of knowledge and databases by incorporating new experiences. With the rapid growth of the IT industry and standardization of software development processes on a global scale, there is a strong foundation for the exchange, replication, and implementation of intelligent technologies that integrate fuzzy logic elements. Given these findings, it is recommended to utilize the fuzzy DSS approach when forecasting the performance of the IT industry in uncertain conditions.

References

1. PwC: Global Top 100 companies by market capitalisation (31 March 2018 update). https://www.pwc.com/gx/en/audit-services/assets/pdf/global-top-100-companies-2018-report.pdf. Accessed 03 July 2021
2. PcC: Global Top 100 companies by market capitalisation. https://www.pwc.com/gx/en/audit-services/publications/assets/global-top-100-companies-june-2020-update.pdf. Accessed 21 Aug 2021
3. Shapiro, S.C., Rapaport, W.J.: The SNePS family. Comput. Math. Appl. **23**, 243–275 (1992).https://doi.org/10.1016/0898-1221(92)90143-6
4. Dudnyk, O., Sokolovska, Z.: Application of fuzzy expert systems in IT project management. In: Project Management - New Trends and Applications [Working Title]. IntechOpen (2022). https://doi.org/10.5772/intechopen.102439
5. Dudnyk, O., Sokolovska, Z.: Forecasting development trends in the information technology industry using fuzzy logic. Eastern-Eur. J. Enterp. Technol. **1**, 74–85 (2023). https://doi.org/10.15587/1729-4061.2023.267906

6. IT Ukraine Report 2021. https://drive.google.com/file/d/1rDOzj3_hKgXfIj8czwIVzP8Ct4lBY5eW/view. Accessed 20 Aug 2022
7. National bank of Ukraine: Financial sector statistics. https://bank.gov.ua/files/ES/BOP_m.xlsx. Accessed 21 Mar 2023
8. Ukraine Refugee Situation. https://data.unhcr.org/en/situations/ukraine. Accessed 13 Nov 2022
9. Ukraine Internal Displacement Report: General Population Survey, Round 6. https://migration.iom.int/sites/g/files/tmzbdl1461/files/reports/IOM_Gen%20Pop%20Report_R6_final%20ENG.pdf. Accessed 13 Nov 2022
10. ILO Brief: The impact of the Ukraine crisis on the world of work: Initial assessments. https://www.ilo.org/wcmsp5/groups/public/---europe/---ro-geneva/documents/briefingnote/wcms_844295.pdf. Accessed 20 Aug 2022
11. The Development Impact of the War in Ukraine: Initial projections. https://www.undp.org/sites/g/files/zskgke326/files/2022-03/Ukraine-Development-Impact-UNDP.pdf. Accessed 13 Nov 2022
12. International Migrant Stock (2020). https://www.un.org/development/desa/pd/content/international-migrant-stock
13. McGrath, J.: Report on Labour Shortages and Surpluses. https://www.ela.europa.eu/sites/default/files/2021-12/2021%20Labour%20shortages%20%20surpluses%20report.pdf. Accessed 13 Nov 2022
14. IT Generation. https://it-generation.gov.ua/. Accessed 13 Nov 2022
15. DOU Open Data: zarplaty razrabotchikov v Ukraine. http://devua.seektable.com. Accessed 21 Mar 2023
16. Komarnytska, E., Supruniuk, I., Toporkov, O., Grzegorczyk, M., Turp-Balazs, C., Wrobel, A.: The country at war: the voice of Ukrainian start-ups. Emerging Europe (2022). https://drive.google.com/file/d/18kV886D29iQ3ENmS9NeIl3AjFgzeBvoV. Accessed 21 Mar 2023
17. World Trade Statistical Review 2022. https://www.wto.org/english/res_e/publications_e/wtsr_2022_e.htm. Accessed 21 Mar 2023

Moving Towards Explainable Artificial Intelligence Using Fuzzy Rule-Based Networks in Decision-Making Process

Farzad Arabikhan[1(✉)], Alexander Gegov[1], Rahim Taheri[1], Negar Akbari[2], and Mohamed Bader-El-Den[1]

[1] School of Computing, University of Portsmouth, Portsmouth, UK
{farzad.arabikhan,alexander.gegov,rahim.taheri, mohamed.bader}@port.ac.uk

[2] Portsmouth Business School, University of Portsmouth, Portsmouth, UK
negar.akbari@port.ac.uk

Abstract. Advanced Machine Learning and Artificial Intelligence techniques are very powerful in predictive tasks and they are getting more popular as decision making tools across many industries and fields. However, they are mostly weak in explaining the inference and internal process and they are referred to as black-box models. Fuzzy Rule Based Network is a powerful white-box technique which maps well the external inputs, intermediate latent variables and outputs a modular approach based on Fuzzy Logic and it is capable of dealing with complexity and linguistic uncertainty in decision making process. To improve the performance of Fuzzy Rule Based Network, it requires to be tuned and optimized to increase its accuracy, transparency and efficiency. In this paper, a method is proposed to tune the Fuzzy Rule Based Network by using Fuzzy C-Mean and Genetic Algorithm for rule reduction and tuning membership functions and also Backward Selection techniques for pruning and input and branch selection. A case study in transport and telecommuting is used to illustrate the performance of the proposed method. The results show the Fuzzy Rule Based Network's ability to explain the internal process of decision making and its capabilities in transparency, interpretability and in moving towards Explainable Artificial Intelligence (XAI).

Keywords: Fuzzy Rule Based Network · Fuzzy Rule Based Network Tuning · Decision Making Process · White-Box Model · Explainable AI

1 Introduction

Decision making process usually incorporate many factors which can be represented as multilayer network with external inputs, intermediate variables and appropriate outputs. The derived conceptual frameworks in decision making illustrate the interactions of the external inputs and intermediate variables and their collective impact on the outputs. This simply highlights the necessity of utilising white-box modelling approaches and enhancing the interpretability and explainability of the developed models.

I. Saad et al. (Eds.): ICIKS 2023, LNBIP 486, pp. 296–306, 2024.
https://doi.org/10.1007/978-3-031-51664-1_21

Fuzzy logic focuses on linguistic variables in natural language and its aim is to provide foundations for approximate reasoning with imprecise propositions [1]. Uncertainty in linguistic information or subjective knowledge can be explained well by fuzzy logic. Fuzzy logic has shown its applicability in uncertain and complex decision-making problems and it is an extremely suitable method to combine subjective and objective knowledge [2] and non-probabilistic uncertainties such as imprecision, incompetents and ambiguity [3].

Concerning the structure of the Standard Fuzzy System (SFS) models, they have black box nature and unable to reflect the interactions between the external inputs and the intermediate variables in comprehensive and multidimensional decision making processes. Chained (CFS) [4] and Hierarchical Fuzzy Systems (HFS) [5] are characterised by white-box nature where external inputs are mapped to outputs using the intermediate variables. The operation of CFS and HFS are based on multiple Fuzzification, Inference and Defuzzification (FIN) sequences for each connection links between intermediate variables which lead to accumulated error terms and end up with low accuracy of the model [6]. It confirms that although transparent modelling techniques is used but in comparison with SFS it performs less accurate and reliable.

1.1 Fuzzy Rule Based Network

Fuzzy Rule Based Network (FRBN) is a Mamdani rule-based network and act as a hybrid fuzzy system which benefits from both SFS and CFS/HFS. In FRBNs, first the whole network is mapped by the external inputs and intermediate nodes which highlights its white-box nature and then an equivalent linguistic rule base is formed using vertical and horizontal merging techniques. Figure 1 shows a fuzzy network with *m* layers and *n* levels which maps the external inputs ($x_1, x_2, \ldots, x_m$), intermediate nodes ($N_{11}, N_{12}, N_{13}, \ldots, N_{1,n}, I_{n-1,1}, \ldots$), their connections ($z_1, z_2, z_3, \ldots, z_n$) and the output (y) [7]. The emerged single node (Fig. 2) acts as an SFS which has the best performance among fuzzy systems in terms of accuracy. Fuzzy network performance is measured using accuracy, efficiency, transparency and interpretability indices [8].

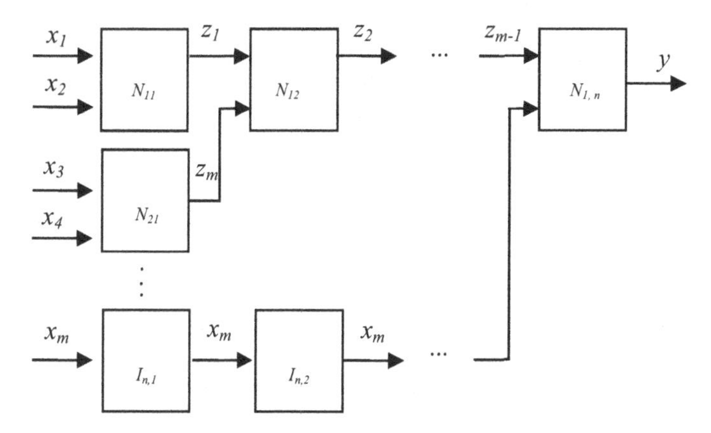

Fig. 1. Fuzzy Rule Based Network

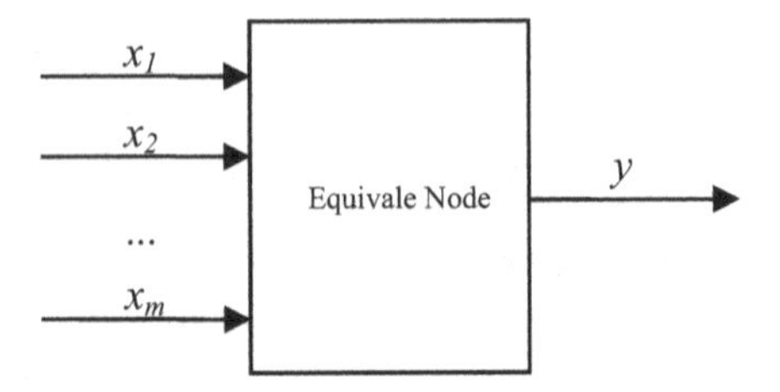

Fig. 2. Equivalent Node of a Fuzzy Rule Based network

Fuzzy Network has shown its powerful capacity in modelling complex problems such as clustering [9] retail pricing [10], transport [6] and software maintenance [11] using relatively small data set and mainly based on expert knowledge in constructing the rules. However, FRBN performance in larger data sets with more external inputs and intermediate nodes, which leads to have exponential rules, is not promising and inefficiently increases the computational cost. On the other side, there is no solid approach to tune fuzzy networks in terms of feature and branch (node) selection as well as data driven rules and rule reduction and also optimising its Membership Functions (MF).

1.2 Contribution

Fuzzy Rule Based Networks was introduced as a modular white-box and transparent modelling tool which can perform well in modelling complex problems. However, it cannot deal with the dimensionality when the number of external inputs and intermediate variables increases and there is no explicit method in the literature to address this issue. The main contribution of this paper is to utilize different techniques and propose a method to tune the FRBNs by pruning the network and to optimize different modelling factors such as membership functions and rule bases. This can pave the way for FRBN to be used extensively in different fields and areas.

1.3 Paper Structure

This paper introduces a method to tune Fuzzy Rule Based Network. The remainder of this paper is as follows. In Sect. 2 the methodology for optimising FRBN is discussed through different steps. In Sect. 3, a case application of proposed method in transportation and suitability of adopting working home is illustrated. Lastly, Sect. 4 presents the conclusion, final remarks and future research.

2 Methodology

This paper In this proposed method, the conceptual framework is used as the road map to develop a white-box FRBN model. In the first step, the hidden and intermediate variables or connection (i.e. z_1 in Fig. 1 for node N_{11}) are quantified by modular FID sequences using the final output y (Fig. 1) and the node's inputs (i.e. x_1, x_2 in Fig. 1). This should start from the first layer nodes to obtain the values of connections and outputs of the corresponding nodes between layer one and two. This will align the nodes in way to predict the final output more efficiently. To obtain the best results, fit a robust model to the data and reduce the error term, the modular models are tuned as explained below.

2.1 Rule Reduction Using Fuzzy C-Mean

To increase the credibility of the model a data driven approach is used to extract the rules and membership functions. Fuzzy C-Mean (FCM) [12] data clustering technique is utilised to identify natural groupings of data from a large data set to produce a concise representation of a system's behavior where each data point belongs to a cluster to some degree that is specified by a membership grade. For a single output of a node, GENFIS3 in Matlab is used to generate a fuzzy inference system for each node. The rule extraction method first uses the FCM function to determine the number of rules and membership functions for the antecedents and consequents [13].

To simplify the problem and reduce the complexity, 3 clusters are considered for each modular rule based systems. In another word, 3 linguistic terms and membership functions are generated for each variable and 3 rules are driven from data for each subsystem. The next step is to tune the initial FIS the same as what ANFIS does for Sugeno fuzzy-type systems. To tune the fuzzy model with having 3 rules for each rule-based subsystem, membership functions should be developed and optimised.

2.2 Tuning Membership Functions Using Genetic Algorithm

To develop membership functions different methods have been proposed in the literature such as Intuition, Inference, Rank Ordering, Neural Network, Genetic Algorithms and Inductive Reasoning [14] among all the existence methods, Genetic Algorithm (GA) has proved itself as a transparent, interpretable and convenient adaptive search technique based on natural selection and genetic rules. GENFIS3 generates Gaussian membership functions for Mamdani fuzzy-type system which is slightly challenging to get optimised. As there is no obligation on the shape of membership functions, for sake of simplicity shape triangular membership function are approximated for this study as Fig. 3 shows.

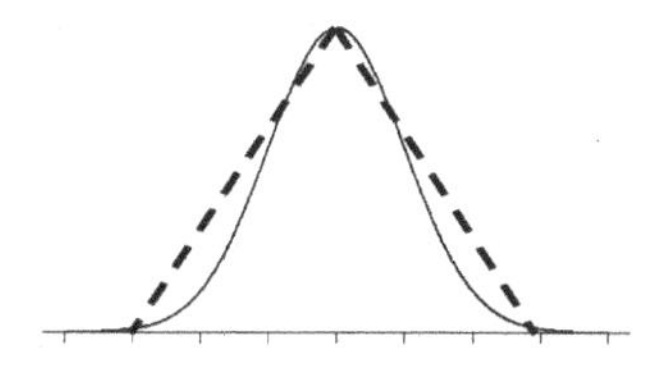

Fig. 3. Approximating Gaussian MF with triangular MF

The first step in GA is constructing an initial population as a potential solution which is generalised randomly or heuristically. The final solution of GA is to coordinate of all vertices for all triangular membership functions. In this research, the initial population size is about *10* * *n* genes where *n* is the number of variables (inputs and output) in each modular rule based. Initial population consist of two types of chromosomes, *n* – *1* are randomly constructed and the other one approximates the Gaussian membership functions (Genfis3 output estimates) with simple triangular membership functions. The reason to put the approximated solution from Genfis3 into the initial population is to provide a chance for search technique to reach to the optimum point faster and speed up the convergence. Also, positions of vertices for each membership function are in binary encoding with the length of 6 bits.

The next step is to assess the performance, or fitness, of individual members of a population. This is done through an objective function that measure of how individuals have performed in the problem domain. The objective function establishes the basis for selection of pairs of individuals that will be mated together during reproduction. The fitness function is normally used to transform the objective function value into a measure of relative fitness.

During the reproduction phase, each individual is assigned a fitness value derived from its performance measure given the objective function. In optimising membership function in this study, the objective function is a minimization problem which tries to minimise the difference between observation and the output of fuzzy inference system. Thus, the lower objective function values correspond to fitter individuals [15]. In this research and using MATLAB Genetic Algorithm Toolbox [16], not to get negative values for minimising problem, values obtain from objective functions are ranked. In linear ranking approach, the fitted individual is given 2 and the least fit value of 0. Then, the relative fitness (probability) of an individual being selected is determined by comparing the given index number $f(x_i)$ with the cumulative sum of indices.

$$F(x_i) = \frac{f(x_i)}{\sum_{i=1} f(x_i)}$$

Selection is the next step. When relative fitness values are obtained, parents are selected. Selection is the process of determining the number of times a particular is chosen for reproduction and, thus, the number of offspring that an individual will produce. Using Stochastic Universal Sampling (SUS) [17], which embedded in MATLAB Genetic Algorithm Toolbox, the selection is taken place and reproduction happens.

By choosing the individuals from the population and recombining them, new chromosomes are produced. The basic operation for producing new chromosomes is Crossover. Like its counterpart in nature, crossover produces new individuals that have some parts of both parents' genetic materials. The simplest form of crossover is single-point crossover which is achieved in three stages. First matching and selecting two individuals randomly, then crossover point is determined in each individual and finally two parts of the individuals are replaced with each other. Crossover is not necessarily performed in all strings in the population and it is applied with a probability when the pairs are chosen for breeding (This probability in this study is 0.7).

For faster convergence, fewer individuals are produced by recombination than the size of the original population. In the other words, usually, the least fit members are replaced by the fit ones. The fractional difference between the new and old population sizes is called the generation gap. Generation gap, in this study, is 0.9 which means 10% of the population is removed in each iteration. In return, most fit individuals are deterministically allowed to propagate through successive generation. This step is called Reinsertion.

Mutation is undertaken for the next step. It is applied to the new individuals with a given probability, which is assumed 0.7 in this study and causes a single bit to changes its state from $0 \rightarrow$ or $1 \rightarrow 0$. Mutation is generally considered to be a background of the operator that ensures that the probability of searching a particular subspace of the problem space is never zero [16].

In the next step, the performance of the population is measured using the objective function. Over a number of iterations, the GA tries to minimise the objective function and produce new membership functions. As the GA is a stochastic search method, it is difficult to specify convergence criteria. So, the termination of GA in this study is defined either for 1000 iterations or getting 50 times the same value for the fittest individual.

2.3 Input and Branch Selection

To select the optimal combination of variables for nonlinear models such as fuzzy models, statistically rigorous criteria for model selection do not exist and must rely on heuristic approaches. In most of the heuristic approaches, new models should be repeatedly developed and evaluated to identify the optimal combination.

In this study a backward selection is used to address the mentioned issues. This method is based on generating only one fuzzy model that takes all the possible input variables and then systematically removing antecedent clauses in the fuzzy rules of this initial model to test the significance of each variable. In other words, this is essentially a backward selection procedure that avoids the need to repeatedly generate new models to test each combination of variables. The idea is to simply remove all antecedent clauses associated with particular input variable from the rules and then evaluate the performance of the model using the checking data.

A model selection criterion is needed to select the best performance. The performance criterion is the model's Root Mean Square (RMS) output error with respect to an independent set of checking data. The data is divided into two groups: training (A) and checking (B) data set. The initial model is generated and tuned using training data (A). The checking error criterion is then defined as:

$$J_c = \sqrt{\frac{\sum_{i=1}^{n_B}(z_i^B - z_i^{BA})^2}{n_B}}$$

where z_i^B is the ith output data in group B and z_i^{BA} is the corresponding predicted output for group B obtained from trained model using A and n_B is the number of data points in group B. This criterion is evaluated repeatedly as each input variable is removed from trained model and choose the variable subset that produced the smallest J_c at each stage [18].

2.4 Linguistic Composition

In order to form the tuned fuzzy network, the first layer (N_{11}, N_{21}, N_{31},..., $N_{1,m}$) needs to be tuned by selecting important external inputs and calibrating the membership functions for all the significant variables. For the next layer (N_{12}, N_{22}, N_{23},..., $N_{2,m}$), the nodes' inputs are the outputs of the layers 1 nodes (z_1,...) and subsequently the explained procedure is duplicated for all these nodes to select the important inputs - which are the branches of the network - and to tune the MFs. This process is propagated for all the layers to select the most important nodes and also tune all the MFs of the entire network. In fact, not only external inputs are selected but also those important branches of the network and intermediate variables are kept in the final model and less important ones are removed to minimise the size and dimension of the rule based network.

The obtained network is now called the tuned network with all the important external inputs, intermediate nodes or variables and also branches. All the rules are data driven and MFs are calibrated by GA. The intermediate variables are also quantified and more analysis can be performed.

The equivalent linguistic rule is obtained using the optimised size of the network, rules for all subsystems and also tuned membership functions. Linguistic composition [6] approaches such as horizontal and vertical merging and also the efficient method of merging modular bases help to achieve the equivalent linguistic rule for the network. Finally, by performing a single Fuzzification, Inference and Defuzzification sequence for the whole model, the output is calculated.

3 Case Study

FRBN is used to develop a model to simulate employees decision making process on how to adopt telecommuting and assess the suitability of working from home. A comprehensive conceptual framework was developed using qualitative approaches which illustrates the internal decision-making process for adopting telecommuting (Fig. 4) [19]. This model reveals the fact that not only external factors have influence on the decision made but also employees group the relevant factors in upper layers and consider those factors together in explaining the decision. That implies the necessity of investigating the role of intermediate factors as well as external ones and a transparent and white-box system should be considered for modelling.

The data was obtained from a questionnaire distributed at seven governmental and semi-governmental organizations in a central business district of Tehran, Iran. Given that dataset, the proposed framework was converted to a network with 37 external inputs with 4 layers: 8 intermediate variables in layer 2, 3 intermediate variables in layer 3 and one output as suitability of working from home in terms of number of days. The discussed method above was used to optimise and tune the fuzzy network. As a result, the network was reduced to 10 external inputs and 8 internal variables in 3 layers (Fig. 5).

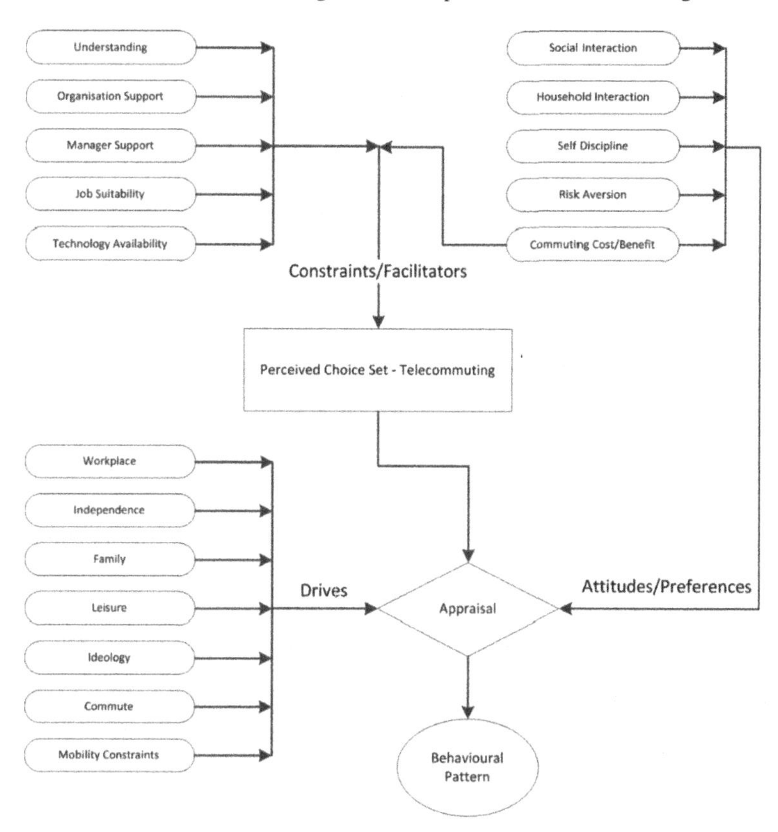

Fig. 4. The conceptual framework of the internal decision-making process for adopting telecommuting [19]

The above optimised network is used and the equivalent rule based is formed using the merging techniques. The model is loaded with the data and the output of the model which is the suitability of the working from home for the users regarding the number of days are compared against the observations (Fig. 6).

In a comparative analysis a Standard Fuzzy System (SFS) is also developed and tuned and the results below was obtained (Fig. 7).

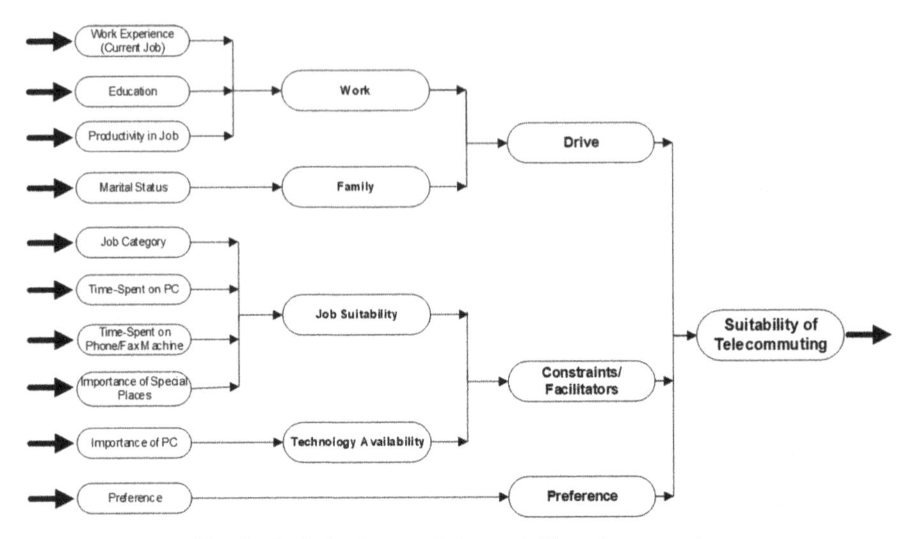

Fig. 5. Optimised network for modelling telecommuting

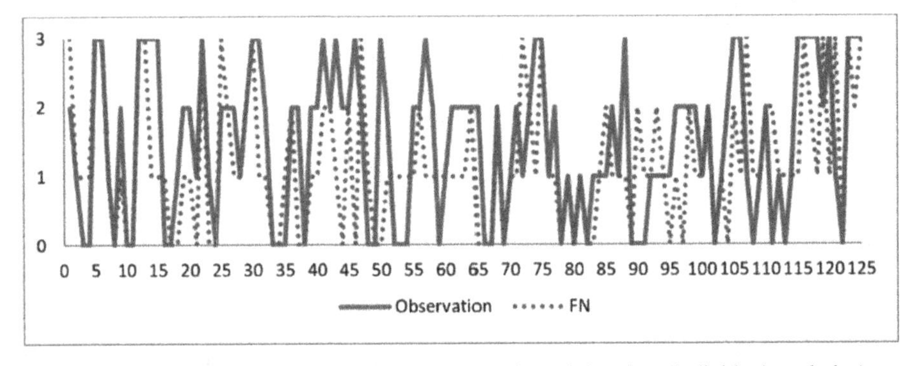

Fig. 6. FRBN model prediction and observation for training data (individual vs choice)

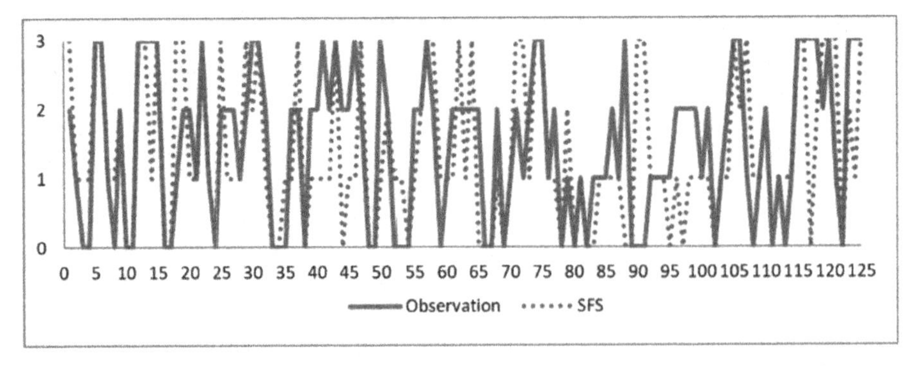

Fig. 7. SFS model prediction and observation for training data (individual vs choice)

The percent correct is terms of the correct choices are used to compare the results above as shown in Table 1 below:

Table 1. Accuracy indicator: Percent correct

	FRBN	SFS
Training data	46	47
Testing data	32	32

As the table above shows, the accuracy of the both models are almost the same. But the FRBN model has great potential in terms of explainability and interpretability. It is less efficient in comparison to SFS but it is more transparent which leads to more XAI.

4 Conclusion

Fuzzy Rule Based Network is a hybrid model between standard and hierarchical/chain fuzzy systems which has proved its power in white-box modelling. In this study, a method of optimising FRBN was proposed and tested using a telecommuting dataset. First using fuzzy c-mean and data driven classification, the rules are obtained and linguistic terms are identified for external inputs and 1st layer. Then the membership fictions are tuned using genetic algorithm based on the obtained Gaussian membership functions but with triangular estimation. Input selection is performed by backward propagation and this process is duplicated for entire network to select the most important external inputs and intermediate nodes. The obtained network is utilised for linguistic composition using vertical and horizontal margining and finally a single fuzzification, inference and defuzzification is performed to get the output. A case study of working from home adoption model was tested using the tuned FRBN and SFS. Results show the FRBN white-box and transparent nature which helps understand in internal decision-making process. Also, FRBN has a very comparable accuracy in comparison SFS with higher power in interpretability and explainability which can be counted as a powerful technique in explainable AI.

References

1. Zadeh, L.A.: Fuzzy sets. Inf. Control. **8**(3), 338–353 (1965)
2. Teodorović, D.: Fuzzy logic systems for transportation engineering: the state of the art. Transp. Res. Part A Policy Pract. **33**(5), 337–364 (1999)
3. Fernández-Caballero, A.: Contribution of fuzziness and uncertainty to modern artificial intelligence. Fuzzy Sets Syst. **160**(2), 129 (2009)
4. Kamthan, S., Singh, H.: Hierarchical fuzzy logic systems. J. Inst. Eng. India Ser. B **103**(4), 1167–1175 (2022). https://doi.org/10.1007/s40031-022-00728-4
5. Lendek, Z., Babuka, R., De Schutter, B.: Stability of cascaded fuzzy systems and observers. IEEE Trans. Fuzzy Syst. **17**(3), 641–653 (2009)

6. Gegov, A., Arabikhan, F., Petrov, N.: Linguistic composition based modelling by fuzzy networks with modular rule bases. Fuzzy Sets Syst. **269**, 1–29 (2015)
7. Gegov, A.: Fuzzy Networks for Complex Systems: A Modular Rule Base Approach. Springer, Berlin (2010). https://doi.org/10.1007/978-3-642-15600-7
8. Arabikhan, F.: Telecommuting choice modelling using fuzzy rule based networks. University of Portsmouth (2017)
9. Wang, X., Chen, Y., Jin, J., Zhang, B.: Fuzzy-clustering and fuzzy network based interpretable fuzzy model for prediction. Sci. Rep. **12**(1), 16279 (2022)
10. Gegov, A., Gegov, E., Treleaven, P.: Advanced modelling of retail pricing by fuzzy networks. In: 9th WSEAS International Conference on Fuzzy Systems, Sofia, Bulgaria, pp. 138–143 (2008)
11. Wang, X., Gegov, A., Arabikhan, F., Chen, Y., Hu, Q.: Fuzzy network based framework for sofware maintainability prediction. Int. J. Uncertain. Fuzziness Knowl.-Based Syst. **27**(5), 841–862 (2019)
12. Bezdek, J.C.: Pattern Recognition with Fuzzy Objective Function Algorithms. Plenum Press, New York (1981)
13. Roger, R.J., Gulley, N.: Fuzzy logic toolbox user's guide. The Mathworks Inc. (1995)
14. Ross, T.J.: Fuzzy Logic with Engineering Applications. Wiley, Chichester (2004)
15. Arslan, A., Kaya, M.: Determination of fuzzy logic membership functions using genetic algorithms. Fuzzy Sets Syst. **118**, 297–306 (2001)
16. Chipperfield, A., Fleming, P., Pohlheim, H., Fonseca, C.: Genetic algorithm toolbox for use with MATLAB (1994)
17. Baker, J.: Reducing bias and inefficiency in the selection algorithm. In: International Conference on Genetic Algorithms, pp. 14–21 (1987)
18. Chiu, S.L.: Selecting Input Variables for Fuzzy Models. J. Intell. Fuzzy Syst. **4**(4), 243–256 (1996)
19. Mokhtarian, P.L., Salomon, I.: Modeling the choice of telecommuting 1: setting the context. Environ Plan A **26**, 749–766 (1994)

Natural Language Processing for Decision Systems

Sentiment Analysis: Effect of Combining BERT as an Embedding Technique with CNN Model for Tunisian Dialect

Seifeddine Mechti[1(✉)], Rim Faiz[2], Nabil Khoufi[3], Shaima Antit[4], and Moez Krichen[3]

[1] LARODEC Laboratory, ISG Tunis, ISSEPS, University of Sfax, Sfax, Tunisia
seif.mechti@isseps.usf.tn

[2] IHEC, LARODEC Laboratory, University of Carthage, Carthage, Tunisia
rim.faiz@ihec.ucar.tn

[3] FSEGS, MIRACL Laboratory, University of Sfax, Sfax, Tunisia
Nabil.khoufi@fsegs.rnu.tn

[4] ENIS, REDCAD Laboratory, University of Sfax, Sfax, Tunisia

Abstract. In this paper, we present an enhanced BERT methodology for sentiment classification of a Tunisian corpus. We introduce a Tunisian optimized BERT model, named TunRoBERTa, which surpasses the performance of Multilingual-BERT, CNN, CNN combined with LSTM, and RoBERTa. Additionally, we incorporate TunRoBERTa as an embedding technique with Convolutional Neural Networks (CNN). The experimental results demonstrate that the combination of TunRoBERTa and CNN yields the highest performance compared to the previous models. Our findings outperform Multilingual-BERT, CNN, and CNN combined with LSTM.

Keywords: Sentiment Analysis · Embedding technique · Deep Learning · BERT · RoBERTa · Tunisian Dialect

1 Introduction

In recent years, the world has witnessed a significant increase in social media users especially after the apparition of Covid-19 pandemic and the lockdown period. Since the emergence of social media platforms, microblogging sites have played a crucial role in people's life. They have gained in popularity as they have represented the only way for people to communicate, connect with each other and express themselves, their opinions, impressions and thoughts. Subsequently, there has been a boost in social media consumption and an increase in the posted comments and tweets. According to Datareportal Global Overview report[1] published on July 2021, 4.48 billion people worldwide use the internet

[1] https://datareportal.com/reports/digital-2021-july-global-statshot.

I. Saad et al. (Eds.): ICIKS 2023, LNBIP 486, pp. 309–320, 2024.
https://doi.org/10.1007/978-3-031-51664-1_22

nowadays. This number represents more than half of the world population. Actually, there has been an increase of social media user numbers by 13% in just one year that is equal to more than half a billion users. Still according to the same report, Facebook remains the most used and popular social media platforms with 2.853 billion monthly active users. These numbers and statistics can only reveal the importance of microblogging sites and how much they could affect our lives in many aspects. In fact, social media networks are becoming an indispensable in people's life, either to keep them up to date or to connect them. Sentiment Analysis (SA) tool has gained more interest and popularity recently. It is defined as an intersection of multiple disciplines such as Natural Language Processing (NLP), Deep Learning (DL) and Artificial Intelligence (AI). This tool permits to predict and classify people's emotions and thoughts based on an unstructured data as people attempt usually to express themselves informally. One of the leading and most effective DL approach is Bidirectional Encoder Representations from Transformers (BERT) a pretrained model proposed by [3], a Google team. BERT [3] has caused a revolution in the NLP field. It has achieved state of the art results for eleven Natural Language Processing tasks including Sentiment Analysis. This could be explain the reasons why BERT is widely adopted especially to analyze, classify and predict sentiments.

In our work, we will be interested in Tunisian SA. When Tunisians interacts on social media, they aims to use Code Switching texts which is composed of Modern Standard Arabic (MSA), French and the Tunisian dialect expressed in a double script arabic and latin. Existing researches in the field of the Tunisian dialect, such as [1,9,10] and [8] are very restricted. This is primarily due to the limitation and lack of free available resources in this dialect. Another main challenge is the multilingualism and the code switching applied in this dialect that makes it more difficult to deal with. In this context, we propose a new pretrained model named Tunisian Robustly Optimized BERT. The contributions of our work are as follows:

- We present a trained Robustly optimized BERT pretraining approach language model for Tunisian dialect, for sentiment analysis.
- We compare the performance of our proposed model with best state of the art method.
- We study the effect of combining TunRoBERTa as an embedding technique with CNN model.

This paper is organized as follows: Sect. 2 provides an overview of the state of art. Our proposed method and the different steps we must take to build our model are presented in Sect. 3. A TunRoBERTa-CNN based sentiment analysis model was raised in Sect. 3.2. Section 4 presents the evaluation section. We conclude in Sect. 5 with a discussion and future works.

2 A Literature Review on Sentiment Analysis

Sentiment Analysis field presents a prominent research topic under Natural Language Processing (NLP) especially with the importance gained by social media networks and its role in our lives. In this context, several studies and researchers works were carried out in this field using various techniques and models. In recent years, various approaches and techniques were applied in the field of sentiment analysis in order to deal with different challenges.

SA techniques are divided into two categories: Traditional approaches (ML approaches, Lexicon-based approaches and Hybrid approaches) and DL approaches.

2.1 Traditional Approaches

According to [2], sentiment analysis approaches could be classified as Machine Learning approach, lexicon based approach and hybrid approach which combines both lexicon and machine learning techniques (Fig. 1).

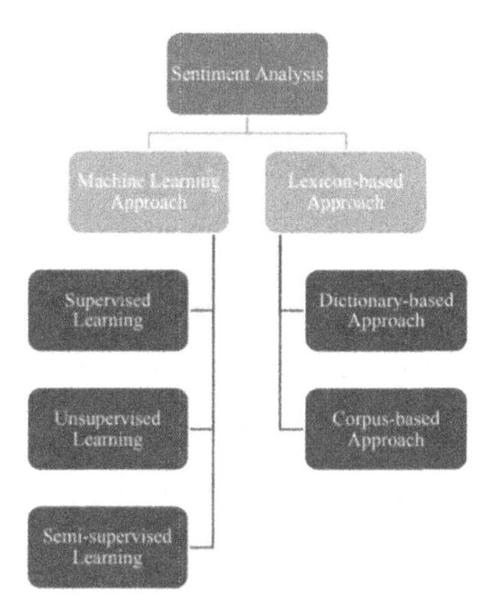

Fig. 1. Traditional Sentiment Analysis Approaches [2]

On the one hand, Machine Learning approaches are based on a learning-by-example paradigm, they require a training and testing set. It is mainly categorized into: supervised, unsupervised and semi-supervised learning methods.

On the other hand, the Lexicon-based approach or Linguistic approach has been widely adopted in the field of sentiment analysis. This approach permits to compute and weigh sentiment related words in the lexicon by splitting sentences into words or tokens, process them and finally classify them. Lexicon-based method uses two main methods in order to establish SA classification which are: corpus based lexicon and dictionary based lexicon (Table 1).

In the study of 'Twitter Sentiment Analysis during COVID19 outbreak' [4], the authors represented a study on how citizens from twelve distinct countries (USA, Italy, Spain, Germany, China, France, UK, Switzerland, Belgium, Netherland, Australia and India)are handling the outbreak. In fact, the authors has

Table 1. Summary of Literature Review Approaches on SA

Authors and date	Corpus language	Approach
Agarwal, Xie, Vovsha, Rambow, & Passonneau, 2011	English	Linguistic: Unigram model (their baseline), Tree kernel model, 100 Senti-features model, Kernel plus Senti-features and Unigram plus Senti-features.
Mukwazvure & Supreethi, 2015	English	Hybrid: Sentiment Dictionary, SVM, KNN
Appel, Chiclana, Carter, & Fujita, 2016	English	Hybrid: a combination of semantic rules, fuzzy sets, unsupervised machine learning techniques and a sentiment lexicon ameliorated with the reinforcement of SentiWordNet
Rathi, Malik, Varshney, Sharma, & Mendiratta, 2018	English	ML: A hybrid model of Decision Tree, Adaboost Decision Treeand SVM
Rahman & Hossen, 2019	English	ML: Bernouilli Naive Bayes, Decision Tree, Support VectorMachine, Maximum Entropy and Multinomial Naive Bayes.
Dubey, 2020	English, Italian, Spanish, German, Chinese, French, Romansh, Dutch, Hindi, Belgium	Linguistic: NRC Emotion lexicon
Horne, Matti, Pourjafar, & Wang, 2020	English	DL: BERT using gated recurrent units (GRUs).
Du, Sun, Wang, Qi, & Liao, 2020	English	DL: BERT-DA, BERT-DAAT, BERT, BERT-AT.
Dong, He, Guo, & Zhang, 2020	English	DL: BERT+CNN
Muller, Salath, & Kummervold, 2020	English	DL: BERT
Wang, Lu, Chow, & Zhu, 2020	Chinese	DL: BERT

collected 5000 tweets from each country. Once they have obtained the dataset, they started the data preprocessing, then applies the NRC Emotion lexicon in order to analyse the tweets on the basis of sentiments and emotions. Finally, they attributed a score of emotions. Subsequently, the authors developed a world cloud for each country from the created corpus. This study represents some limits as it only uses tweets in English language and the NRC lexicon does not take into consideration sarcasm and irony.

2.2 Sentiment Analysis for Tunisian Dialect

Tunisia has witnessed an increase in the use of social media platforms especially since 2011. For example, Facebook, the most popular micro-blogging site in Tunisia, accounted for 8270000 users in Tunisia in January 2021 which represents 68.6% of its entire population[2]. In fact, Tunisian users attempt to generate

[2] https://napoleoncat.com/stats/facebook-users-in-tunisia/2021/01.

content on social networks using an informal language which mainly include: Modern Standard Arabic (MSA), Tunisian Dialect in Arabic script and Latin script and French. Tunisian Dialect Sentiment Analysis works could be divided into two types: Non-Code-Switched text and Code-Switched text.

- Based on Non-Code-Switched text: An emotional dictionary for Tunisian Sentiment Analysis was proposed in [1]. The authors collected 60.000 political comments from Tunisian Facebook pages from 1 January 2010 to 31 December 2013 in order to create a new method to construct an emotional dictionary for sentiment analysis based on emoticons. In order to establish the given method they applied two steps: initial construction of emotional dictionaries then emotional dictionaries enrichment. As a result, they obtained 9 dictionaries: satisfied, happy, gleeful, romantic, disappointed, sad, angry, disgusted and surprised state. To evaluate their method, a test corpus containing 755 words manually labelled by three experts was applied.
  [9] proposed two deep learning models in order to establish sentiment analysis on Tunisian Romanized script. In the first model, they applied Word2vec or frWaC as initial representation. Then, they used distinct classifiers which are CNN and Bi-LSTM followed by a fully connected layer with a softmax activation function for prediction. As for the second model, they adopted BERT multilingual followed by CNN or Bi-LSTM and a fully connected layer with a softmax activation function for prediction. As a result, Multilingual BERT embeddings combined with CNN have outperformed other models.
- Based on Code-Switched text:
  [10] presented a new manually annotated dataset collected from twitter from October,4th 2014 to December, 23rd 2014. The provided dataset contains 5541 tweets in the context of tunisian election where 3760 tweets are written in Modern Standard Arabic and 1754 are in Tunisian Dialect. The authors performed feature selection using Information Gain in order to use it while training six distinct classifiers for a binary classification and a multi-class classification. The experimental results showed that the SVM outperforms other classifiers with the set of one-gram, two-grams and three-grams as features. In addition, an increase in the accuracy of different classifiers is observed when they are trained only with Tunisian Dialect features or when dealing with binary classification (Table 2).

Also, we notice that latest works and researches based on Deep Learning approaches are achieving better results and performance compared to the traditional approaches and sometimes demonstrate state of the art results.

As for the Sentiment Analysis of the Tunisian dialect, the literature reviews highlight many challenges which explains the restricted number of works dealing with this problematic. Another limitation for the SA Tunisian Dialect is the lack of free available resources in this dialect. Actually, the majority of proposed works in this context are only considering Tunisian dialect either written using latin script only or arabic scipt such as [9] and [1]. We were able to identify no more than two works [8,10] which tackles SA of code switched Tunisian Dialect that

Table 2. Summary of Literature Review Approaches on Tunisian SA

Authors and Date	Corpus Language	Approach
Ameur, Jamoussi, & Hamadou, 2016	Tunisian Dialect	Linguistic: An emotional dictionary based on emoticon
Sayadi, Liwicki, Ingold, & Bui, 2016	Tunisian Dialect+ Modern Standard Arabic (MSA)	ML (using one gram, two grams and three grams): NB, SVM, KNN, C4.5, Random Forest
Mdhaffar, Bougares, Esteve, & Hadrich-Belguith, 2017	Tunisian Dialect	ML: SVM + NB; DL: Multi-Layer Perceptron (MLP)
Messaoudi, Haddad, Ben HajHmida, Fourati, & Ben Hamida, 2020	Tunisian Romanized script	DL: Word2vec/ frWaC (as initial representation)+CNN, Word2vec/frWac+Bi-LSTM, Multilingual BERT+CNN, Multilingual BERT+Bi-LSTM

is composed of: Modern Standard Arabic (MSA), French and Tunisian dialect expressed using both latin and arabic script. Although, available works dealing with Tunisian dialect are basically based on classical approaches.

In this context, we propose a Tunisian Robustly Optimized BERT approach for the Sentiment Analysis of Tunisian code-switched dialect. Then we will compared it to some baseline models and we will study the effect of combining our proposed model with a CNN model.

3 Proposed Method

In our work, we present a Tunisian Robustly Optimized BERT Approach (TunRoBERTa). In simple words, we will build a Tunisian pretrained model based on the RoBERTa model [7] that is itself an improvement of BERT model [3].

In order to create our model, we start with building our tokenizer as it represents a key component in our process. A tokenizer allows to split strings sequences into tokens which are generally words, keywords or even phrases.

In our implementation, we have used a byte-level BPE tokenizer [5], a data compression algorithm, with a vocabulary of 50,265 subword units. BPE tokenizer generates new subword according to the next highest frequency pair. The subword vocabulary is built from an alphabet of single bytes so all words will be split into tokens. As a result, our proposed model will be able to encode any input without getting "unknown" tokens.

To train our tokenizer on our unlabelled tunisian dataset corpus, we must set our vocabulary size and the min frequency for a token to be included to 2. Once training is finished, an optimized tokenizer for our specific vocabulary is obtained. The tokenizer is saved in order to be applied to create our model.

Two phases will be applied in our work in order to create our model a pretrained phase followed by a smaller fine-tuning phase (see Fig. 2).

Our model will be pretrained following the RoBERTa training-regime. The pretrained phase occurs in an unsupervised way by providing a corpora of text

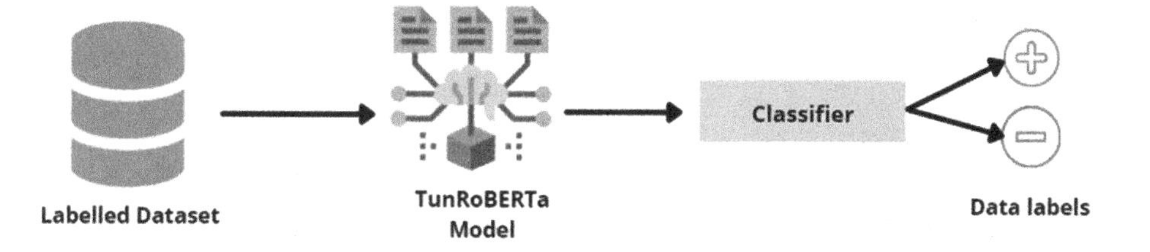

Fig. 2. Fine-tuning phase

int he Tunisian language. Due to the limitation of resources in Tunisian dialect, TunRoBERTa was pretrained on a collection of tunisian datasets with a total size 231k. TunRoBERTa model is trained from scratch on a collection of tunisian datasets using our trained tokenizer. This tokenizer will preprocess, tokenize our corpus and train our model on dynamic Masked Language Modeling task. We will mask 15% of our tokens. The Masked Language Modeling task offer to our model the possibility to memorize textual patterns from our tunisian unlabelled datasets. Then our model will be capble of predicting the masked tokens. A data collator object is used to form input data batches in a specific form so the language model could be trained on.

We encode and tokenize the input data using our created tokenizer. Each input will be encoded using token embeddings, position embeddings and sentence embeddings. The token embeddings is defined as a standard word embedding. For the sentence embeddings, it returns the sentence that contains the token. The position embedding permits to return the position of a token into a specified sequence.

Our proposed TunRoBERTa model architecture is composed of 12 self-attention layers with 12 heads including 117M trainable parameters since it has the same architecture of RoBERTa [7] but we have modified some of the original hyperparameters. Once the pretraining is finished, the second phase fine tuning starts. Fine tuning applies Transfer Learning that is based on transferring knowledge gained while solving a problem to a similar new problem. More precisely, fine tuning consists in applying a pretrained model for one specific task and then tunes it in order to make it perform a second similar task.

3.1 Convolutional Neural Network (CNN)

Convolutional Neural Network (CNN) is a class of Deep Neural Networks. It is basically applied in classification and image or object recognition. CNN model is made up of two parts: a convolution tool that allows to extract features and a fully connected layer which permits to predict the class of the object,[3] CNN model consists of 3 layers:

[3] https://www.upgrad.com/blog/basic-cnn-architecture/.

- **Convolutional Layer:** this is the first layer of our model, it is applied to extract several features from the input. It is represented by the following formula [6]:

$$z^l = h^{l-1} * W^l \tag{1}$$

- **Pooling Layer:** mainly this layer follow the convolutional layer, it allows to reduce the size of the convolved feature map which is the output of the first layer to decrease the computational costs. In our case, we will define a max pooling layer represented as follow [6]:

$$h^l_{xy} = max_{i=0,\ldots,s,j=0,\ldots,s} h^{l-1}_{(x+i)(y+j)} \tag{2}$$

- **Fully connected Layer:** this is the last layer of our model, it allows to classify data and predict their classes. This layer applies the following formula [6]:

$$z_l = W^l h^{l-1} \tag{3}$$

  l: layer z^l: pre-activation output of l
  h^l: activation of l
  $\star$: discrete convolution operator
  W: learnable parameter

Furthermore to these 3 layers, CNN model consists of two important parameters which are Dropout layer and the activation function. Actually, the Droput layer is used to overcome overfitting problem in the training data when all the features are connected to the Fully Connected layer.

The activation function permits to add non-linearity to the network, it is applied in order to decide if a neuron should be activated or not. In our case, as we will be dealing with a binary and multi-class classification we will choose softmax as an activation function.

Softmax activation function is calculated as shown below:

$$Softmax(z_i) = \frac{e^{z_i}}{\sum_j e^{z_j}} \tag{4}$$

An example of softmax activation function[4]:

3.2 Combination: TunRoBERTa and CNN Model

In this section we present a TunRoBERTa-CNN model which is a combination of two model TunRoBERTa as an embedding technique and Convolutional Neural Network (CNN).

Our model TunRoBERTa will be used to convert comments to word embedding, will be applied as a contextual embedding method. Our proposed model will map words to their indexes and representations in the embedding matrix

[4] https://towardsdatascience.com/softmax-activation-function-explained-a7e1bc3ad60.

depending on their context. Then, we will feed these representations to the CNN model in order to establish classification. In other words, our proposed model will be applied only as an encoder that reads the text input (contrary to Sect. 3.3 where we have applied it as an encoder and a decoder) and CNN will be the decoder that make predictions for our task sentiment analysis classification.

3.3 Combination: CNN and LSTM Model

In this section, we present a combination between CNN and LSTM model. CNN model will be applied in order to extract features from input data. For the LSTM model, it will be applied for the sequence learning in order to provide sequence prediction. First, we will apply CNN model to extract local features of text inputs. Then, the LSTM model will save the historical information and start extracting dependencies of text. The feature vector output by CNN will be applied as the input of LSTM. Finally, in order to establish text classification the Softmax activation function is used.

4 Experiments and Evaluation

All experiments were trained and tested on the datasets presented below (Table 3):

Table 3. Overview of the evaluation datasets

Dataset	Size	Classes	Positive labels	Negative labels
Tunisian Dialect Corpus	20k	2	9006	11326
TSAC dataset	17k	2	8854	8199

The first dataset Tunisian Dialect Corpus (TDC)[5] contains 20k comments on total written in multilingual Tunisian dialect of which 9006 are labelled as positive and 11326 labelled as negative. This corpus was collected from YouTube and Facebook public pages from June 2020 to October 2020 and then annotated manually. It contains both comments written using arabic and latin script and some comments that mix both scripts.

The second dataset TSAC dataset[6] consist of 17k comments manually annotated to positive and negative polarities. It contains 8854 positive labels and 8199 negative labels. This dataset was scrapped from Facebook public pages related to Tunisian radios and TV channels fom January 2015 until June 2016. Comments are written in Modern Standard Arabic, Tunisian Dialect in arabic script, Tunisian Dialect in latin script and French.

[5] https://github.com/BoulahiaAhmed/Tunisian-Dialect-Corpus.
[6] https://github.com/fbougares/TSAC.

All datasets were split into training set, a test set and a validation set. The training set is a group of data applied to train the model and make it capable of learning hidden features of the data. For the validation set, it is used to validate the model performance while the training phase. Finally, the test set which permits to evaluate the model after training.

In order to evaluate and compare the different results of our proposed models and the baseline models, different measures are applied. Accuracy, F1, Precision and Recall metrics are calculated.

We experimented several models which include our proposed model TunRoBERTa, CNN model, LSTM model, TunRoBERTa+CNN model, CNN+LSTM model, BERT, RoBERTa and m-BERT. Each dataset will be tested using the different models.

4.1 Results of the Experiments

Experiments results of Sentiment Classification using both datasets Tunisian Dialect Corpus dataset and TSAC dataset are displayed below:

- **Tunisian Dialect Corpus dataset:**
  Table 4 reviews the sentiment classification performances for Tunisian Dialect Corpus dataset. Our proposed model TunRoBERta achieves an accuracy of 79.1% and outperforms CNN model (62.2%), RoBERTa model (71.6%), CNN-LSTM model (75.3%) and multilingual BERT (78%).
  Therefore, TunRoBERTa embeddings combined with CNN demonstrates state of the art results and performs better than other models for all performance measures with an accuracy of 80.6%, a precision equal to 83.9%, a recall metrics of 80.6% and F1 measure of 81.1%.
- **TSAC dataset:** To confirm previous classification results, experiments were also performed on the TSAC dataset. Table 5 shows the results of the different proposed models on the TSAC dataset. We observe that our proposed models CNN combined with LSTM, Multilingual-BERT, TunRoBERTa and TunRoBERTA combined with CNN outperforms the results of the existing research of Medhaffar et al. models [8].
  Also, we can notice that TunRoBERTa embeddings combined with CNN models presents the best accuracy with a 90.8% compared to 89.2% scored by TunRoBERTa, 88.7% recorded by the multilingual BERT model, CNN-LSTM with an accuracy of 81.6%, CNN with a 63.8% of accuracy and last 51% scored by RoBERTa model. This is also the case for the F1 metric performances.
  For the precision measure, the best performance was recorded by CNN-LSTM model with 95.9% followed by TunRoBERTa-CNN (91.3%), TunRoBERTa (90.6%), Multilingual-BERT (90.4%), CNN (36.6%) and finally RoBERTa (0%). Best recall performance was presented by TunRoBERTa-CNN with a value of 90.8% proceedings ahead TunRoBERTa (88.4%), Multilingual-BERT (87.7%), CNN (82.3%) CNN-LSTM model (75.3%) and RoBERTa (0%).

Table 4. Tunisian Dialect Corpus dataset Results

Proposed models	Accuracy	Precision	Recall	F1-measure
CNN	0.622	0.414	0.539	0.486
RoBERTa	0.716	0.639	0.685	0.661
CNN+LSTM	0.753	0.759	0.822	0.789
m-BERT	0.780	0.710	0.767	0.737
TunRoBERTa	0.791	0.732	0.768	0.749
TunRoBERTa+CNN	0.806	0.839	0.806	0.822

Table 5. TSAC dataset Results

[8]		Proposed models				
Model	Accuracy	Model	Accuracy	Precision	Recall	F1 measure
SVM	0.77	RoBERTa	0.51	0	0	0
MLP	0.78	CNN	0.638	0.366	0.823	0.506
		CNN-LSTM	0.816	0.959	0.753	0.834
		Multilingual BERT	0.887	0.904	0.877	0.890
		TunRoBERTa	0.892	0.906	0.884	0.854
		TunRoBERTa-CNN	0.908	0.913	0.908	0.910

4.2 Discussion

Table 4 reviews the sentiment classification performances for Tunisian Dialect Corpus dataset. Our proposed model TunRoBERta achieves an accuracy of 79.9% and outperforms CNN model (62.2%), RoBERTa model (71.6%), CNN-LSTM model (77.1%) and multilingual BERT (78%). Therefore, TunRoBERTa embeddings combined with CNN demonstrates state of the art and performs better than other models for all performance measures with an accuracy of 80.6%, a precision equal to 83.9%, a recall metrics of 80.6% and F1 measure of 81.1%. Table 5 shows the results of the different proposed models on the TSAC dataset. We observe that our proposed models CNN combined with LSTM. Also, we can notice that TunRoBERTa embeddings combined with CNN models presents the best accuracy with a 90.8% compared to 89.2% scored by TunRoBERTa, 88.7% recorded by the multilingual BERT model, CNN-LSTM with an accuracy of 81.6%, CNN with a 63.8% of accuracy and last 50% scored by RoBERTa model. This is also the case for the F1 metric performances. For the precision measure, the best performnace was recorded by CNN-LSTM model with 92.8% followed by TunRoBERTa-CNN (91.3%), Multilingual-BERT (89.7%), TunRoBERTa (89.2%), CNN (58.4%) and finally RoBERTa (0%). Best recall performance was presented by CNN model with a value of 91.8% proceedings ahead TunRoBERTa-CNN (90.8%), TunRoBERTa (89.2%), Multilingual-BERT (87%), CNN-LSTM model (67.9%) and RoBERTa (0%).

5 Conclusion

In this work, we have tackled sentiment analysis task on code-switched tunisian dialect. We propose a Tunisian Robustly optimized BERT approach model called TunRoBERTa which has outperformed Multilingual-BERT, CNN, CNN combined with LSTM and RoBERTa. Furthermore, we combined our proposed model TunRoBERTa as an embedding technique with Convolutional Neural Networks (CNN). Subsequently, results showed that TunRoBERTa combined with CNN achieved the best results compared to previous models. In future work, we will try to apply our model to others NLP task such as Dialect Identification, Next Sentence Prediction, Reading Comprehension Question-Answering.

References

1. Ameur, H., Jamoussi, S., Ben Hamadou, A.: Exploiting emoticons to generate emotional dictionaries from Facebook pages. In: Czarnowski, I., Caballero, A.M., Howlett, R.J., Jain, L.C. (eds.) Intelligent Decision Technologies 2016. SIST, vol. 57, pp. 39–49. Springer, Cham (2016). https://doi.org/10.1007/978-3-319-39627-9_4
2. Aydoğan, E., Akcayol, M.A.: A comprehensive survey for sentiment analysis tasks using machine learning techniques. In: 2016 International Symposium on INnovations in Intelligent SysTems and Applications (INISTA), pp. 1–7. IEEE (2016)
3. Devlin, J., Chang, M.W., Lee, K., Toutanova, K.: Bert: pre-training of deep bidirectional transformers for language understanding. arXiv preprint arXiv:1810.04805 (2018)
4. Dubey, A.D.: Twitter sentiment analysis during covid-19 outbreak. Available at SSRN 3572023 (2020)
5. Gage, P.: A new algorithm for data compression. C Users J. **12**(2), 23–38 (1994)
6. Hafemann, L., Sabourin, R., Oliveira, L.: Learning features for offline handwritten signature verification using deep convolutional neural networks. Pattern Recogn. **70**, 163–176 (2017). https://doi.org/10.1016/j.patcog.2017.05.012
7. Liu, Y., et al.: Roberta: a robustly optimized bert pretraining approach. arXiv preprint arXiv:1907.11692 (2019)
8. Mdhaffar, S., Bougares, F., Esteve, Y., Hadrich-Belguith, L.: Sentiment analysis of tunisian dialects: linguistic ressources and experiments. In: Third Arabic Natural Language Processing Workshop (WANLP), pp. 55–61 (2017)
9. Messaoudi, A., Haddad, H., Ben HajHmida, M., Fourati, C., Ben Hamida, A.: Learning word representations for Tunisian sentiment analysis. In: Djeddi, C., Kessentini, Y., Siddiqi, I., Jmaiel, M. (eds.) MedPRAI 2020. CCIS, vol. 1322, pp. 329–340. Springer, Cham (2021). https://doi.org/10.1007/978-3-030-71804-6_24
10. Sayadi, K., Liwicki, M., Ingold, R., Bui, M.: Tunisian dialect and modern standard Arabic dataset for sentiment analysis: Tunisian election context. In: Second International Conference on Arabic Computational Linguistics, ACLING, pp. 35–53 (2016)

An Enhanced Machine Learning-Based Analysis of Teaching and Learning Process for Higher Education System

Majed Alsafyani(✉)

Department of Computer Science, College of Computers and Information Technology, Taif University, P. O. Box 11099, Taif 21944, Saudi Arabia
alsufyani@tu.edu.sa

Abstract. As the use of AI in colleges and universities becomes increasingly common, more studies on its impact on classroom instruction are likely to be conducted. However, this study highlights the lack of research on AI in pre-service teacher HES programs. The ML-based neural network model used in this study was an improvement on the standard classification neural network, and it successfully solved a multi-classification problem with multiple constraints as a supervised ML model. Further empirical research is needed to explore the use of AI by pre-service teachers, as it could facilitate the implementation of AI-based teaching in future classrooms if pre-service teachers have greater awareness and competency in AI. Our analysis suggests that the development of technically and pedagogically capable AI systems that can enhance quality education in various TLM environments is still a work in progress. Future work should focus on incorporating two equally important technologies into the model: blockchain technology, which can protect the institution's and students' data and business processes, and the internet of things, which can include devices that collect data to enable ongoing improvement of the TLM.

Keywords: Machine Learning · Higher Education Systems · Teaching and Learning Methods · Accuracy · Active Learning

1 Introduction

Introducing Artificial Intelligence (AI) into our daily lives has had a significant impact on our society. The development of knowledge-based economies and emerging technologies are studied systematically. In today's technologically advanced information era, a wealth of knowledge is widely available through various technological tools. [1] As technology becomes more prevalent in education, it is important for teachers to modify their Teaching and Learning Materials (TLM) to better suit their students' social environment. This means taking into account the digital tools and resources that are available to students, as well as the ways in which they use technology in their daily lives. By adapting their TLM to these realities, teachers can provide more effective instruction that is better suited to their students' needs and learning styles. However, the divide between digital immigrants

I. Saad et al. (Eds.): ICIKS 2023, LNBIP 486, pp. 321–332, 2024.
https://doi.org/10.1007/978-3-031-51664-1_23

and natives (Digital immigrants are people who were born before the widespread use of digital technology and have had to adapt to it later in life, while digital natives are people who were born into a world where digital technology was already prevalent) is one of the principal factors of the modern digital revolution and its effects on human lifestyles. One approach [2] to investigate student participation in HE in the significantly more extensive HES is by using a TLM. Educators must create methods that increase student participation and encourage a sense of community to enable an interesting online program.

Innovative Teaching Practices (ITP) [3] can motivate students and improve the HES. However, the lecture-based teaching method is the primary focus of learning in a traditional classroom or learning environment [4]. "Smart Teaching" is how things happen in a smart classroom. Students can access online lectures, YouTube clips, or readings that contain educational programming that they can review at their own pace. With the teacher's help, students apply what they have learned through problem-solving, case-based scenarios, or interactive discussion and debate [5]. Active Learning (AL) is a way for students to improve their thinking and creativity by considering their research and its purpose. AL calls each student to participate, contribute significantly, and pay more attention [5].

The TLM should be surveyed by course instructors in HE institutions to involve students more in the learning process [7]. The goal of TLM in HE institutions should be to develop a community of learners who can draw conclusions and solve problems. This paper aims to highlight the significance of digitalization for HE in the initial research after the revolution of AI's rise. It examines the new era's teaching skills problems and their application in actual teaching activities by combining AI application criteria, teachers' standards of practice, and other ideas. A novel direction for the long-term development of HE is recommended, as can methods to improve and improve HES.

The improvement of an online learning mode and subsequent enhancement of student learning is proposed in this work by integrating AI, data analysis, and TLM. The model includes an AI system like ML that interacts with and manages student performance [7]. The study used a question-by-question analysis of the most recent research on the subject to investigate the perspective and roles of teachers in AI-based research. However, there are some errors in the text that need to be corrected, such as the incorrect reference number and missing articles. For instance, the correct reference for the Innovative Teaching Practices is [3], not [2]. Also, the correct sentence should be "Smart Teaching" is how things happen in a smart classroom, and not "Smart Teaching" is how things happen in a smart classroom.

2 Related Works

In the field of ML, it holds a key position. In-depth research is being done by an increasing number of researchers in the market today, causing the technical field to advance quickly [9]. According to researchers working in this area, educational ML is a new field created by integrating ML and learning science. Through observation and comprehension of the TLM, educational ML supports student learning [10]. According to relevant researchers, educational ML primarily uses computers and TLM to help AL with good-quality in teaching, help teachers teach more successfully, and reinforce AI with technology in the future [11].

More sensors monitor TLM behavior. Online learning data is analyzed and predicted by the AL-TLM [12]. The priority of the works about using AI in LMSs is to assist teachers in creating more successful TLM for use in these environments. Specialized AI techniques are used to interact with users and learn from their actions, which is a key point [12].

The development of this work has prospered from these dependable models. Based on the review, the suggested work varies from others in integrating AI and data analysis technologies in one environment. A virtual assistant that initially manages each student's data and is in charge of automated and individualized monitoring is developed by coordinating all educational administrative tasks into a single platform [13]. Data analysis and user interaction are available to the assistant. Data from other sources must be adapted into the analysis in order to fully understand each student's desires and requirements, which means that it is not just limited to data from the TLM. ML algorithm allows Decision-Making System adaptability (DMS) [14]. AI can make decisions quickly and efficiently about student performance through this integration. Analysis of the sensor-free effects detection problem using ML. The ML outcomes were compared with previous results achieved using conventional ML algorithms. ML models had better AUC, but Cohen's kappa values did not improve [15].

3 Proposed Improved Machine Learning Model for Higher Education System

3.1 Data Analysis

Due to the large dataset predicted to be extracted and the data type that will be unified into the analysis, data analysis is crucial in this work. Students' activity-related data and data gleaned from their interactions with the TLM are structured in their database. In any case, granularity in the analysis is not accomplished if these data alone are considered. And the data will be decomposed further so that you understand what you did in each task. In other words, this does not guarantee accurate data about each student's progress in class. Since universities store students' economic and educational data and sometimes educational outcomes data from primary HES, the analysis system needs more data. This data enables the identification of significant power trends in the composition of the student population and their modes of learning. This study collects student data from social media and other sources.

3.2 Methodology

Before choosing a technology, address the business goal. Knowing what to optimize is crucial and not proportional gain the goal. Figure 1 shows how the different phases of the ML process interact. The goals may be to enhance conversions, decrease revenues and profits, or increase user satisfaction.

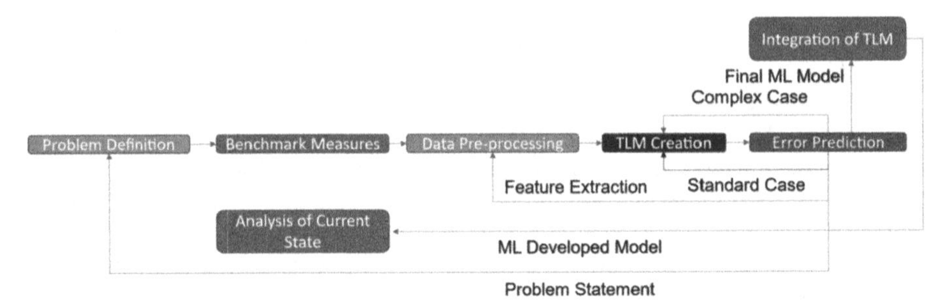

Fig. 1. Step-by-Step Process of Proposed Improved ML model

Step 1. Problem Definition: To solve the problem, we must understand it. Suppose computation time is high, mainly if the problem happens from a low-data sector. Collaborating with problem experts is fundamental during this phase.

Step 2. Data Sharing: Exploratory data analysis is typical to get acquainted with it. The exploratory analysis uses descriptive statistics, correlations, and graphs to understand the data better. It estimates whether there is enough data to build a model.

Step 3. Benchmark Measures: Regression and classification use root-mean-square error and cross-entropy. Two-class classification problems use accuracy and completeness.

Step 4. Analysis of Current State: There is no need to use ML to fix the issue. Better results are the reason to use ML to solve this problem. Automatically attaining similar outcomes is another common motivation. The ML model's performance is compared to the current solutions. A simple, easy-to-implement solution is defined without a current solution.

Step 5. Data Pre-processing: The big data turn allows this process, but the ML phases require specifics. ML's data preparation requires more work. Incomplete data is the main problem. The ideal data for ML is not always available. An online survey predicts students who are more likely to enter an STS. Few will have completed all fields. However, partial data is better than zero data, and there are different methods for preparing, deleting, imputing a reasonable value, using an ML, or doing nothing that processes incomplete data. Data from databases, spreadsheets, files, and other sources may be combined. ML algorithms need to combine data. Instead of data, relevant features work better for ML algorithms. For simulation purposes, measuring temperature is more accessible than in μg of mercury.

Step 6. Deploy the ML Model: ML model building is simple after data preparation. The prepared historical data will teach the ML algorithm to get the right results.

Step 7. Error Prediction: Understanding how to improve ML results is crucial in this phase. Ensure the model can generalize during error analysis. By using this method, good results are usually easy to achieve. However, we must repeat the previous phases to get good results. It helps design more relevant features and reduces generalization errors. Users can choose the best ML method with a better understanding.

Step 8. System Model for Integration: Error-adjusted ML models are integrated into TLMs. Integrating ML models into systems requires more effort. The ML model must communicate with other system parts to automatically repeat the data preparation phases and use its results. Errors are monitored automatically. If model errors increase, the model informs to rebuild the ML model manually or automatically with new data. Data interfaces are needed so the model can automatically get data, and the system can use its prediction.

3.3 Supports for STS

In environments where teachers collaborate, STS occurs more frequently. STS scores typically are higher in classrooms where teachers report working together more frequently to improve TLM (Fig. 2). This model learned from interviews with educators that collaboration is a powerful means by which educators learn from one another and work together to enhance their classroom practices.

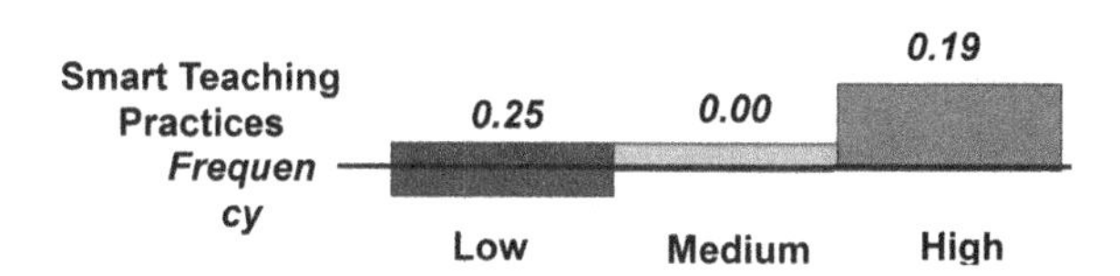

Fig. 2. More frequent Collaboration about Teaching Predicts HES

When teachers have access to suitable applications for career development, STS is more likely to happen. Design and intensity are key factors in professional development. The survey found that teachers who have performed TLM and done experiments have more innovative TLM than those who have only identified sessions and listened to lectures (Fig. 3).

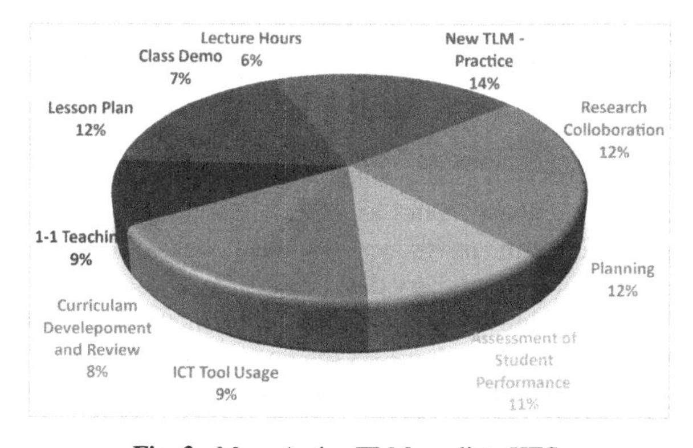

Fig. 3. More Active TLM predicts HES.

Students are more likely to engage in STS that uses ICT in HE, providing easy access. According to the survey results, ICT integration is a major influencing factor of

TLM. Students' ability to enter classrooms is a crucial element in promoting integration. According to surveys, teacher and student access to computers in the classroom is a better predictor of ICT integration than access in public spaces like computer labs or libraries. ICT integration is also more strongly associated with student access to classroom computers than teacher access. Lack of student access in classrooms proved to be the major deterrent mentioned by teachers when asked about the greatest challenges to using ICT in their teaching (Fig. 4).

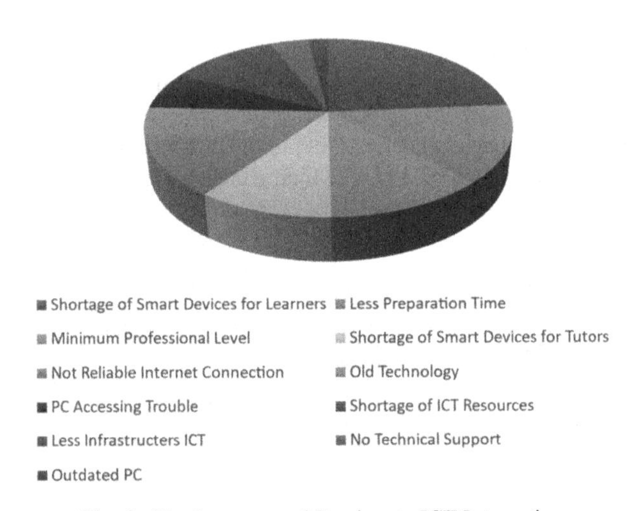

Fig. 4. Teacher reported Barriers to ICT Integration

4 Neural Network Model for Classification

Every node in Conditional Class has a value of either 0 or 1 (binary units), regardless of whether it is in the visible or hidden layer. Obtain this model to accurate prediction score because the threshold data indicates the lower bound criteria, and the network data is obtained from the absolute number, not the binary representation. The best method for processing continuous data with binary units is to initially scale the data to the closed interval (0, 1) and then use these values as the likelihood that the associated unit will be activated. Simulated experiments showed that the model's effect is not ideal, but it is still a method for communicating binary data [16].

Since continuous data representation is a core problem, this work extends the model to address it. In the improved model, we set the conditional and input layer node units to gaussian. The NN model includes the following energy equation: the visible layer is Gaussian, and the hidden layer is binary:

$$Test(x, y, z) = \varepsilon_1^2 \bullet \left|p_i - q_j\right| + r_j \bullet s_i - w_{i,j} \bullet h_j \bullet u_i \bullet \varepsilon_1^2 + t_y$$

Where ε is the standard deviation, i is the visible layer; k is the conditional layer.

Parameter solving uses GD. Adjusting each step's calculation will be more complicated. Therefore, in many NN model applications, a simpler preferred system is normalizing the original input data before bringing it into the CD algorithm training. Each node unit's data is normalized to 0 mean and unit variance so that β in the above formula equals 1. In the conceptual framework and test phase, the accurate CD algorithm will ensure process completeness and result accuracy. Joint probability density of x,y,z in the binary-binary model:

$$P_{BC}(x, y, z) = R_{BC}^{-1} \bullet \varepsilon^{-E_{BC}(x,y,z)}$$

Where BC is the partition equation, ensuring the probability density distribution.

5 Results and Discussion

HE institutions must find novel methods that can adapt to the needs of people because of the new normality that humanity now experiences. The goal of this paper is to enhance an STS by taking this into account. Technology performance becomes the foundation for improving education and tracking student performance. Online HE models are expected to continue TLM. This work is used to improve the design and building infrastructure of the university that took part in the study. Focusing on the ML model's design after the proportion of the required infrastructure has been set up is seen as a benefit. Any time a modern architecture layer needs to be changed, it is updated without incurring additional financial, technical, or human costs. Integrating these technologies improves teacher-based student performance monitoring. This model eliminates the need for human vision, allows for continuous analysis of each student under the supervision of the systems, and even uses ML to identify cases with the most significant potential for poor academic performance. When the comprehensive model is detected early, projections can be created based on the student's past. Students who suffered in the emergence of mathematics class, for example, the case, are blocklisted by the system as types of cases for problems in advanced math I and other courses that require it. The system can even identify a valid case of repetition by examining the key points considering a subject, though this analysis can be very preliminary.

ML recommends learning resources based on student activity performance. The decision is based on the student's best performance. Some students don't like type activities and accurate/false assessments. The model suggests tasks for these groups. ML recommends these learning activities to students based on their needs. When solving problems, intelligence fields are combined. This proposed HES must implement more complete courses to meet students' learning and intelligence needs.

Simulation of HES Module

The course's final module uses the most resources. As the new course incorporates the HE curriculum system and module coding, many HE-based modules have been compiled to meet the performance standards of each HE institution. Users can add, remove, or change modules whenever needed. Thus, all introductory education courses—language and literature, mathematical skills, social and humanities sciences, biology, future technologies, art, sports and health, data analysis learning, community-based,

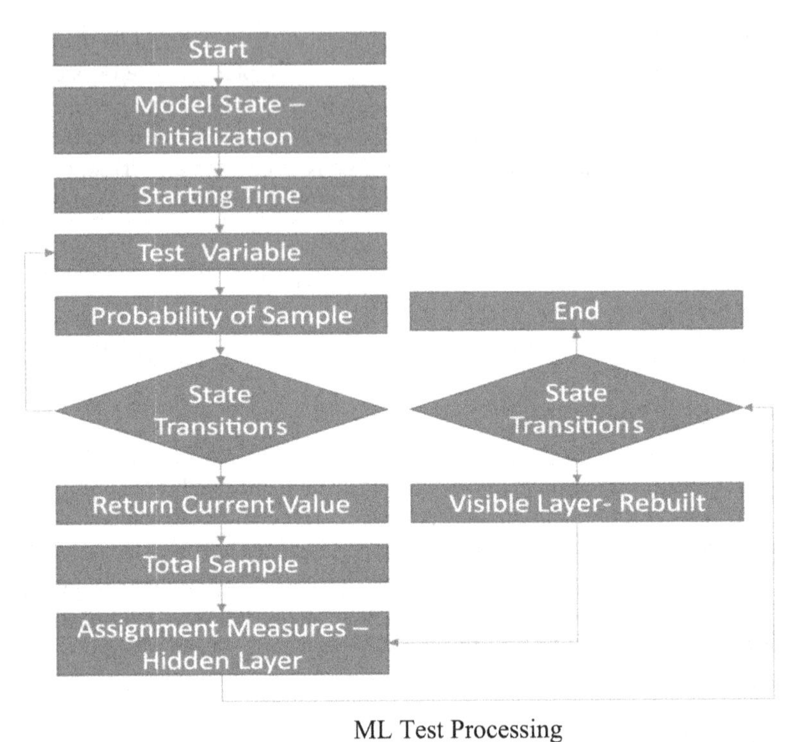

Fig. 5. ML Test Processing

and social practice—are coded. The three categories of HE, local, and school-based courses are split into each field. According to the improved ML model, Fig. 7 displays the time needed to query the HES module of a college education. Statistics show that the few smart fields of multiple intelligence theory represent a great deal of the HE-based curriculum. The distribution of course prediction rate using the upgraded ML model is displayed in Fig. 8.

In this ML model, it is possible to identify the specific AI field each course belongs to by looking at its unique name. It is possible to enter all of the courses into the computer, and after that, we can label them to the intelligent codes associated with the individual courses. HE administrators can easily design and set up courses because it is simple to determine how many courses there are in each intelligent field.

Figure 9 shows the improved ML model's university teachers' satisfaction with HES information management. Several exercises evaluated the model using two administrative career parallels. Each parallel had 500 students for 64 weeks, one academic year. The variables include the degree, subjects passed, lecture hours (1 and 20), sex, and the students' ages between 18 and 25. Repudiation, defined as repeated failure, is the issue addressed. The first exercise, MySQL logs of teacher and student activities, requires extensive data. Hadoop found student patterns after processing and transforming data from multiple sources (Fig. 6).

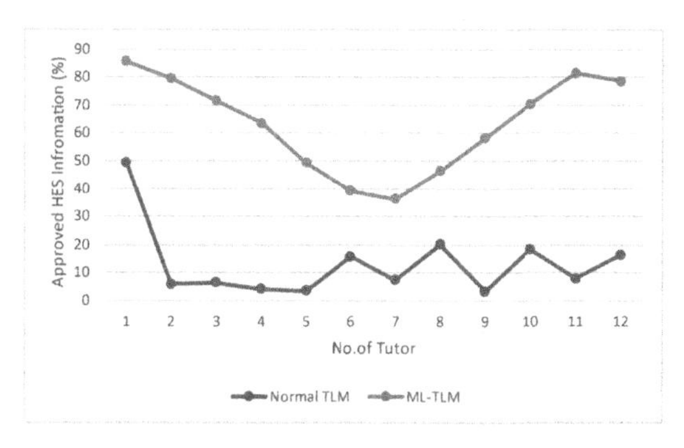

Fig. 6. HES information of College Tutors using normal vs. ML model.

Figure 5 shows the first exercise's patterns, which show the students' activities over the set time. Learning-guaranteed performance assessment procedures are shown to influence student performance. From 1 to 5, these test scores are presented. The minimum learning criteria are met by a grade of six, denoted as an acceptable grade on this axis. The forums usually have a high learning level, and low grades are usually due to students not registering their participation and subjective contributions. The task's mean values, calculated using Bloom's taxonomy, indicate that some students have achieved the necessary level of mastery.

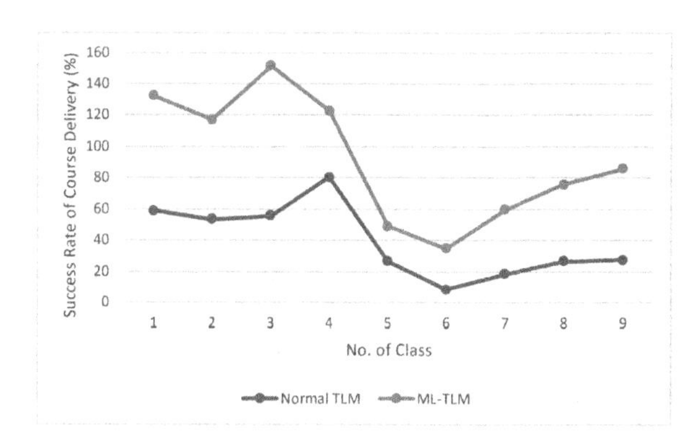

Fig. 7. Success Rate of Course Delivery using Normal vs. ML Model.

The evaluations in the form of questionnaires are closest to 1. In 30 min, students must answer 20 questions on these evaluations. This activity's values are extremely low.

The outcome of data is input to ML, which uses this data to learn and make decisions. The AI model included data from the TLM on students' dedication to reading the teacher's resource and a study on students' response times. The ML algorithm was used on the analysis's data, and the outcomes are shown in Table 1.

The results of this evaluation are presented in Table 1, which provides a detailed breakdown of the findings. The table likely includes various metrics and statistics that were used to evaluate the performance of the algorithm on the dataset. By analyzing the

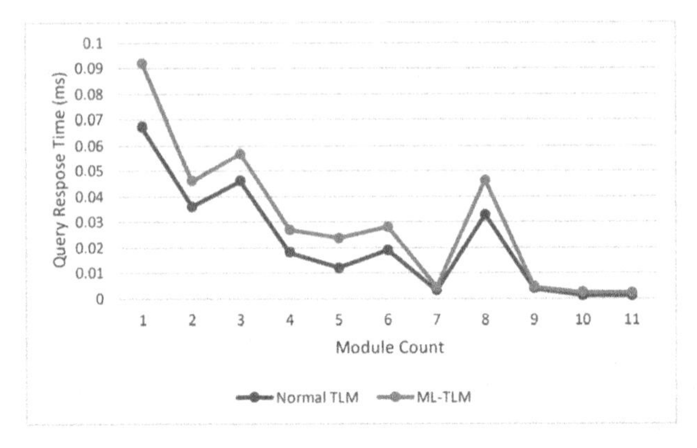

Fig. 8. The Query Response Time for information model of HES using Normal vs. ML model.

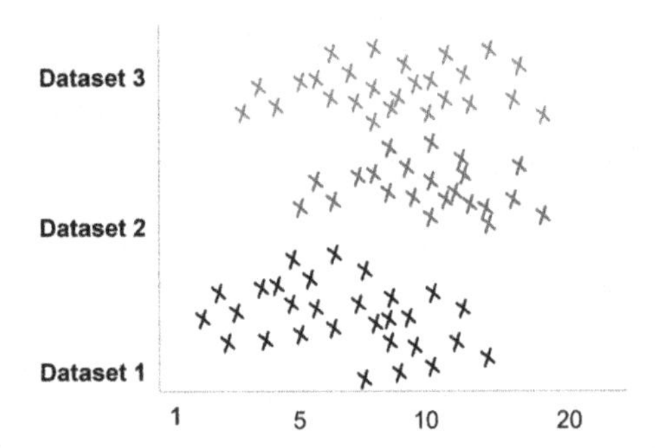

Fig. 9. Performance analysis of TLM activates of STS using ML Model

Table 1. Ranked Cross-validation.

Description	Value
Corrected Class Sample	53
Error Class Sample	5
Test Date Statics	0.92173
Mean Error	0.0582
RMSE	0.18332
RAE	11.2382%
RRSE	38.129%
Total Sample	100

data presented in the table, researchers can gain a deeper understanding of the factors that contributed to the substandard performance of the dataset.

Overall, this type of evaluation is crucial for identifying and addressing performance issues in datasets. By using algorithms and statistical analyses, researchers can gain insights into the underlying causes of poor performance and make informed decisions about how to improve the dataset's quality. Ultimately, this can lead to better results and more accurate predictions in a wide range of applications.

With a sample size of 100, the algorithm determined the causes of the substandard performance shown by the evaluation scores. 100 instances were examined, and 68 had classifications of the correctness of 96.79%. This value was assumed to be true to assume the analysis's decision. The findings are displayed in Table 1, which is found below.

6 Conclusion and Future Work

More studies on the impact of AI on classroom instruction are sure to be conducted as the use of AI in colleges and universities becomes progressively common. This study demonstrates that there has been little interest in researching AI in pre-service teacher HES programs. The ML-based neural network model was improved from the standard classification of neural network. It also solved the multi-classification problem with multiple constraints as a supervised ML model. The use of AI by pre-service teachers thus merits further empirical research, which we advise. It may be easier for AI-based teaching to be implemented in future classrooms if pre-service teachers have greater AI awareness and competency. Our analysis leads us to conclude that developing AI systems that are technically and pedagogically capable of enhancing quality education in many TLM environments is still a work in progress.

Future work are suggested to incorporate two equally important technologies into this model. The blockchain is one of these technologies, and it aims to protect the institution's and students' data and business processes. Also important will be the inclusion of the internet of things. It would be ideal for including devices in this modality that can collect data to enable ongoing improvement of the TLM.

References

1. Janson, A.: Integrating digital learning objects in the classroom: a need for educational leadership. In: Proceedings of the World Conference on Educational Multimedia, Hypermedia and Telecommunications (EDMEDIA), pp. 3341–3349 (2008)
2. McGee, P.: Planning for the digital classroom and distributed learning policies and planning for online instructional resources. In: Proceedings of the Annual Meeting of the Society for College and University Planning (SCUP), pp. 55–61 (2005)
3. Ruan, S.: Research and analysis on the correlation between college students' mental health education, psychological counseling and moral education. In: Proceedings of the International Conference on Food Research and Development (ICFRD), pp. 233–234 (2021)
4. Yang, L., Ding, J., Hu, T., Liu, Y.: Research status of artificial intelligence in psychology. In: Proceedings of the International Conference on Computer Knowledge and Technology (ICCKT), pp. 1–6 (2022)
5. Deng, X.: Innovative application of artificial intelligence in self-diagnosis and treatment of psychological problems and health education in colleges and universities. In: Proceedings of the Annual Conference of Jilin Medical College (ACJMC), pp. 427–428 (2021)

6. Claudiu, C., Laurent, G., Luiza, M., Carmen, S.: Online teaching and learning in higher education during the coronavirus pandemic: students' perspective. Sustainability **12**(5), 1827 (2020)
7. Akanksha, J., Joe Arun, C.: Potential of artificial intelligence for transformation of the education system in India. In: Proceedings of the International Conference on Education and Development using Information and Communication Technology (ICEDICT), pp. 142–158 (2021)
8. Kim, T., Lim, J.: Designing an efficient cloud management architecture for sustainable online lifelong education. Sustainability **11**(5), 1523 (2019)
9. Villegas-Ch, W., Palacios-Pacheco, X., Buenaño-Fernandez, D., Luján-Mora, S.: Comprehensive learning system based on the analysis of data and the recommendation of activities in a distance education environment. Int. J. Eng. Educ. **35**(5), 1316–1325 (2019)
10. Mahmud, M.S., Huang, Z., Salloum, S., Emara, T.Z., Sadatdiynov, K.: A survey of data partitioning and sampling methods to support big data analysis. Big Data Min. Anal. **3**(2), 85–101 (2020)
11. McHugh, J., Cuddihy, P.E., Williams, J.W., Aggour, K.S., Vijay, S.K., Mulwad, V.: Integrated access to big data polystores through a knowledge-driven framework. In: Proceedings of IEEE International Conference on Big Data, pp. 1494–1503 (2018)
12. Doleck, T., Poitras, E., Lajoie, S.: Assessing the utility of deep learning: using learner-system interaction data from BioWorld. In: Bastiaens, J.T. (ed.) Proceedings of EdMedia + Innovate Learning, pp. 734–738. AACE, Amsterdam (2019)
13. Xiao, C., Choi, E., Sun, J.: Opportunities and challenges in developing deep learning models using electronic health records data: a systematic review. J. Am. Med. Inform. Assoc. **25**(10), 1419–1428 (2018)
14. Winzenried, A., Dalgarno, B., Tinkler, J.: The interactive whiteboard: a transitional technology supporting diverse teaching practices. Australas. J. Educ. Technol. **26**(4), 534–552 (2010)
15. Kokku, R., Sundararajan, S., Dey, P., Sindhgatta, R., Nitta, S., Sengupta, B.: Augmenting classrooms with AI for personalized education. In: Proceedings of the 2018 IEEE International Conference on Acoustics, Speech and Signal Processing (ICASSP), Calgary, AB, Canada, pp. 6976–6980 (2018)
16. Maghsudi, S., Lan, A., Xu, J., van Der Schaar, M.: Personalized education in the artificial intelligence era: what to expect next. IEEE Signal Process. Mag. **38**, 37–50 (2021)

Topic Modelling of Legal Texts Using Bidirectional Encoder Representations from Sentence Transformers

Eya Hammami[1(✉)] and Rim Faiz[1,2]

[1] LARODEC Laboratory, University of Tunis, Tunis, Tunisia
hammami.eya@isg.u-tunis.tn
[2] IHEC, University of Carthage, Carthage, Tunisia
rim.faiz@ihec.ucar.tn

Abstract. Topic Modeling of legal texts is a challenging task because of its complicated language structures, and technical features. Recently, there has been a big boost in the number of legislative documents, which makes it very difficult for law experts to keep up with legislation like implementing acts and analyzing cases. The importance of topics is affected by the processing and the presentation of law texts in some contexts. The aim of this work is to figure out the legal opinions from cases seen by the supreme court of the United States and the legal judgments from cases seen by the supreme court of India. In this study we used different Language Models to create sentence embeddings from those legal texts datasets. This paper employs BERTopic technique and a baseline approach in order to discover significant topics from legal opinions and legal judgment.

Keywords: NLP · Topic Modeling · Sentence Transformers · Clustering · Language Models · BERTopic · LDA

1 Introduction

law texts are expected to be used in a specific way. They have certain methods about how they should be written and organized. In legal texts, a group of words is called a clause that has a subject and is part of complex sentences [1]. Thus, natural language processing (NLP), simultaneously with machine learning (ML), is an essential part for recognition and analyzing of legal cases.

The detection of words from the topics present in a text of data is called Topic Modelling [2], which is the most widely used approach in NLP for text understanding, such as Latent Semantic Analysis (LSA) and Latent Dirichlet Allocation (LDA) [3].

LSA is a statistical technique for defining and extracting the contextual meaning of words from corpus [4]. The hidden theory of a particular corpus is collected by LSA using singular value decomposition (SVD) [5]. It is also useful for information retrieval and it works efficiently if the corpus is made up of documents that are deliberately related [6].

I. Saad et al. (Eds.): ICIKS 2023, LNBIP 486, pp. 333–343, 2024.
https://doi.org/10.1007/978-3-031-51664-1_24

LDA is a famous topic modeling technique for recognizing a set of hidden subjects associated with a corpus of documents. [7]. In LDA, each document is represented as a bag-of-words, with each topic represented as a distribution of words [8].

However, these existing topic modeling techniques like LDA and LSA have many limitations. For example, LDA fails to determine any relationship between topics because it uses bag-of-words (BoW), which makes the assumption of word exchangeability without taking into account sentence structure. And as LSA is a linear model, it is not applicable to datasets having non-linear dependencies. Also LSA uses SVD, which requires a lot of work by the way it is difficult to update it when new data becomes available.

In our study we consider that the legal judgements and opinions process is mainly dependent on textual information. Therefore, understanding legal cases is an example of time-consuming task that can gain a lot of advantages from NLP approaches to be automated and saving time.

Thus, the objective of this paper is to get an abstract definition of legal cases by exploring the use of a stochastic topic modeling method on legal opinions and legal judgments datasets. Accordingly, we use BERTopic [7], which is a neural topic modeling approach that presents text from embeddings of BERT and its variants, with the assumption of avoiding the limitations of classic Topic modeling techniques. Particularly, the proposed approach represents legal documents in content form using different pre-trained sentence transformer models along with different clustering algorithms. Our main contributions regard:

- First, use different pretrained sentence transformers model to generate Embeddings.
- Second, implement clustering algorithm that reduce outliers to generate topics form our legal cases datasets.

This paper is organized as follows: Sect. 2 provides an overview of some related work. The proposed Topic Modeling technique is described in Sect. 3. The experimental results are presented in Sect. 4.

2 Related Work

Recently the implementation of ML and NLP methods have been used to analyze legal text documents. To figure out a solution to unstructured data in [9], the web-based text analytic is used to produce organized data. In [10], the classical application of document clustering was merged with the topic modeling method. With this unified approach, it is possible to see the pattern. In [11] and [12], LDA has been mostly used to model legal corpus. In [13] the authors analyzed and compared the classification of scientific unstructured e-books using LDA and LSA. Besides, The work in [14] concentrates on making it easier to navigate and identify legal topics and their associated sets of specific topic terminology by analyzing the performance of topic models to summarize British statute. The work done in [15] analyzes also the usage of LDA in obtaining accurate and meaningful topics in case legal documents to figure out the possibility of detecting themes in the documents related to case legal documents.

Furthermore, we find that in [16] the researchers also use LDA to group Indian court decisions and they use cosine similarity as a distance metric between documents.

According to the power of distributed representations to detect the semantics of words and texts is gaining celebrity [17]. The LDA has stayed the favored technique for Topic Modelling task until now regardless of its omnipresence, LDA has a number of weaknesses. To get better results from the LDA model, there should be a good number of topics. Moreover, the LDA method employs a bag-of-words model of words to represent the texts, which ignores word order, semantics and contextual meaning.

Thus, in order to have token level context specific word embedding, the researchers used generic context-specific language models like GPT-2 [18], BERT [19], and RoBERTa. Then, a new Topic modeling technique was introduced called BERTopic. This revolutionary technique in the area of topic generation gains a lot of advantages from transformer-based models to get reliable word representation [20] which yields good results that could better understand the contextual meaning of Topics especially in the legal domain.

However, despite its effectiveness, this model generates a lot of outliers. Therefore, in this work we tackle this limit by changing the default BERTopic clustering algorithm (HDBSCAN) with another algorithm in order to reduce the generation of outliers on our legal opinions and judgments datasets.

3 Proposed Topic Modeling Method

An unsupervised learning method that defines the distribution of topics in a corpus is introduced as topic modelling, where subjects are known as a recurring pattern of terms [21]. The objective of topic modelling is to extract the words that carry the document's concept. The method considered in this work uses BERTopic technique to identify document topics. BERTopic is a topic-modelling method that designs concentrated collections using bidirectional encoder transformers (BERT embedding and its variant) and class-based TF-IDF. Figure 1 shows the architecture of topic modeling via BERTopic. This technique contains three main steps.

The first step, consist of using embedding methods like BERT to represent document embeddings. In our case we used different pretrained Sentence Transformers models to generate sentence embeddings from our legal cases dataset.

The second step cope with the constructing of clusters. It employs uniform manifold approximation and projection (UMAP) to reduce embedding dimensionality and HDBSCAN algorithm to cluster decreased embeddings and build semantically comparable document clusters. In this study we tested the use also of Kmeans clustering algorithm instead of HDBSCAN in order to reduce the generation of outliers. Then we compare the efficiently of Topic Modelling by using these two different clustering algorithms. The final step is to use class-based TF-IDF to generate and extract reduced topics where interesting words

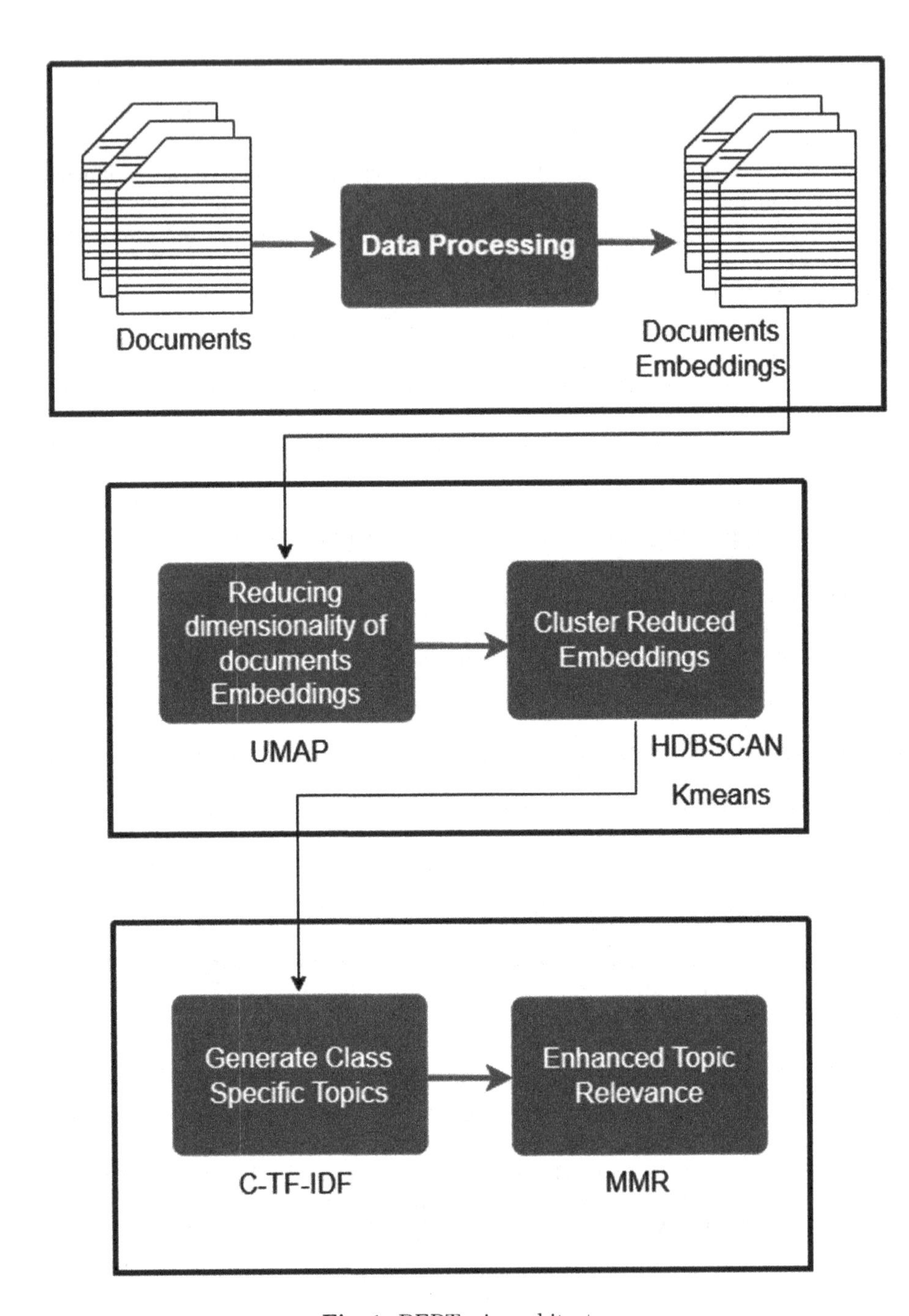

Fig. 1. BERTopic architecture

can be found in clusters of documents rather than per document. C-TF-IDF is a TF-IDF formula that has been used to multiple classes by combining all documents in each class. Therefore, rather than a set of documents, each class is transformed into a single document. For each class I the frequency of words t is calculated and divided by the number of total words W.

$$W_{x,c} = TF_{x,c} \times log(1 + \frac{A}{fx})$$

where $TF_{x,c}$ denotes the frequency of word x in class c, fx denotes the frequency of word x across all classes. A stands for average number of words per class. Then the last thing is to use Maximal Marginal Relevance (MMR) to improve word coherence.

4 Experiments

For the purpose of evaluating the performance of BERTopic approach, we exploit experiments on US legal opinions and on Indian legal judgments datasets referring to three evaluation metrics. Therefore, we conducted a comparison with three different pretrained sentence transformers models along with two different clustering algorithms. We compared also our obtained results from BERTopic method with the results that are obtained using LDA technique which is considered as a baseline approach.

4.1 Dataset

We applied our model on US legal opinions dataset collected from Courtlistener.com[1], which is a free legal research website containing millions of legal opinions from US federal and state courts. The dataset includes 10991 opinions from 1970 through 2020. Figure 2 shows the distribution frequency of number of words per opinion of US legal opinions dataset. According to this figure the longest text opinion's contains 30589 words and the majority of them have less than 5000 words.

We applied also our model on Indian legal judgment dataset crawled from Thomson Reuters Westlaw India[2], which is open source and available on github[3]. This dataset includes in total a set of 50 court case documents judged in the Supreme Court of India, where each sentence has been annotated with its rhetorical role by law student [24], it covers approximately 9380 sentences. Figure 3 shows also the distribution frequency of number of words per sentence of the Indian legal judgment dataset. According to this figure the longest sentence contains 145 words and the majority of them have less than 40 words.

[1] https://www.courtlistener.com/.
[2] http://www.westlawindia.com.
[3] https://github.com/Law-AI/semantic-segmentation.

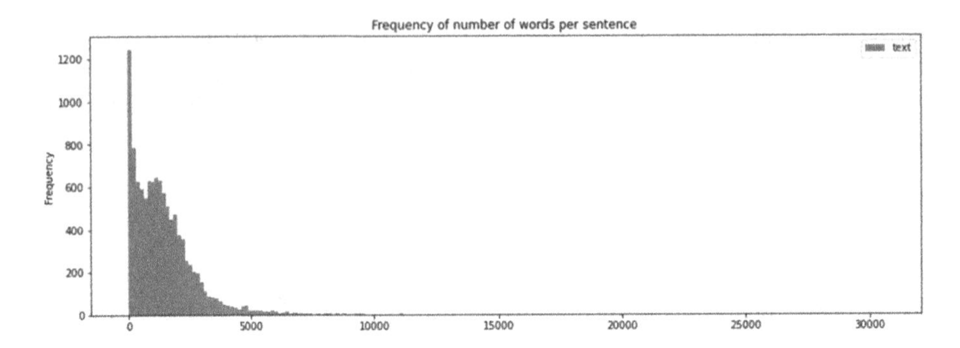

Fig. 2. Analyse of US legal opinions texts

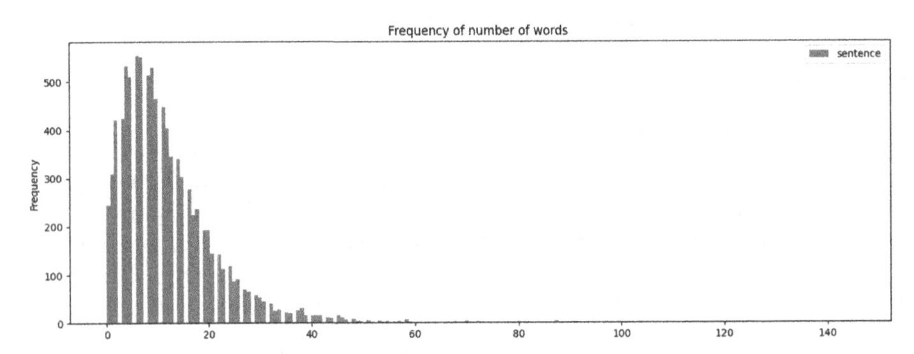

Fig. 3. Analyse of Indian Legal Judgment texts

4.2 Evaluation Metrics

Topic Coherence measures score a single topic by measuring the degree of semantic similarity between high scoring words in the topic. These measurements help distinguish between topics that are semantically interpretable and topics that are artifacts of statistical inference. For our evaluations, we consider three evaluation measures designed for BERTopic, which have been shown to match well with human judgements of topic quality:

Coherence C_{umass} Metric: this metric defines the score to be based on document co-occurrence. It computes document co-occurrence counts over the original corpus used to train the topic models, rather than an external corpus. This metric is more intrinsic in nature. It attempts to confirm that the models learned data known to be in the corpus. The range score of this measure is between -14 and 14, such that values closer to 14 represent good quality of topics, however those which are closer to -14 indicate bad quality of Topics results [22]. It is defined by the following formula:

$$C_{umass}(v_i, v_j, \epsilon) = log\frac{D(v_i, v_j) + \epsilon}{D(v_j)} \tag{1}$$

where:

- $D(v_i, v_j)$ counts the number of documents containing words v_i and v_j
- $D(v_j)$ counts the number of documents containing v_j.

Coherence C_v Metric: This metric is based on a sliding window, one-set segmentation of the top words and an indirect confirmation measure that uses normalized pointwise mutual information (NPMI) or the cosine similarity. The range score of C_v is between 0 and 1, such that values closer to 1 represent good quality of topics, however those which are closer to 0 indicate bad quality of topics results [22]. It is defined by the following formula:

$$C_v = \frac{1}{T} \sum_{t=1}^{T} \frac{1}{|D_t|} \sum_{w \in D_t} NPMI(w, D_t) \tag{2}$$

where:

- T is the number of topics
- D_t is the set of words in topic t
- w is a word in topic t
- $NPMI(w, D_t)$ is the normalized pointwise mutual information between word w and topic t

Silhouette Metric: it measures how similar an object is to its own cluster compared to other clusters. Its score ranges from -1 to 1 where a high value indicates that the object is well matched to its own cluster and poorly matched to neighboring clusters. If most objects have a high value, then the clustering configuration is appropriate. If many points have a low or negative value, then the clustering configuration may have too many or too few clusters [23].

$$Silhouette = \frac{b - a}{max(a, b)} \tag{3}$$

where:

- b is the distance between the data point and the nearest cluster that it is not a part of.
- a is the average distance between the data point and all other points in its cluster

4.3 Parameters and Settings

The default parameters for the different embedding techniques that we used in our BERTopic model are described in Table 1, where we present the Max Sequence Length (MSL) values, the Dimensions values, the Normalized Embeddings (NE) and we mention the original pretrained model that are used for each of the pretrained sentence transformers models that we utilized[4]. Then, for the clustering algorithm we used first the default clustering algorithm of BERTopic model which is HDBSCAN by it's default parameters. After that we replaced HDBSCAN with the k-means clustering algorithm by fixing the k parameter to 36 which belong to the number of Topics that are generated first by HDBSCAN on US legal opinions dataset and to 25 for the Indian legal judgment dataset. As a baseline we used LDA approach from Gensim library[5].

Table 1. Parameter of embedding models

Pretrained SBERT Models	MSL	Dimensions	NE	Original Model
all-MiniLM-L6-v2	256	384	True	MiniLM-L6-H384-uncased
all-distilroberta-v1	512	768	True	distilroberta-base
all-mpnet-base-v1	512	768	True	mpnet-base

4.4 Results and Discussion

The model is initialized with the parameter verbose set to true so that the model's stages can be observed. By running the model first with the default clustering algorithm (HDBSCAN) by its default parameters along with all-MiniLM-L6-v2 pretrained sentence transformers model as documents embedding technique, we found 38 topics for the US legal opinions dataset and 25 topics for the Indian legal judgment dataset. Table 2 and Table 3 show the results over CV, $UMASS$ and silhouette evaluation metrics on our datasets. Second we keep the same embedding model but we used Kmeans clustering algorithm by fixing the parameter k to 36 for US legal opinions dataset and 25 for Indian legal judgment dataset. Here according to the evaluation metrics CV and Silhouette seen in Table 2 and Table 3 we notice that the results are improved compared to the results of the first implementation where we used HDBSCAN as clustering model.

We did the same thing again but we changed the embedding model. We used all-distilroberta-v1 first with the default HDBSCAN algorithm and then with Kmeans algorithm. As a result we see that the use of Kmeans clustering provides better results than using HDBSCAN over CV and Silhouette evaluation metrics when using all-distilroberta-v1 as an embedding model. We also conduct our last test by using all-mpnet-base-v1 as another pretrained embedding model with HDBSCAN and with Kmeans clustering algorithms. Consequently, Kmeans offer better results than HDBSCAN. Following the results of these experiments we can assume that the use of Kmeans instead of HDBSCAN which reduces the generation of outliers, outperforms the default clustering algorithm of BERTopic

[4] https://www.sbert.net/.

[5] https://radimrehurek.com/gensim/index.html.

independently from the pretrained language models that are used in the embedding documents phase, especially over *CV* and silhouette evaluation metrics. Moreover, regarding the results of the evaluation with LDA technique we can say that the use of BERTopic approach in order to detect topics outperforms baseline approach especially in our specific domain knowledge.

Table 2. Evaluation Results on US Legal Opinions Dataset

Topic Model	*CV*	*UMASS*	Silhouette
Bertopic all-MiniLM-L6-v2 HDBSCAN	0.6468	−0.0816	0.4508
Bertopic all-MiniLM-L6-v2 Kmeans	**0.7122**	−0.1625	**0.6262**
Bertopic all-distilroberta-v1 HDBSCAN	0.6044	−0.0843	0.4364
Bertopic all-distilroberta-v1 Kmeans	**0.6456**	−0.0940	**0.6250**
Bertopic all-mpnet-base-v1 HDBSCAN	0.6236	−0.0627	0.4494
Bertopic all-mpnet-base-v1 Kmeans	**0.7015**	−0.1334	**0.6306**
LDA (baseline-approach)	0.3151	−0.5406	-

Table 3. Evaluation Results on Indian Legal Judgments Dataset

Topic Model	*CV*	*UMASS*	Silhouette
Bertopic all-MiniLM-L6-v2 HDBSCAN	0.6274	−0.6229	0.4329
Bertopic all-MiniLM-L6-v2 Kmeans	**0.7729**	−0.6939	**0.6043**
Bertopic all-distilroberta-v1 HDBSCAN	0.6994	−0.5634	0.4575
Bertopic all-distilroberta-v1 Kmeans	**0.7915**	−0.6552	**0.4758**
Bertopic all-mpnet-base-v1 HDBSCAN	0.6543	−0.6394	0.4302
Bertopic all-mpnet-base-v1 Kmeans	**0.8056**	−0.5898	**0.4430**
LDA (baseline-approach)	0.4425	−3.8274	-

5 Conclusion

In this paper, we have presented the use of BERTopic algorithm for topic modelling of US legal opinions and Indian legal judgment datasets. In terms of qualitative evaluation, the approach offer good results, by picturing topics that are coherent with the subject of the document, especially when using Kmeans algorithm instead of HDBSCAN which avoid the generation of outliers. This work can be considered as an initial method for future studies. As well, the performance of BERTopic can be compared with other topic modelling methods in other specific domains knowledge.

Acknowledgments. This work was supported by Google PhD Fellowships program and by Google Cloud Platform (GCP).

References

1. Nogales, A., Täks, E., Taveter, K.: Ontology modeling of the estonian traffic act for self-driving buses. In: Lossio-Ventura, J.A., Muñante, D., Alatrista-Salas, H. (eds.) SIMBig 2018. CCIS, vol. 898, pp. 249–256. Springer, Cham (2019). https://doi.org/10.1007/978-3-030-11680-4_24
2. Ruhl, J.B., Nay, J., Gilligan, J.: Topic modeling the president: conventional and computational methods. Geo. Wash. L. Rev. **86**, 1243 (2018)
3. Dieng, A.B., Ruiz, F.J.R., Blei, D.M.: Topic modeling in embedding spaces. Trans. Assoc. Comput. Linguist. **8**, 439–453 (2020)
4. Ray, S.K., Ahmad, A., Kumar, C.A.: Review and implementation of topic modeling in Hindi. Appl. Artif. Intell. **33**(11), 979–1007 (2019)
5. Pilato, G., Vassallo, G.: TSVD as a statistical estimator in the latent semantic analysis paradigm. IEEE Trans. Emerg. Top. Comput. **3**(2), 185–192 (2014)
6. Rajandeep, K., Manpreet, K.: Latent semantic analysis: searching technique for text documents. Int. J. Eng. Dev. Res. **3**(2), 803–806 (2015)
7. Blei, D.M., Ng, A.Y., Jordan, M.I.: Latent dirichlet allocation. J. Mach. Learn. Res. **3**(Jan), 993–1022 (2003)
8. Mu, W., Lim, K.H., Liu, J., Karunasekera, S., Falzon, L., Harwood, A.: A clustering-based topic model using word networks and word embeddings. J. Big Data **9**(1), 1–38 (2022)
9. Kadir, N.H.M., Aliman, S.: Text analysis on health product reviews using R approach. Indones. J. Electr. Eng. Comput. Sci. (IJEECS) **18**(3), 1303–1310 (2020)
10. Mangsor, N.S.M.N., Nasir, S.A.M., Yaacob, W.F.W., Ismail, Z., Rahman, S.A.: Analysing corporate social responsibility reports using document clustering and topic modeling techniques. Indones. J. Electr. Eng. Comput. Sci. **26**(3), 1546–1555 (2022)
11. Remmits, Y.: Finding the topics of case law: latent dirichlet allocation on supreme court decisions (2017)
12. Luz De Araujo, P.H., De Campos, T.: Topic modelling brazilian supreme court lawsuits. In: Legal Knowledge and Information Systems, pp. 113–122. IOS Press (2020)
13. Mohammed, S.H., Al-augby, S.: LSA & LDA topic modeling classification: comparison study on e-books. Indones. J. Electr. Eng. Comput. Sci. **19**(1), 353–362 (2020)
14. O'Neill, J., Robin, C., O'Brien, L., Buitelaar, P.: An analysis of topic modelling for legislative texts. In: CEUR Workshop Proceedings (2016)
15. Angelov, D.: Top2vec: distributed representations of topics. arXiv preprint arXiv:2008.09470 (2020)
16. Rawat, A.J., Ghildiyal, S., Dixit, A.K.: Topic modelling of legal documents using NLP and bidirectional encoder representations from transformers. Indones. J. Electr. Eng. Comput. Sci. **28**(3), 1749–1755 (2022)
17. Silveira, R., Fernandes, C., Neto, J.A.M., Furtado, V., Pimentel Filho, J.E.: Topic modelling of legal documents via LEGAL-BERT. In: Proceedings http://ceur-ws org ISSN 1613 0073 (2021)
18. Grootendorst, M.: BERTopic: neural topic modeling with a class-based TF-IDF procedure. arXiv preprint arXiv:2203.05794 (2022)

19. Gunjan, V.K., Zurada, J.M.: Modern Approaches in Machine Learning & Cognitive Science: A Walkthrough. Springer, Cham (2022). https://doi.org/10.1007/978-3-030-96634-8
20. Blei, D.M.: Probabilistic topic models. Commun. ACM **55**(4), 77–84 (2012)
21. Abuzayed, A., Al-Khalifa, H.: BERT for Arabic topic modeling: an experimental study on BERTopic technique. Procedia Comput. Sci. **189**, 191–194 (2021)
22. Röder, M., Both, A., Hinneburg, A.: Exploring the space of topic coherence measures. In: Proceedings of the Eighth ACM International Conference on Web Search and Data Mining, pp. 399–408 (2015)
23. Thinsungnoena, T., Kaoungkub, N., Durongdumronchaib, P., Kerdprasopb, K., Kerdprasopb, N.: The clustering validity with silhouette and sum of squared errors. Learning **3**(7) (2015)
24. Ghosh, S., Wyner, A.: Identification of rhetorical roles of sentences in Indian legal judgments. In: Legal Knowledge and Information Systems: JURIX 2019: The Thirty-second Annual Conference, vol. 322. IOS Press (2019)

Author Index

I. Saad et al. (Eds.): ICIKS 2023, LNBIP 486, pp. 345–346, 2024.
https://doi.org/10.1007/978-3-031-51664-1

Printed in the United States
by Baker & Taylor Publisher Services